V · O · L · U · M · E 1

METHODS IN
CHEMICAL
ECOLOGY

CHEMICAL METHODS

D1745015

90 0382739 5

WITHDRAWN
FROM
UNIVERSITY OF PLYMOUTH
LIBRARY

SEVEN DAY LOAN

This book is to be returned on
or before the date stamped below

UNIVERSITY OF PLYMOUTH

PLYMOUTH LIBRARY

Tel: (01752) 232323
This book is subject to recall if required by another reader
Books may be renewed by phone
CHARGES WILL BE MADE FOR OVERDUE BOOKS

V · O · L · U · M · E 1

METHODS IN
CHEMICAL
ECOLOGY

CHEMICAL METHODS

EDITED BY

JOCELYN G. MILLAR
University of California, Department of Entomology, Riverside, CA

KENNETH F. HAYNES
University of Kentucky, Department of Entomology, Lexington, KY

CHAPMAN & HALL

Distributors for North, Central and South America:
Kluwer Academic Publishers
101 Philip Drive
Assinippi Park
Norwell, Massachusetts 02061 USA

Distributors for all other countries:
Kluwer Academic Publishers
Distribution Centre
Post Office Box 322
3300 AH Dordrecht, THE NETHERLANDS

Library of Congress Cataloging-in-Publication Data

Methods in chemical ecology
      p. cm.
    Includes bibliographical references and index.
    Contents: v. 1. Chemical methods / edited by Jocelyn G. Millar and
Kenneth F. Haynes-- v. 2. Bioassay methods / edited by Kenneth F. Haynes
and Jocelyn G. Millar.
      ISBN 0-412-08071-0 (alk. paper : v. 1). -- ISBN 0-412-08041-9
(alk. paper : v. 2)
    1. Chemical ecology--Methodology. I. Millar, Jocelyn G.
      II. Haynes, K.F.
QH541.15.C44M48 1998
577'.01'54--dc21   97-39820
                                                                    CIP

Copyright © 1998 by Kluwer Academic Publishers

All rights reserved. No part of this publication may be reproduced, stored in a
retrieval system or transmitted in any form or by any means, mechanical, photo-
copying, recording, or otherwise, without the prior written permission of the
publisher, Kluwer Academic Publishers, 101 Philip Drive, Assinippi Park, Norwell,
Massachusetts 02061

Printed on acid-free paper.

Printed in the United States of America

Dedication

This volume is dedicated to the memory of Audrey Bingham Millar, 1924–1995.

UNIVERSITY OF PLYMOUTH	
Item No.	9oo **382739**S
Date	2 5 NOV 1998 S
Class No.	*5 74. 5 MET*
Contl. No.	O4I2O8O7IO
LIBRARY SERVICES	

UNIVERSITY OF PLYMOUTH

Item No.	900 1521535
Date	2 5 NOV 1998
Class No.	574.5 MET
Contl. No.	0412080710

LIBRARY SERVICES

Contents

Preface

A working definition of the discipline of chemical ecology might be "the study of the structure, function, origin, and significance of naturally occurring compounds that mediate inter- and intraspecific interactions between organisms." In particular, chemical ecology focuses on determining the role of semiochemicals and related compounds in their natural contexts. Thus, chemical ecology is distinct from disciplines such as pharmacology, in which compounds are screened for uses outside their natural context, for example in the screening of natural products for use as drugs. Superficially, many of the methods used in the various branches of natural products chemistry, such as pharmacology and chemical ecology, are very similar, but each branch has developed its own set of specialized methods for dealing with the problems characteristic of that discipline. For example, in chemical ecology, many semiochemicals are isolated and identified using only a few micrograms or less of material. Although the same general chromatographic and spectroscopic techniques are used as would be used with the identification of most organic compounds, specialized techniques have been developed for handling these very small quantities, allowing the maximum amount of information to be recovered from the minimum amount of sample. These microscale techniques, and the problems unique to working with very small amounts of sample, are rarely covered in detail in reference books on the isolation and identification of biologically active natural chemicals. Thus, one of our main objectives in this volume is to provide a sense of what can and cannot be done with the various separation and identification techniques. We have tried to present assessments of the advantages and limitations of each technique, including the information content, the sensitivity, and the ease of use, to serve as a guide to practitioners when faced with the task of identifying small and very valuable samples, which may have taken weeks or even years of effort to accumulate and isolate.

With this volume, and the companion volume on bioassay methods, we have also attempted to bring together a large number of techniques and references, which are scattered through the literature of several disciplines over a period of several decades. Those of us who have been in the field for a number of years have assembled personal collections of references, but there has been no single source that might serve as a general reference manual for chemical ecology methods. With these two volumes, we hope to remedy this situation.

We would also stress the inseparable link between the chemical methods used to fractionate and purify biologically active compounds and the bioassays required to demonstrate the chemically mediated phenomenon under study. Normally, unknown compounds are isolated via bioassay-driven fractionation, with the results of each bioassay cycle determining, in part, the choice of the next separation step. Thus, the division of these volumes into a book on chemical methods and a second on bioassays is done solely for organizational reasons; in practice, most projects proceed by the sequential application of a fractionation step followed by a bioassay step. We cannot overemphasize the importance of a robust, reliable, and reproducible bioassay in a chemical ecology project. Conversely, the astute application of efficient fractionation steps leading to the isolation of compounds in a minimum number of steps, and the careful use of chemical identification methods that use the minimum amount of sample, may easily determine the success or failure of a project. In short, chemistry and bioassays are inextricably interdependent in studies in chemical ecology.

We have attempted to take the mystery out of some of the more specialized techniques, which can seem so daunting to the uninitiated; that is, many authors of papers in specialized journals assume, quite naturally, that they are writing for experts, so that the methods and materials sections are often short and lacking in detail, to the point of being perfunctory. Furthermore, the rationale for using various steps in various sequences is often not given. This can make repetition of the methodology difficult, particularly by nonexperts. It also hinders understanding of the general processes and procedures used, and the circumstances in which specific procedures are most appropriate.

We as authors and editors have also been faced with the task of preparing volumes to serve two somewhat different audiences; that is, we have attempted to produce chapters that will be straightforward enough to be used directly by newcomers to the field, while at the same time having enough detailed information and compilations of references to specific techniques to be of value to seasoned veterans. It is also obvious that one volume could not turn readers into experts on separation and spectroscopic techniques, nor was that our aim. In some chapters, such as the chapter on nuclear magnetic resonance (NMR) spectrometry, we expect readers to have a basic working knowledge of the NMR experiment, which can be found in any standard undergraduate text on spectroscopic methods, and so in some cases the basics have not been covered again here. Instead, our aim has been to provide readers with an understanding of the capabilities and general operating characteristics of various techniques, along with citations to key in-depth references for those desiring more detailed information. To borrow a word used by one of our reviewers, we would like this volume to be "enabling."

The most successful chemical ecology projects and research groups are often collaborations between chemists and biologists/ecologists, each with their distinct and different training and backgrounds. Consequently, we have also tackled the challenge of making each of the two facets of chemical ecology, the chemistry and the biology/ecology, comprehensible to the various partners in chemical ecology studies, on the basis that greater understanding should lead to greater insight. From the viewpoint of a chemist (J.G.M.), the companion volume on bioassay methodology has been fascinating reading, illustrating as it does the cleverness and ingenuity embodied in the designs of bioassays for different organisms. Conversely, from the perspective of a biologist (K.F.H.), this volume on chemical methods has resulted in increased appreciation for the resourceful solutions that chemists have developed to problems in microscale analyses and identifications.

This volume focuses primarily on methods of isolating and identifying unknown compounds, rather than methods of analyzing and quantifying known compounds, which could encompass a volume in itself. This focus was chosen deliberately because the isolation and identification of an unknown compound is a major technical and intellectual challenge, requiring a researcher to make careful and reasoned choices from a wide range of separation and identification methods. In some cases, separation steps do not work, or in the worst case, actually destroy the sample, and a researcher must be prepared to assess what went wrong and to work out contingency plans to circumvent the problem. Furthermore, almost all spectroscopic and other identification techniques do not provide molecular structures per se, but rather they provide more or less cryptic information from which the structure is eventually deduced, usually on the basis of integrating fragments of information from several sources. In contrast, the methodology for the accurate analysis and quantification of known compounds encompasses a different set of requirements and skills, the starting point for which is usually the careful duplication or modification of literature procedures.

This volume reflects the background of the editors and contributors, many of whom work primarily with insects and/or plants. Consequently, many of the examples given have an entomological bias. However, we anticipate that this bias will have minimal effect on the practical information content of the chapters because to a large degree, the instruments and techniques used are independent of the origin of the sample. A gas chromatograph or an NMR spectrometer works equally well with samples from elephants or insects! What is generally more important is the gross characteristics of the sample; for example, very oily or fatty samples, be they of plant or animal origin, may require particular cleanup steps to remove most of the contaminating material.

We have laid out this volume in a logical sequence, beginning with a chapter on the handling and extraction of crude samples, including some cleanup steps. The next two chapters cover the main liquid and gas chromatography techniques used in the separation and isolation of the comparatively small organic molecules (molecular weights < 1000 Da) commonly used as semiochemicals by many organisms. Due to space limitations, we have focused on the most generally useful methods, with brief mention, or even no mention at all, of more specialized techniques such as countercurrent chromatography or capillary electrophoresis. The next group of four chapters focuses on identification methods, some of which can be used even with crude samples. They are arranged in the order in which they are most commonly applied, beginning with mass spectrometry. These chapters assume some basic knowledge of the techniques being described, and we have tried to concentrate instead on the information to be gained from each method, and its advantages and limitations. Thus, for example, the chapter on NMR focuses on the application of the most useful and valuable NMR techniques to a single molecule, quinine, illustrating how the results of the various experiments are integrated to arrive at an unambiguous determination of the structure. The next chapter, on infrared (IR) and ultraviolet (UV) methods, focuses primarily on the coupled gas chromatography-Fourier transform (GC-FTIR) technique, which has proven so useful in the identification of very small quantities (a few tens of nanograms) of volatile semiochemicals. The last of these four chapters, the long and detailed chapter on microchemical methods, describes a wide variety of reactions that can be used to provide critical information on the presence or absence of specific functional groups, often with nanogram to subnanogram quantities.

The final chapters cover two specialized areas. The first, on the separation of enantiomers, the determination of enantiomeric purity, and the determination of absolute configurations, provides comprehensive tables of chiral insect pheromones and the methods used to resolve them. Although restricted to insect pheromones, the ideas and methods presented are of general applicability throughout chemical ecology and vividly illustrate the critically important relationship between chirality and biological activity. The last chapter focuses on an area that has been most widely developed and used in entomology; that is, the use of electrophysiological techniques such as electroantennography in the identification and understanding of the function of inter- and intraspecific semiochemicals. Although this chapter is restricted to techniques used in the study of insects, the extraordinary role that these techniques have played in the rapid advances in insect chemical ecology of the past several decades merits its inclusion in a general volume on chemical ecology methods.

In summary, we hope that these volumes will serve as useful and well-used compendia of methods and references for practitioners of chemical ecology. We further hope that in writing and editing these volumes, we have performed a service for the chemical ecology community and for those contemplating studies in our discipline. Our understanding of the plethora of chemical signals and their functions in the world around us is in its infancy, and expanded studies in chemical ecology can only be encouraged by making the techniques of chemical ecology both more understandable and more accessible.

Contributors

Athula B. Attygalle
Department of Chemistry,
Cornell University
Ithaca, NY 14853, USA

Louis B. Bjostad
Bioagricultural Sciences and
  Pest Management Department
Colorado State University,
Fort Collins, CO 80523, USA

Barbara D. Deuben
Center for Medical, Agricultural, and
  Veterinary Entomology
Agricultural Research Service, United
  States Department of Agriculture
Gainesville, FL 32608, USA

Robert R. Heath
Center for Medical, Agricultural and
  Veterinary Entomology
Agricultural Research Service, United
  States Department of Agriculture
Gainesville, FL 32608, USA

David J. Kiemle
Department of Chemistry,
1 Forestry Drive
SUNY ESF, Syracuse, NY 13210, USA

Walter Soares Leal
Laboratory of Chemical Prospecting
National Institute of Sericultural and
  Entomological Science
1-2 Ohwashi, Tsukuba-city 305 Japan

Jocelyn G. Millar
Department of Entomology,
University of California
Riverside, CA 92521, USA

Kenji Mori
Department of Chemistry, Faculty of
  Science, Science University of Tokyo
Kagurazaka 1-3, Shinjuku-ku,
Tokyo 162 Japan

James J. Sims
Department of Plant Pathology,
  University of California
Riverside, CA 92521, USA

Francis X. Webster
Department of Chemistry,
1 Forestry Drive
SUNY ESF, Syracuse, NY 13210, USA

Acknowledgments

The editors would like to acknowledge the contributions of numerous people to this volume, beginning with our editors at Chapman & Hall, Greg Payne, with whom we initiated the project, and Henry Flesh, who helped us finish it. We thank our authors, many of whom undertook major revisions without complaint, and Kris Gilbert and J. Steven McElfresh for their assistance with editorial chores and figures. We also acknowledge the contributions of colleagues who volunteered to review drafts for us. Finally, no list of acknowledgments would be complete without an acknowledgment of gratitude to our spouses, Dorothy E. Hartley and Elizabeth W. Haynes, for whose patience and support during the vicissitudes of this project we are deeply grateful.

1

Preparation, Cleanup, and Preliminary Fractionation of Extracts

Jocelyn G. Millar and James J. Sims

1.1. Introduction

The preparation of a biologically active extract represents the first important milestone in the identification of a semiochemical. It is a task that should not be underestimated and careful consideration of methods and conditions can save an immense amount of time and effort. There are several points to consider. First, natural products vary widely in stability; a number of semiochemicals have been shown to degrade easily (e.g., Gries et al. 1997) and special methods may be required to minimize degradation. Second, it can be difficult, if not impossible, to recover all the activity in a single extract. A case in point is the corn rootworm host attractant blend of CO_2 (a gas) and relatively nonvolatile dimboa (Hibbard and Bjostad 1988, 1989), in which no single extract contained representative amounts of both compounds. Third, extraction efficiency should be carefully considered. An efficient extraction is one in which most or all of the desired compound is extracted while the amounts of extraneous material are minimized; in these terms, the solvent that extracts the most material overall may not be the best choice. Finally, care must be exercised to minimize contamination of extracts with materials from sources as varied as fingerprints, dissolved Parafilm®, or stopcock grease.

When testing fractions in bioassays, there are also several potential pitfalls. Activity may decrease as you fractionate, and correct identification of the cause is important. If full activity is recovered on recombination of all fractions, then it is likely that compounds with additive or synergistic activity have been separated into different fractions. If the activity is not recovered, then degradation or other loss of material has probably occurred. In contrast, activity may also increase as fractionation progresses due to the separation of antagonistic compounds. For example, the mixture of four components, or any mixture of three components of the volatiles produced by the rice bug, *Leptocorisa chinensis,* were not attractive to male bugs, whereas a blend of two of the components was highly attractive (Leal et al. 1996). As a general rule, fractions should always be tested versus the starting material for that fractionation step to monitor sudden activity changes. If synergism is suspected, it is best to test compounds at a marginally active dose of each component, because the synergistic effect is usually more obvious. That is, the difference between 20 and 50% response is easier to demonstrate than the difference between 80 and 90% response.

A thorough search of both the chemical and biological literature using species and genus names is strongly recommended before embarking on a large-scale fractionation and isolation project. Such searches are now easy to do with computerized databases (e.g. Chemical Abstracts, Biosis and Current Contents) and they can reveal all manner of useful information, such as compounds characteristic of or unique to the organism under study, useful fractionation or isolation procedures, and not infrequently, the active compound under study. Identification then simply becomes a matter of matching assorted spectra from the isolated compound with those of the previously identified compound in the literature.

Conversely, for organisms containing compounds of known biological activity, there is always a danger of assuming that the activity seen in preliminary bioassays is due to these compounds. This assumption can be false, and the bioassay data, rather than preconceived notions, should always guide all fractionation steps (e.g. Meade et al. 1994).

In addition to a specific search, simply browsing through some of the journals dedicated to isolation of natural products may prove useful. These include the *Journal of Natural Products, Phytochemistry, Phytochemical Analysis,* the *Journal of Antibiotics,* and the *Journal of Agricultural and Food Chemistry.*

Pilot-scale extractions are strongly recommended to find solvents and methods that maximize recovery while minimizing coextractants. They also enable estimation of the amount of raw material needed for the isolation sequence, based on the activity obtained per gram of material and the complexity of the extract (i.e., few or many fractionation and bioassay steps required, and the amount of material consumed in each bioassay). Pilot-scale fractionations are also invaluable for testing how well a fractionation will work in preventing inadvertent destruction of active compounds and in highlighting problems such as emulsion formation, incomplete extraction or partitioning, or inadvertent loss of volatiles (e.g., during concentration *in vacuo* or freeze drying). Overall, the solubility and volatility characteristics identified on a small scale will guide the choice of separation and analytical techniques.

Once ready to begin the full-scale extraction, enough material should be harvested to carry out the whole isolation sequence because it can be difficult or impossible to get an identical sample, for example, due to seasonal changes or stresses that can qualitatively and quantitatively alter chemical profiles. Ideally, a considerable excess should be harvested and frozen. There may be a conflict between the amount that can be comfortably handled and the amount needed, so batchwise extraction may be necessary. The solvent used in batchwise extraction can usually be recovered and used for the next batch. When doing extractions, several sequential extractions with smaller volumes of solvent are usually more efficient than a single extraction with the same total volume. Overall, the volume of extracting solvent should be minimized because it has to be removed again after extraction.

For multistep fractionations, a flowchart is useful for keeping track of fractions and their weights. Reference samples of each fraction should also be saved in case a fractionation has to be repeated. The fractionation sequence should first use simple, large-scale fractionation methods such as solvent partitioning to reduce the volume of material before going on to smaller scale, high-resolution chromatographic methods.

Microscale extractions present special problems, and appropriate adjustments to methods and equipment need to be made. The number of handling steps should be reduced to minimize losses and contamination. Often only a single cleanup step, or no cleanup step at all for relatively clean extracts, is used before going directly to high-resolution chromatographic methods. Microscale equipment, such as conical-bottomed vials or vial inserts (or homemade equivalents) should be used, with small amounts of solvent. For long-term storage, extracts are best stored in sealed glass ampoules rather than screw-cap vials. Solvents are removed by blowing down under nitrogen, rather than by rotary evaporation. Microscale equipment is described in detail in chapter 7.

In the following sections, we first discuss contamination and degradation, and how to avoid them; followed by descriptions of apparatus and methods for collecting or extracting volatile and nonvolatile chemicals; and conclude with prechromatography, crude fractionation, and cleanup steps.

1.2. Contamination

Contamination can be a nagging and insidious problem, particularly when isolating trace amounts of a new compound with unknown characteristics. The problem is exacerbated when compounds have to be carried through a number of steps with potential for contamination at each stage. The importance of contamination varies, depending on the scale of the extraction (i.e., more important in submilligram- than gram-scale isolations) and extract utilization. For example, the chance of random contamination affecting bioassay results is slight, but not unknown. However, contamination can complicate identifications. It is less of a problem with methods such as gas chromatography-mass spectrometry (GC-MS) or coupled gas chromatography-Fourier transform infrared (GC-FTIR), where the method incorporates a separation step. However, methods such as nuclear magnetic resonance (NMR), direct-insertion-probe MS, or optical rotation, where the spectra obtained are characteristic of the whole sample, can be severely compromised. If contamination is suspected, the sample may need to be repurified, and/or the suspected contaminants can be identified by consulting a compilation of spectra of artifacts (e.g., Middleditch 1989). Many labs maintain their own collections of artifact spectra for this purpose. Generally, rigorous cleanliness and constant vigilance minimize most contamination.

Potential sources of contamination are numerous, particularly for inexperienced workers, who may be unaware of them. Common sources of contamination and how to avoid them, are listed in Table 1.1.

Contaminants in solvents and water deserve special discussion. One liter of

Table 1.1.   Common Contaminants, Sources of Contamination, and How to Avoid Them

Contaminant/source	Solution
Plasticizers and related substances leached out of tubing, squeeze bottles, stoppers, plastic bags, or containers	Use only glass, metal, or Teflon® equipment and vials. Use Teflon capliners and Teflon® tubing for gases.
Sample collection	Clean off all dirt and debris. Discard all diseased, parasitized, or damaged parts.
Contaminants in solvents and water	See text.
Lab glassware; rotary evaporators	Thoroughly clean all glassware, even disposable glassware. See text for cleaning methods.
Gases, compressed air	Compressed air in particular can be heavily contaminated. Clean by passage through charcoal scrubbers, made from, e.g., charcoal granules used for fish tank air sources. Lines and fittings should be metal, glass, or Teflon®.
Syringes and other sample introduction devices	Rinse plungers with solvent, suck solvent through syringe barrel with Teflon® spaghetti tubing connected to aspirator vacuum.
Chromatographic contamination	Filter with syringe filter to remove particulates from liquid chromatography (LC). GC column bleed cannot be prevented; possibly use a different, nonbleeding GC phase, or use LC.
Parafilm	Never use Parafilm®, a mixture of hydrocarbons readily soluble in organic solvents. Use aluminum foil for short-term or glassware with ground glass joints for longer term closures, with Teflon® joint sleeves if necessary.
Stopcock grease	Avoid wherever possible. Use separatory funnels with Teflon® instead of ground glass stopcocks. Use minimal or no grease for vacuum work. Remove grease by wiping off visible amounts, then wipe with hexane-soaked tissue and rinse with hexane.
Careless handling	One fingerprint contains many micrograms of skin oils etc. (Biedermann & Grob 1991; Grenacher & Guerin 1994). Never touch syringe needles or plungers, cap liners or vial necks, thin end of pipettes, etc. Use forceps for small parts such as cap liners.

solvent with 1 ppm contamination contains 1 mg of contaminant, which may be more material than the active compound being isolated. Solvent purity is readily checked by concentrating a few hundred milliliters to a few microliters, and screening for contaminants by GC, thin-layer chromatography (TLC), or spectral methods such as NMR. Use only high-grade solvent, or even better, distill solvents through a fractionating column before use. It is worth mentioning that careless attempts to further purify high-grade solvents may actually introduce more contaminants during handling than they remove!

It is not usually necessary to dry solvents used for extractions. Use due care when distilling very volatile and highly flammable solvents like ether and pentane. Distilled diethyl ether or tetrahydrofuran must be used rapidly because of their propensity to form potentially explosive peroxides over time. Similarly, chloroform should be used rapidly because of its tendency to form traces of phosgene and HCl; more stable methylene chloride can usually be substituted. Water, even distilled or highly purified water, has a nasty habit of developing microbial growth, and should be used quickly once purified. Purification of water by distillation may be problematic because of concurrent steam distillation of volatile impurities. It may be better to use water purified by passage through charcoal filters, as is done with high-pressure liquid chromatography (HPLC) water generators or MilliQ® type water purification systems (Waters). Once distilled, most solvents can be stored in brown glass bottles with Teflon® cap liners.

Above all, plan ahead before starting extractions and handling. Have enough (even extra!) clean glassware of appropriate size on hand, and minimize handling and transfers, because each transfer increases the chance of contamination.

1.2.1. Decontamination and Cleaning

Glassware for routine work can be adequately cleaned by soaking overnight with dilute residue-free detergent (e.g., Micro® liquid, Baxter Scientific), rinsing with distilled water and clean acetone or ethanol, and finally baking overnight at ~125°C. Stubborn deposits may require soaking in Chromerge® or similar acid baths, followed by the complete wash procedure. Sintered glass funnels and other fritted glassware can be cleaned by soaking in bleach, followed by thorough rinsing (H.G. Cutler, personal communication, 1997). As a final precaution, glassware can be rinsed immediately before use with the solvent to be used in extractions. In cases where extreme cleanliness is required, glassware can be baked overnight in a glassblower's annealing oven, which burns off all traces of organic material. Clean apparatuses can be wrapped in aluminum foil, and smaller pieces can be kept in sealed wide-mouthed jars.

1.3. Degradation

Many organic chemicals are readily degraded by environmental conditions or living organisms, and samples should be handled and stored accordingly. At the least, extracts should be stored sealed in a freezer. Potential degradation routes are numerous, depending on the structure and functional groups present. Some of the more common degradation routes and methods of minimizing them are as follows.

Some classes of compounds are subject to oxidation by air, often catalyzed by light. Common examples are the polymerization of phenolics (e.g., the browning of cut fruit surfaces), oxidation of aldehydes to acids, and oxidation of alkenes (e.g., fats turning rancid). In one interesting case, the parasitic wasp *Macrocentris grandii* apparently relies on air oxidation of cuticular hydrocarbons to form a pheromone component (**Z**4-tridecenal; Swedenborg & Jones 1992). Oxidation in extracts can be inhibited by addition of traces (0.1–1% of active compound) of stabilizers such as butylated hydroxytoluene (BHT) or butylated hydroxyanisole (BHA) to nonpolar extracts, or ascorbic acid (Bone and Bottjer 1984) to aqueous extracts. Extracts should be stored under nitrogen.

Thermal degradation can be minimized by avoiding heating during extraction or concentration. Degradation routes can include dehydration (e.g., of 3° alcohols), rearrangements (e.g., terpenes such as germacrenes and farnesenes, Hamilton et al. 1996; conjugated trienes, Doolittle et al. 1990), or polymerization. Problems are minimized by using more volatile solvents, working in cold rooms, and using alternative fractionation methods that do not require heat (e.g., LC instead of GC).

Problems with acid or base sensitivity can be circumvented by working at neutral pH, using buffers if necessary. Esters are particularly base sensitive, while acid sensitivity may be a problem with semiochemicals with extensive conjugation (Doolittle et al. 1990), or that undergo acid-catalyzed rearrangements (e.g., Hamilton et al. 1996). Watch out for small amounts of acid (e.g., acetic or trifluoroacetic acids) in recipes of solvent blends used for extraction and chromatography. Furthermore, silica gel is both slightly acidic and a dessicant, and there are numerous reports of acid-sensitive compounds degrading on silica (e.g., Hamilton et al. 1996). In fact, degradation of chromatographically purified compounds during storage is a common problem, and traces of silica are almost certainly a contributing factor.

Enzymatic degradation frequently occurs when tissues are disrupted. For example, nonvolatile glucosinolates are converted to volatile isothiocyanates when *Brassica* tissues are damaged. In this and other cases, enzymatic degradation can be prevented by grinding and extracting the sample at subzero temperatures (e.g., by grinding the tissues with liquid nitrogen, followed by extraction of the still-frozen powder with solvent), or by denaturing the enzymes, for example, by plunging the sample into boiling alcohol. Alternatively, dried sample can be extracted, since many enzymes require an aqueous environment to function.

Cis-trans isomerization of isolated or conjugated double bonds is the most common manifestation of sensitivity to light, carried to extremes in the visual pigment retinal. There are also a number of chemicals loosely classified as phototoxins that form reactive intermediates and act as photosensitizers on exposure to light (Heitz & Downum 1995). Examples include acetylenic thiophenes from Composite plants, and furanocoumarins from the Apiaceae. If light sensitivity is suspected, work should be carried out under dim red light (e.g., photographers safe light) using dark red or brown glassware.

Microbial degradation is usually a problem only with aqueous extracts, which, after all, may be rich in nutrients. Growth of microorganisms can be prevented by storing samples frozen or freeze-dried, or with a thin layer of organic solvent (e.g., toluene), provided the compounds of interest do not partition into the solvent. Heat or filter sterilization may be useful, but requires that all subsequent sample handling be done under sterile conditions.

Although not strictly degradation, very volatile semiochemicals may be lost through evaporation during handling and concentration. Losses can be minimized by keeping extracts cold, using volatile solvents such as pentane and ether, and removing solvent by fractional distillation rather than rotary evaporation.

As a general rule, extracts stored for extended periods should be monitored for decreases in activity and physical changes such as deposits or flocculence. All are signs that more stringent procedures and precautions may be required.

1.4. Sampling and Collection of Volatile Chemicals

Volatile semiochemicals constitute one of the most important methods of communication between terrestrial organisms. Odor signals are involved, for example, in development, reproduction, location of food and/or hosts, oviposition site location, territory marking, and defense. Historically, volatile semiochemicals were extracted by the same methods used for extraction of semi- and nonvolatiles, such as solvent extraction or steam distillation. These methods have the advantage that they provide relatively large quantities of material to work with, and extraction is relatively complete. However, they have a number of major disadvantages. First, the extracts are complex mixtures of volatile and nonvolatile compounds, and numerous fractionation steps may be required to isolate pure compounds. Second, the profile of volatiles obtained is often not representative of the blend released by the intact living organism (e.g., Teranishi et al. 1993; Lorbeer et al. 1984; Takeoka et al. 1988; Tollsten & Bergstrom 1988; Heath & Manukian 1992). Enzymes released by tissue maceration can rapidly change the chemical profile. For example, a characteristic odor chemical of garlic, allicin, is produced by enzymatic degradation of nonvolatile alliin. More generally, the "green leaf volatiles" Z3-hexenal, Z3-hexenol, and Z3-hexenyl acetate are produced by oxidation of fatty acids in disrupted plant tissues (e.g., Charron et al. 1996).

Third, heating (e.g., in steam distillation or Soxhlet extraction) can lead to degradation or rearrangements (see section 1.3). Finally, solvent removes compounds that may normally be sequestered by the intact organism, such as defensive compounds. These compounds can mask the activity of the compounds under study, and make bioassays difficult to interpret. Furthermore, solvent extraction often results in a substantial volume of solvent, which must be removed, and care must be taken to minimize loss of volatile semiochemicals during concentration.

Because of these drawbacks, most recent work has focused on collection of volatiles emitted into the airspace around a living organism, and a number of devices have been designed for this purpose. Organisms are usually isolated in an "aeration chamber" to minimize contamination, but field collections have also been carried out (e.g., collecting volatiles from flowers *in situ,* see below). Collection of volatiles by "aeration" has several advantages. First, it provides a more representative sample of the volatiles given off by an organism than solvent or other extracts. Second, the resulting extract is less complex than a solvent extract, and "clean" in that it contains only volatile components, and so can be analyzed directly by GC or GC-MS with no cleanup steps required. Third, the extract is recovered in a relatively small volume of solvent, so concentration is easier.

Several general considerations are pertinent when choosing volatiles sampling or collection methods. First, the amounts of material obtained are small, depending as they do on rate of release, volatility, and flow rate in the aeration setup. Normally, a few nanograms to milligrams of material are obtained per aeration. If only a few nanograms of material are required, for example, for quantification of a known compound, then straightforward sampling methods are satisfactory. However, if larger amounts of material are required, for example, for fractionation and bioassay or NMR analysis, then repeated collections may be required.

The method will also depend on the characteristics of the bioactive blend. If the blend of volatiles consists of nonpolar to intermediate-polarity compounds of roughly similar volatilities, most of the sampling or large-scale aeration setups described below will suffice. However, if the blend consists of compounds of widely differing polarities or volatilities, then the collection and analysis of truly representative samples becomes much more difficult. That is, commonly used adsorbents such as Porapak Q®, Tenax®, or activated charcoal, while excellent for trapping low- to medium-polarity compounds of intermediate size, are poor to useless for collection of low-molecular-weight and/or polar compounds such as ethanol, ammonia, and short-chain amines. This problem is discussed in more detail below. The fact that the blend of volatiles produced by most sampling and collection methods is quantitatively and qualitatively biased has been largely ignored and is undoubtedly partly responsible for slow progress in areas such as the attraction of insects to host plants, where very complex blends of volatiles may be involved.

Furthermore, volatile semiochemicals may be obscured under solvent peaks during GC analyses of extracts prepared by elution of trapping media. This problem may be circumvented by thermal desorbtion of the trapped volatiles (section 1.4.1.2). However, thermal desorbtion is also not without problems, including degradation of fragile compounds and production of artifacts from the trapping media.

A final word of caution: matrix effects on release of volatiles are critically important (Ebeler et al. 1988). When a volatiles blend is reconstructed based on results of an aeration analysis, and the reconstructed blend is put into/onto a release substrate (e.g., filter paper, into water, etc.), the profile of volatiles released may be substantially different from that measured initially due to interactions of the analytes with the release substrate. For example, when reconstructing aqueous blends of volatiles involved in attraction of mosquitoes to oviposition sites, we found that polar, water-soluble compounds were released much more slowly than anticipated, and nonpolar compounds were released up to 50 times faster than anticipated, so that the profile of the reconstructed blend bore little relationship to the true blend (Y. Du & J. G. Millar, unpublished data). The problem was readily corrected by several iterative adjustments. However, release rate adjustments will be different for different matrices and semiochemicals.

Several excellent reviews of methods and apparatuses for the collection of volatile semiochemicals have been published (e.g., Golub & Weatherston 1984; Blight 1990; Dobson 1991). In the discussion that follows, the focus is on detailed practical descriptions needed to set up and operate micro-, medium-, and large-scale aeration units, rather than on attempting a comprehensive survey of aeration methods.

1.4.1. Sampling Methods

In general, sampling methods use small closed systems rather than systems with airflow. The constraints and materials to use for construction of sample containment chambers are essentially the same as for aeration and collection systems (section 1.4.2).

1.4.1.1. Headspace Analysis

The simplest sampling method consists of direct sampling of headspace volatiles of a sample equilibrated, and sometimes heated, in a closed container. An aliquot of the headspace gases (~100–1000 µl or more) is removed with a gas-tight syringe and slowly injected directly into the GC, using splitless or on-column injection so that most of the volatiles are transferred to the column. Cryogenic cooling may be required to focus the injected volatiles as a narrow band at the head of the column. This method is practical only for compounds whose concentration and vapor pressure is sufficient to provide at least nanogram quantities in the sample aliquot. Direct headspace sampling has been used, for example, to determine concentrations of very volatile compounds such as ethylene (e.g., Kendall & Bjostad 1990; Campos et al. 1994) and CO_2 in headspace samples (L. Bjostad, personal communication, 1997). Commercial headspace autosamplers and applications literature are available (e.g., Hewlett-Packard).

1.4.1.2. Thermal Desorbtion and Purge and Trap

Simple modifications to a gas chromatograph (GC) allow the heating of small tissue samples (e.g., insect glands) in the GC injector, with the volatiles being swept onto the column with the carrier gas (review, Attygalle & Morgan 1988).

Alternatively, materials can be coated on a solids injection syringe (Deml & Dettner 1994). A refinement of this technique consists of collecting (or thermally desorbing) volatiles onto an adsorbent cartridge by aeration. The cartridge is then purged with dry inert gas to remove water, and thermally desorbed directly into the GC (short-path thermal desorbtion). A number of commercial devices are available (e.g., Supelco, Scientific Instrument Services, Perkin-Elmer, Carlo Erba, Tekmar), including sample tubes with specialized adsorbents for different applications. Extensive technical and applications literature is available from the suppliers. When the volatiles collection and the thermal desorbtion units are combined into a single instrument coupled to a GC or GC-MS system, the method is known as purge-and-trap analysis. It has been used extensively in water and food analysis, but applications to chemical ecology appear to be rare. It has been used, for example, to analyze attractants from ripe or fermenting fruit (Cossé et al. 1995; Cossé et al. 1994; Phelan & Lin 1991, using home-made equipment). Commercial units are available from the suppliers listed above.

Sample degradation and generation of artifacts need to be considered with all methods of thermal desorbtion. Artifacts can be generated by degradation of the sample, possibly catalyzed by the adsorbent being used, or from the adsorbent itself. Numerous studies have addressed these problems (e.g., Patt et al. 1988; Rothweiler et al. 1991; Teranishi et al. 1993; Calogirou et al. 1996). Artifacts produced from common adsorbents such as Porapak® and Tenax® are discussed in section 1.4.2.3.

1.4.1.3. Solid-Phase Microextraction

This recently developed sampling technique involves trapping volatiles on adsorbent-coated fibers followed by thermal desorbtion of the volatiles by insertion of the fiber directly into a GC injector. The apparatus is simple, consisting of a fused silica fiber, coated with various adsorbents, mounted on a modified GC syringe (Fig. 1.1). The sample (plants, live animals, aqueous solutions, etc.) is held in a container with a septum closure. The needle of the device pierces the septum, and the adsorbent fiber is extended, either into the headspace or, in

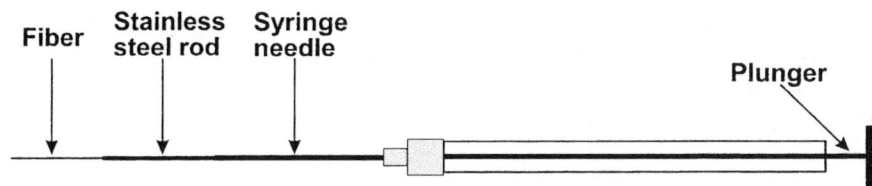

Figure 1.1. Solid-phase microextraction device. The adsorbent fiber is attached to the end of the stainless steel rod, which retracts into the syringe needle. The needle is used to pierce the septum of the sample vial or the GC septum, and the fiber is extended into the sample for adsorption or into the GC injection port for desorption, respectively.

the case of solutions, directly into the solution (both volatiles and nonvolatiles will be adsorbed from solutions). After an adsorbtion period of minutes to hours, the fiber is withdrawn into its needle sheath, the needle is inserted through the GC septum, and the loaded fiber is extended into the hot injector for thermal desorbtion of the adsorbed volatiles. Each fiber is reusable many times.

Solid-phase microextraction (SPME) devices are available (Supelco, Bellefonte, PA) with interchangeable fibers coated with different thicknesses (7–100 µm) of several adsorbents (polydimethylsiloxane, polyacrylate, Carbowax/divinylbenzene, Carboxen™-polydimethylsiloxane) similar to GC stationary phases. The phases differ in their adsorbtion of different classes of compounds, enabling the efficient sampling of compounds of differing polarities and volatilities. New 100-µm polydimethylsiloxane fibers are efficiently cleaned by desorbtion at 200°C for 5 h in a clean inert gas stream (Bartelt 1997); cleaning of other fiber types is described in the product literature. Technical bulletins describing use and applications of SPME are available (Supelco).

Several studies have delineated advantages and limitations of the SPME method (Bartelt 1997; Matich et al. 1996; Schäfer et al. 1995; Yang & Peppard 1995), as follows. First, the fiber must be exposed to headspace volatiles long enough to reach equilibrium to obtain reproducible quantitative results, with equilibration time inversely proportional to volatility. Second, adsorbtion of compounds is critically dependent on molecular size and polarity or functional group(s). For polydimethylsiloxane, larger molecules are adsorbed several orders of magnitude more effectively than small, very volatile compounds. Conversely, at the time of writing, a new type of fiber (Carboxen™-polydimethylsiloxane) designed specifically for sampling of gases and other low-molecular-weight analytes had just been released. Thus, calibration with appropriate standard mixtures covering the entire range of interest is crucial for quantitative work.

Third, adsorbtion is inversely dependent on temperature. When sampling the headspace above a sample, however, some heating may be required to increase the headspace concentrations of the volatiles being sampled. Finally, there appears to be a slight dependence of adsorbtion on concentration, with lower concentrations of analytes being adsorbed slightly better than higher concentrations (Bartelt 1997).

It also must be realized that SPME provides only enough material for GC, GC-MS, and possibly GC-FTIR, and when attempting to analyze trace components, the sensitivity may not be good enough even for some GC-based methods. Thus, SPME's primary value is as a sampling tool for compounds of known structure, or for compounds with structures simple enough to be determined by GC-linked methods.

Despite these limitations, SPME has several major attractions. It is rapid and simple, enabling sampling of virtually any odor source. Sequential samples can be taken continuously with a pair of devices, with the proviso that compounds that are not being produced continuously may become depleted (Bartelt 1997).

Furthermore, samples are free of solvent, so that volatile compounds are not obscured under a solvent front. SPME can also be used to sample compounds spanning a wide range of polarities and volatilities, apparently even including compounds such as ammonia and volatile amines (Robacker & Flath 1995; Robacker & Bartelt 1996), which can be difficult to sample reliably by other methods.

Several techniques may be required to obtain good chromatographic resolution and symmetrical peaks, particularly as there is no solvent effect to concentrate desorbed compounds at the head of the column (see chapter 2). Focusing of the analytes at the column head can be accomplished only by temperature and stationary phase effects (Schäfer et al. 1995). Useful techniques include cryogenic focusing of desorbed volatiles at the head of the column; use of narrow bore, small-volume injector liners (0.75 mm i.d.; Supelco); and use of thick-film capillary columns, particularly for more volatile compounds (Langenfeld et al. 1996).

In addition to numerous applications in food, water, and environmental analyses, several applications of SPME to chemical ecology have appeared. For example, SPME was used to sample ammonia and volatile amines from fruit fly attractants (Robacker & Flath 1995; Robacker & Bartelt 1996). SPME has also been used to quantify the daily rhythms of pheromone emission from sugarcane weevils (Malosse et al. 1995), in headspace sampling of plant volatiles (Matich et al. 1996; Schäfer et al. 1995), and in the identification of lepidopteran sex pheromones (Borg-Karlson & Mozuraitis 1996; Mozuraitis et al. 1996). In the latter case, SPME collected volatiles from one calling female equaled the amount obtained from an extract of 50 pheromone glands. Pheromones of brown algae have also been sampled from seawater by SPME (Maier et al. 1996). Overall, there is no question that the use of SPME will increase, particularly as the sensitivity of analytical methods such as GC-MS continues to improve. However, the reader is again cautioned that, as with most other methods of sampling volatiles, the profile of adsorbed volatiles does not represent the true proportions of the compounds in the headspace, and calibration is crucial.

1.4.2. Volatiles Collection Apparatus

Aeration setups should be prepared whenever possible from only glass, metal, Teflon®, or in some cases polycarbonate or Plexiglas® pieces. A typical setup is shown in Figure 1.2. Rubber, plastics, glues and adhesives, wood, and other building materials should be avoided since they bleed and/or retain volatiles. All parts should be cleaned, and metal and glass parts oven baked overnight (~125°C) before use. Connections can be made with ground-glass joints (ball and socket joints are particularly good because they allow some motion), firmly clamped. Tubes can be connected with brass Swagelok® unions with Teflon® ferrules (from most valve and fittings companies). Connections through solid surfaces such as screw-cap lids can be made by drilling holes of appropriate sizes to accommodate

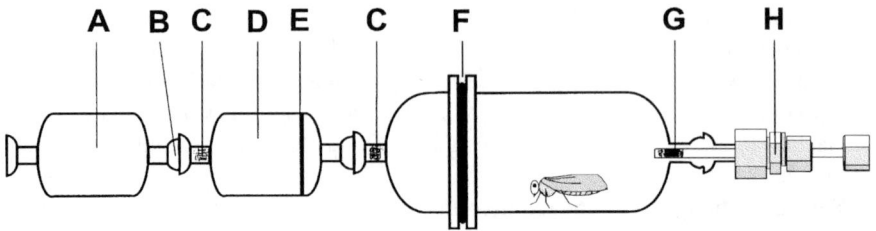

A B C D E   C   F   G   H

Figure 1.2. Typical apparatus for aeration of small organisms: *A;* glass humidifier filled with water-soaked glass wool; *B;* ground glass ball and socket joints; *C;* glass wool plug; *D;* air cleaner packed with activated charcoal granules; *E;* coarse glass frit; *F;* two-part glass aeration chamber with O-ring seal in center. The two pieces are clamped together; G, volatiles collector, made from a glass tube with a short bed of charcoal, Super Q,® or Tenax® adsorbent, held in place with glass wool plugs; H; brass Swagelock reducing union, with Teflon® ferrules.

Swagelok® bulk-head unions. For small setups, Teflon® tubing connected to syringe needles with Teflon® heat-shrink tubing can be used. If necessary, Teflon®-faced septa punctured by syringe needles may be used, but watch for traces of septum components. Rubber or plastic tubing may be used downstream of the collection apparatus, for example, to connect the apparatus to a vacuum source (pump or water aspirator). With a little ingenuity, apparatuses of appropriate sizes can often be assembled from commonly available metal or glass containers. If large numbers of aerations are anticipated, it may be worthwhile to have some pieces custom built. A more detailed description of parts follows.

1.4.2.1. Air or Nitrogen Supplies

Clean air is usually used, although nitrogen or other inert gases can be used if anaerobic conditions are required, bearing in mind that volatiles produced under aerobic and anaerobic conditions may differ (e.g., Hamilton-Kemp et al. 1989). Suitable air sources include higher grades of compressed air (e.g., so-called "medical air"), or even room air supplied by an oil-free pump (e.g., a diaphragm aquarium pump, from pet shops). Alternatively, air can be pulled through the system by vacuum; in "push-pull" systems, both pressure and vacuum are used (e.g., Heath & Manukian 1994). In any case, care should be taken to prevent overpressuring the system or creating a partial vacuum, particularly when aerating live organisms. Either case is readily created by the resistance to air flow created by the charcoal air scrubber on the inlet side, or the volatiles collection trap on the outlet end (see below).

More recently, "open" systems for the aeration of undisturbed intact plants have been developed (Heath & Manukian 1992; Loughrin et al. 1994). In these systems, a large volume of clean air is passed over a plant in an open-ended sleeve, with a small portion of the airstream being sampled.

Inlet air from any source must be cleaned with an activated charcoal filter. For the large scale, a piece of metal pipe or an old metal solvent can is filled with activated charcoal granules (e.g. for cleaning aquarium air supplies, or Alfa Products, Danvers, MA). Medium- to microsized filters are prepared from glass tubes, with the charcoal held in place with small wool plugs, or coarse glass frits or metal screens permanently sealed in place. For example, for aerations of individual insects in 20-ml vials, we use a filter made from a disposable pipette filled with 5 cm of charcoal granules.

For extended periods of aeration, particularly for organisms that desiccate rapidly, inlet air should be passed through a humidifier such as a water-filled gas-washing bottle (from any glassware supplier) or a tube loosely packed with water-soaked glass wool, placed in front of the charcoal scrubber. In both cases, watch for microbial growth in the water.

1.4.2.2. Aeration Chambers

Aeration setups vary in size from small vials (e.g., multiple simultaneous aerations, Chang et al. 1989) to large chambers. Small chambers are usually glass (polycarbonate also may be used; H.D. Pierce, Jr., personal communication, 1996). Typically, wide-mouthed containers with ground-glass joints (e.g., vacuum desiccators, wide-mouthed Erlenmeyer flasks, etc.) or Teflon-lined screw-on lids are used. Custom-made cylindrical flow-through chambers that open in the middle or at one end, with a securely clamped, O-ring-sealed middle joint, and ground-glass ball and socket joints at either end for connections (Fig. 1.2) are also excellent. Animals often like to have something to climb or perch on, so clean screening can be inserted.

Larger chambers can be somewhat problematic. Chambers can be made from steam-cleaned, solvent-washed 55-gal drums (Browne et al. 1979; Millar et al. 1986), with the original outlet adapted to accept a charcoal filter. A large hole, closed with a bolted-on metal plate sealed with a Teflon® sheet gasket, with an outlet for the collection tube, is cut in the bottom for loading.

More recently, sophisticated setups for the aeration of intact plants have been developed, incorporating techniques for purification of large volumes of inlet air, and computer-controlled sequential sampling and monitoring of environmental conditions (Heath & Manukian 1992, 1994; Heath et al. 1992). These systems are now commercially available in custom designs (A. Manukian, Analytical Research Systems, Gainesville, FL).

Sampling live organisms in the field has also been carried out on small and large scales. For example, polyvinyl fluoride (Pham-Delegue et al. 1989) or polyacetate cooking bags (Dobson et al. 1987; Dobson 1991) have been used to bag flower heads for sampling. On a larger scale, aluminum (Browne et al. 1979) or polyvinyl fluoride sheets may be used (e.g., we collected volatiles from trunks of living trees by "bagging" sections with Tedlar® fabric, from Dupont; J.G.

Millar, L.M. Hanks, and T.D. Paine, unpublished data). Due to difficulties in sealing these large bags, a push-pull system may be used in which clean air is pumped in and only a portion of the output is sampled, so the system is under slight positive pressure.

Field sampling of odor sources has also been carried out in the open air, for example, with flowers (Burger et al. 1988). Small, battery-operated personal air sampling pumps with flows of several liters per minute (e.g., Aldrich Safety Products, Sensidyne), used for air sampling in industrial settings, are convenient for field applications. This can be a considerable advantage because volatiles released by organisms in their natural settings may differ from those released in laboratory settings. Obviously, background volatiles may be a problem, and samples should be taken close to the source.

1.4.2.3. Adsorbent Traps for Collecting Volatiles

When nanogram to multimilligram samples are required for bioassays, fractionations, or other manipulations, volatiles are collected and extracted, usually by elution of an adsorbent with solvent rather than thermal desorbtion. Traps usually consist of glass or metal tubes filled with a bed of granulated adsorbent (~40–80 mesh, to compromise between high surface area and resistance to air flow), held in place with glass wool plugs. A variety of adsorbents have been used, but this discussion will focus on three of the most common and versatile, Porapak Q® (and its refined version, Super Q®; Waters, Supelco), Tenax GC® (refined version, Tenax TA®; Chrompack, Alltech), and activated charcoal (most large lab suppliers).

Porapak® is an ethylvinylbenzene-divinylbenzene copolymer, and Tenax® is a 2,6-diphenyl-p-phenylene oxide polymer. For many applications, either should work equally well. Both Porapak® and Tenax® have a low affinity for water, and a high affinity for lipophylic organic compounds. Furthermore, when working with humid samples, residual water may be removed from traps by a short flush with dry N_2 before elution (Helmig & Vierling 1995). Tenax® has higher thermal stability than Porapak®, but this may not be an issue if traps are not thermally desorbed. Porapak® is claimed to have a higher capacity for small molecules than Tenax® (Williams et al. 1978), but for most applications, either adsorbent can probably be used interchangeably (Schaefer 1981).

Porapak® and Tenax® adsorbents need conditioning before use. The problem is particularly acute with Porapak Q® and Tenax GC®, and it is strongly recommended that the newer, cleaner versions, Super Q® and Tenax TA®, be used. Porapak® is typically conditioned by heating in a glass tube at 200°C overnight under a slow flow of clean N_2, followed by Soxhlet extraction of the cooled adsorbent with pentane, methylene chloride, or ether (Byrne et al. 1975; Bjostad et al. 1980). The extracted material is then packed into sampling tubes, and warmed to 60°C for 2 h while flushing with clean N_2 to remove residual solvent.

Alternative reported treatments include flushing with hot solvent (CAUTION!; Chang et al. 1989). Tenax® is conditioned by heating as described above, except that somewhat higher temperatures (270–340°C for 2–24 h; Lewis et al. 1988; MacLeod & Ames 1986; Rothweiler et al. 1991) are used in the initial step. Detailed conditioning instructions are provided in the Alltech and Scientific Instrument Services catalogs, which also recommend that Tenax® not be conditioned with chlorinated or aromatic solvents. Once conditioned, both adsorbents should be stored in sealed tubes and protected from light until used.

Aromatic ketones, alcohols, and other artifacts have been identified from solvent elution of Porapak® (Sturaro et al. 1992; Lewis and Williams 1980), and particularly by thermal desorbtion (Krumperman 1972; Lewis & Williams 1980). Similar artifacts from Tenax® adsorbents have also been identified (MacLeod and Ames 1986), including artifacts produced by oxidation and light (Pellizzari and Krost 1984; Peters et al. 1994). Running a system blank (i.e., aeration of an empty setup) is always recommended to identify artifacts.

Trapped volatiles are recovered from either adsorbent by elution (5–10 ml/g adsorbent) or Soxhlet extraction with low-boiling solvents such as pentane or ether (methylene chloride may be used with Porapak $Q^®$/Super $Q^®$). CS_2 may be required to elute polar compounds efficiently, but it is not generally recommended because of its toxicity and instability. Extracts may be concentrated by passive evaporation from open vials in a fumehood, blow-down under a gentle stream of clean N_2, or for larger volumes, distillation of most of the solvent with a fractionating column packed with glass beads or helices, or a Snyder or Dufton column. Eluted sample tubes are reconditioned by further rinsing with clean solvent, while Soxhlet-extracted material is usually ready for reuse after a second extraction with ether. Residual solvent is flushed off with N_2, as described above.

The amount of adsorbent to use depends on factors such as the polarity and volatility of analytes, flow rate, volume of air to be sampled, and temperature. However, virtually all collections are made at ambient conditions, so temperature is not usually an issue. It often is not appreciated that volatile and/or less hydrophobic compounds are retained relatively poorly and can break through traps quite quickly, resulting in erroneous quantitation. Breakthrough volumes per gram of adsorbent for compounds of different classes have been compiled (e.g., Manura 1994; Maier and Fieber 1988) and are available from adsorbent suppliers (e.g., Scientific Instrument Services). In practice, the simplest way to check for breakthrough is to put two traps in series; if the second trap collects any compounds, breakthrough is occurring.

For many applications, volatiles from tens to hundreds of liters of air can be trapped on a few milligrams to a gram or so of adsorbent packed in tubes, in beds of ~5–50 mm in length, held in place with small glass wool plugs; longer beds result in greater resistance to flow. The amount of adsorbent should be kept small to minimize artifacts and the volume of solvent required for elution. For higher flow rates or larger sampling volumes (times), trap sizes should be increased accordingly. With any setup, to be safe, breakthrough experiments should be performed.

Activated carbon (activated coconut charcoal) is also an excellent adsorbent, and may be the adsorbent of choice for many applications. It is highly retentive, so very small beds can be used, has a high capacity, is easily cleaned, and unlike Super Q® and Tenax TA®, it is cheap (Aldrich, Acros Organics). Disposable traps (to eliminate chances of cross-contamination) are readily fabricated from glass tubes filled with a short bed of a few milligrams. A wide, thin bed of a few milligrams held in a stainless steel Luerlock filter holder has also been used (Aldrich et al. 1994, and references therein). In our experience, charcoal is readily cleaned in bulk by heating under clean N_2 overnight at 250°C, after which further solvent cleanup is rarely necessary. Other researchers have recommended conditioning with exhaustive solvent stripping with, e.g., hot EtOH, CH_2Cl_2, and CS_2 (Matile & Altenburger 1988), or CH_2Cl_2 and pentane (Grob & Zurcher 1976). Loaded traps are eluted with pentane, methylene chloride, or CS_2 (~5–20 volumes). Alternatively, small reusable traps can be fabricated by permanently sealing a few milligrams of charcoal between stainless steel screens in glass tubes (Heath et al. 1991; Tumlinson et al. 1982; Grob and Zurcher 1976; Hollingdale-Smith 1975). These are also commercially available (Brechbühler AG, Urdorf Switzerland; U.S. distributor, Chromapon Inc., Whittier, CA).

1.4.3. Miscellaneous Methods

1.4.3.1. Trapping on Glass

Small amounts of insect pheromones have been efficiently collected on a few glass beads (Charlton & Carde 1982) or small plugs of glass wool (Haynes & Hunt 1990; Haynes et al. 1983; Baker et al. 1981), in devices such as that shown in Figure 1.3; in our modification, all of the parts of this device are commercially available. Volatiles have also been trapped in glass capillaries (Witzgall et al. 1996: Shani 1990; Witzgall & Frérot 1989). With all three methods, compounds are eluted with a minimum volume of solvent, and the cleaned glass materials give minimal background.

1.4.3.2. Closed-Loop Stripping

In addition to open, flow-through systems, closed-loop-stripping systems, adapted from water-sampling methods (Grob & Zurcher 1976) have been used. Air is circulated around a closed loop and through a volatiles trap with a diaphragm pump. These systems have been used in studies of plant metabolism (Boland et al. 1984) and pheromone release (Bestmann et al. 1988). These systems have the advantage that because the air is recirculated, chances of contamination during a single run are decreased. However, care should be taken to eliminate "memory" effects due to adsorbtion of compounds to system components.

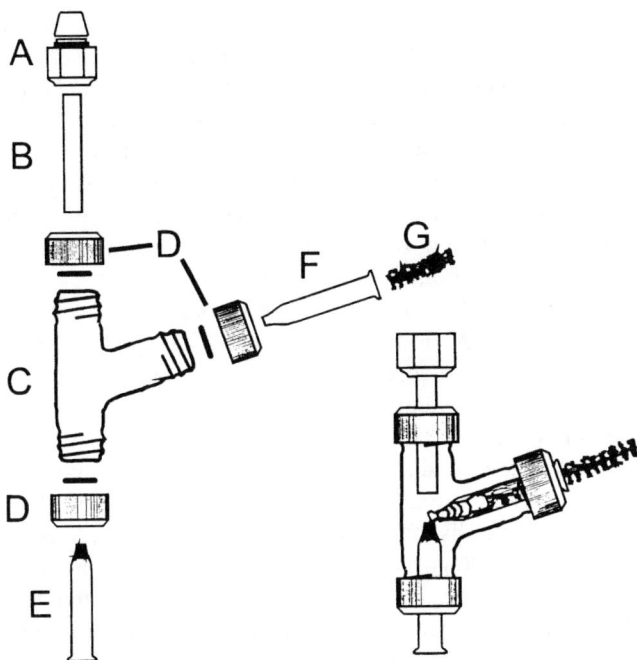

Figure 1.3. Device for aeration of extruded pheromone glands of individual moths. All parts are commercially available. *A* and *B*, connection to clean gas supply; *C*, connecting adaptor (Kimble-Kontes # 747120-0000); *D*, connector with O-ring seal; *E*, conical vial insert with bottom cut off, packed with small plug of glass wool; *F*, conical vial insert with bottom cutoff; moth is inserted abdomen-first, and pushed into place with pipecleaner *G*.

1.4.3.3. Cryogenic Trapping

Cryogenic trapping of volatiles, for example, by condensing volatiles-laden air in liquid-nitrogen-cooled traps, has found occasional use (e.g., Browne et al. 1979). The trap is warmed slightly to distill off the condensed air, leaving a volatiles concentrate. This method, unlike many others, can trap compounds with widely differing volatilities and polarities, including gases like ethylene and CO_2. However, it is not generally recommended as the method of first choice because of the large volumes of water also trapped, from which volatiles must be subsequently extracted or otherwise separated. Formation of ice plugs can also cause hazards. This method has been used successfully for trapping and assaying very volatile compounds such as CO_2 (Hibbard & Bjostad 1988, 1989).

1.5. General Methods of Extraction

Because of the diverse range of polarities and solubilities possible with unknown compounds, there is no perfect solvent or extraction method that can be used for all situations. Unfortunately, few papers describing isolation of new natural products discuss why a particular method of extraction was used, and details of the extraction and subsequent fractionation are often sketchy, at best. In addition, many isolations are a one time effort to obtain enough compound to identify or screen for biological activity, with minimal subsequent effort to maximize yield. Method optimization is often put off until an analytical method for multiple samples is required. This can have adverse consequences; for example, Nakanishi's (1989) determination of the structure of azadirachtin was hindered for 10 years by an improper isolation method.

To get an impression of the methods most commonly used, techniques published in the *Journal of Natural Products* were surveyed for 1 year. Of the 243 papers examined, 62% concerned plants, 24% concerned animals, and the remainder covered microorganisms. Sample pretreatment favored drying and powdering (60%), while 27% of the papers reported using fresh or frozen materials, and 13% used freeze-dried samples. Choice of solvent heavily favored alcohols, with or without added water or $CHCl_3/CH_2Cl_2$ (62%), with 16% using two to four solvents sequentially. Percolation or soaking (88%) was the preferred method of extraction, while postextraction treatments were split equally between solvent partitioning and going directly to some form of chromatography.

Despite these general trends, each case should be considered on its own merits, and the choice of solvent and extraction conditions should be based, at least in part, on the information available about the chemical and biological properties of the material of interest. For example, if evidence suggests that the compounds of interest appear to be in leaf cuticular waxes, then a quick dip in nonpolar solvent may prove to be a more appropriate extraction technique than exhaustive extraction of dried and ground leaf material. The pros and cons of various techniques are the subject of the following sections.

It may be useful to take the NMR spectrum of the crude extract and of each fraction as the fractionation progresses, for several reasons. First, signals characteristic of various functional groups may be recognizable, giving a general feel for which types of compounds are present. Second, the appearance in a fraction spectrum of peaks that were not present in the crude extract spectrum can alert you to possible degradation or contamination. Third, comparison of each fraction with the starting material for a fractionation step provides some idea as to how well the separation is proceeding.

1.5.1. Pretreatment of Sample

Every effort should be made to collect a single species or part of an organism (e.g., leaves, stems, roots, fruit, etc.) in the field, rather than trying to separate unwanted material in the lab. Specimens should be free from dirt and debris, and diseased, parasitized, or other potentially contaminated parts should be discarded, because pathogens both alter the metabolism of their hosts and produce their own unique compounds.

It is often useful to dry or freeze-dry materials before extraction, for several reasons. First, tissues can consist of 90% or more water, so that drying cuts the mass of material to be extracted considerably. Second, dried material can be ground to a powder for more efficient extraction. Third, blending of fresh tissues with solvent often results in intractable emulsions, which are difficult to process expeditiously.

Plant material can usually be air-dried or oven-dried at low temperatures (50–60°C, to prevent degradation, or loss of volatile compounds). To be sure, extracts of samples of fresh and dried material can be compared for biological activity and/or any quantitative or qualitative differences. This problem is obviously most likely in the study of volatile compounds such as monoterpenes. Animals and animal tissues with their greater propensity to decompose rather than dry, are usually chilled or frozen in the field, then homogenized and extracted, or freeze-dried. Whenever freeze-drying is used, the trap contents should be saved until it is certain that the condensate contains no active material.

Organisms such as fungi, bacteria, or algae cultured on solid or semisolid media are usually extracted in their entirety. When cultured in liquid medium, the culture medium and the tissue should be extracted separately because compounds may be selectively excreted into the culture medium or retained in the tissues. Liquid cultures can be centrifuged to separate the cells from the medium, or for larger organisms, the tissues are separated by filtration through coarse media such as cheese cloth. The liquid medium is directly extracted, or extracted after concentration by rotary evaporation. The tissues can be homogenized and extracted, or as described above, freeze-dried to make extraction easier.

If looking specifically for alkaloids, the material is often pretreated with base (e.g. 0.1-M NH_4OH or Na_2CO_3) to ensure that the alkaloids are present in their organic-soluble, neutral forms. However, even these mild bases can cause degradation (e.g., Laus & Kepplinger 1994).

1.5.2. Extraction of Dry and/or Powdered Samples

When starting the isolation of a compound with unknown properties, material is often extracted with a series of solvents, usually of increasing polarity, so that compounds are roughly segregated into solubility classes. For example, material can be extracted sequentially with hexane to extract non-polar compounds such as waxes, fats, and hydrocarbons, followed by CH_2Cl_2, ethyl acetate or acetone, ethanol or acetonitrile, and finally water to remove the most polar and ionic compounds. This list of solvents is by no means hard and fast; solvents are chosen both for their solvating power and for other desirable properties, such as low boiling points (to facilitate removal), ease of purification, low reactivity (no production of artifacts by reacting with sample compounds), cost, and not least, safety. Thus, solvents such as benzene and chloroform are used less nowadays because of health hazards. Diethyl and other dialkyl ethers are often avoided because of their extreme flammability and their tendency to form explosive peroxides and are replaced by the less hazardous *t*-butyl methyl ether. If alcohols are used in extractions, they will occasionally react with compounds; if alkoxy substituents show up on the molecules of interest this can be investigated.

Materials can be extracted by soaking or percolation of solvent through them, or by continuous extraction in a Soxhlet extractor. Soaking and percolation are usually done at room temperature or cooler, or by fast extraction with hot solvent; for extended periods of soaking, protection from light is desirable. Material is often first "defatted" by soaking with a nonpolar solvent to remove nonpolar lipids, before changing to a polar solvent like methanol to finish the extraction. Both extracts must be checked for activity.

Soxhlet extraction, in which refluxing solvent is leached many times through a sample, is highly efficient (because the sample is extracted many times) and provides an extract in a relatively small volume of solvent. The apparatus can run unattended for days, under an inert atmosphere if necessary to avoid oxidation. Because the concentrated extract is boiled for extended periods, however, thermal degradation may be a significant problem, and high-boiling-point solvents should be used with caution. For example, Queiroz et al. (1996) described a class of compounds extracted by soaking that were destroyed by Soxhlet extraction.

If a single-step extraction is desired, instead of sequentially extracting with solvents of increasing polarity, blends of solvents such as chloroform-methanol (9:1 to 2:1) are often used (e.g., Hemming & Hawthorne 1996; Certic et al. 1996). This combination disrupts membranes and releases chloroform-soluble compounds from inside the cells. *sec*-Butanol has also been recommended (Martinson & Plumley 1995). Honeycutt et al. (1995) and Brooks et al. (1996) have compared several extraction methods.

1.5.3. Extraction of Fresh or Frozen Samples

Fresh or frozen samples are used with tissues that do not dry well (e.g., animal tissues), where chemical changes may occur as the tissue dies or dries, or with small samples. Samples may be extracted by dipping, soaking, homogenizing in solvent, or Soxhlet extraction. The method chosen depends on several factors. For example, dipping (in hexane, chloroform, or ethyl acetate, for example) is best for selective removal of surface components such as cuticular chemicals from whole organisms or parts (e.g., leaves) providing an extract contaminated with few internal tissue constituents (e.g., Stammitti et al. 1996). The other three methods involve more thorough extraction, which has both advantages and disadvantages; more thorough extractions may provide higher yields, but they also provide larger amounts of extraneous material that must be separated from the compound(s) of interest in later steps. A judicious compromise is often useful.

For example, small tissue samples or glands can often be extracted adequately by soaking for a few minutes to hours, without extracting too much lipid material, while whole organisms may need soaking for much longer (days to months). Furthermore, the extraction is solvent dependent, with solvents such as alcohols penetrating memebranes more effectively than less polar solvents and consequently resulting in greater extraction of cellular contents.

Small samples can be homogenized with ground glass tissue grinders, or quick-frozen in liquid nitrogen followed by grinding in a chilled mortar and pestle. Larger samples are chopped or homogenized with solvent in blenders or motorized homogenizers, using nonflammable solvents unless the apparatus is explosion-proof. The aqueous homogenate can be intractable to work with due to its thick, soupy texture and the tendency to form emulsions, and water may be added to thin it out. Centrifugation or filtration may be required to separate the extract from the residues.

When emulsions do form, it is usually pointless to add more solvent in hopes of the emulsion breaking; only a larger volume of emulsion results. Methods of breaking emulsions are listed in Table 1.2. In general, prevention is the best approach. The extraction should first be tried on a small scale. Also, when starting an extraction in a separatory funnel, the mixture should be swished gently, rather than shaken vigorously. If the mixture shows signs of emulsifying, it should be first extracted gently by swishing with several aliquots of solvent which often extracts the emulsifying agents, allowing the extraction to proceed more vigorously. Chilling the mixture to 2–4°C during extraction may also be helpful. Alternatively, a different solvent can be tried.

Table 1.2. Methods for Breaking Emulsions

Method	Comments
Wait	Emulsions sometimes break; let stand 1 h; if signs of breaking, let stand overnight.
Add polar solvent	A small amount of methanol may help break emulsion. If not, do not keep adding methanol.
Salt out	Add saturated brine. This changes density and polarity of aqueous phase, and partition ratio will be changed.
Filter	Filter through sintered glass funnel or plug of Cellite® in a Buchner funnel, with vacuum.
Freeze	Chilling may induce separation, or it may simply freeze the emulsion like ice cream.
Evaporate	Rotary evaporate off solvent, watching for bumping. Try another solvent.
Centrifuge	
Alter pH	Partitioning may change with pH, as compounds change from ionic to neutral or vice versa.
Sonicate	Immerse flask for a few minutes in sonication bath. Sonicatation for long periods may cause reactions.

1.5.4. Extraction of Aqueous Samples and Culture Broths

Marine organisms, or organisms grown in aqueous culture media, often exude biologically active compounds into their aqueous environment. These compounds can be extracted in several ways. Relatively concentrated samples such as culture filtrates can be extracted with solvents of increasing polarity in separatory funnels. The pH of the filtrate may need to be adjusted before starting to ensure that the desired classes of compounds are present in neutral form and will be partitioned into the extracting solvent. If the properties of the biologically active material are not known, sequential adjustments of pH followed by extraction may be necessary. For example, the pH can first be adjusted to 3 with 1-M HCl, followed by extraction with a water-immiscible organic solvent to remove a fraction containing neutral, acidic, and phenolic compounds. Alternatively, the aqueous phase can be sequentially extracted with a series of solvents of increasing polarity, such as hexane, ether, ethyl acetate, and n-butanol. The pH of the aqueous residue is then adjusted to 11 to 12 with NH_4OH or NaOH, and the alkaloids and other amines are extracted as their neutral forms with one or more organic solvents as described above. The aqueous residue contains polar, water-soluble compounds such as sugars and inorganic salts.

For dilute aqueous solutions, where solvent extraction may be impractical due to the large volume requiring extraction, hydrophobic polystyrene-divinyl benzene resin beads, such as the Amberlite® XAD and related resins (Supelco; applications bibliographies available) or the Bio-Beads® SM resins (Biorad, Richmond, CA) may be used to adsorb and concentrate the organic compounds. Both these materials may be preferable to activated charcoal for this purpose because recovery of compounds from charcoal can be problematic. The resins have a high capacity for lipophilic compounds (>100 mg/g dry resin), so hundreds of liters of dilute solution can be extracted with quite small volumes of resin. Before use, the resins must be stripped of impurities (wash several times with MeOH, then Soxhlet extract sequentially for 8 h each with MeOH, acetonitrile, and ether; Dressler 1979; Loomis et al. 1979). The cleaned resin is then washed several times with water before use, or it can be stored in MeOH. Precleaned Amberlite® XAD resin is now available under the names Supelpak® 2 and Supelpak®-2B (Supelco; literature available).

The aqueous solution to be extracted is flowed or pumped through a bed (5–25 cm) of the resin packed in water (or slurried with resin for batch processes), and the bed is rinsed with distilled water. The adsorbtion of compounds with marginal lipophilicity can be increased by addition of salt (1–5%) to the solution before running through the column (Colegate & Molyneux 1993). For example, filtered seawater or aquarium water can be pumped through a resin bed for long periods, with the trace lipophilic analytes being continuously trapped (Rittschof et al. 1984). The trapped compounds are eluted with water, mixtures of water with solvent (e.g., methanol, ethanol, acetone, tetrahydrofuran [THF]) or the pure solvents, depending on the lipophilicity of the compounds of interest. During elution, the column must not run dry, and degassed water and solvents should be used to minimize the formation of air bubbles, which disrupt the resin bed, leading to uneven elution. Elution with stepwise gradients of increasing amounts of organic solvents will give some fractionation; alternatively, the solutes can be eluted rapidly in a small volume by switching directly to 100% organic solvent. In addition, the resins swell on exposure to organic solvents, so they must not be tightly confined in glass columns during abrupt changes in eluting solvent (going directly from water to 100% MeOH).

For some applications with limited sample volumes, it may be possible to concentrate analytes from aqueous solutions with recently developed adsorbent disks (e.g., Empore® disks, 3M). These are described in more detail in section 1.6.6.

If the compounds of interest are acids or bases, ion-exchange columns may be used to extract and fractionate them (section 2.6).

1.5.5. Drying Agents

Crude extracts, even in hydrocarbon solvents, may contain trace amounts of water, which can be detrimental to chromatography. Large-scale extracts are normally dried by treatment with anhydrous Na_2SO_4 or $MgSO_4$ granules or powder. The former is preferred because it is less reactive and has a higher capacity, even though it is not as efficient a drying agent. More polar solvents such as ether and ethyl acetate dissolve significant amounts of water, and extracts can be backwashed with saturated brine to remove the bulk of the water before adding drying agent. Extracts should not be stored over drying agents for extended periods. Other drying agents, such as calcium chloride and molecular sieves, are not recommended.

1.5.6. Concentration of Extracts

Medium- to large-volume solvent extracts of semi- or nonvolatile compounds are routinely concentrated on a rotary evaporator, using water aspirator vacuum and the minimum amount of warming required to remove the solvent (<50°C). Extracts of volatiles are more problematic. They can be concentrated by removal of most of the solvent by distillation through a fractionating column (e.g., Vigreux, Snyder, Dufton columns, or column packed with glass Raushig rings; most glassware suppliers). It is advantageous to use a low-boiling solvent like pentane or ether, which makes removal easier, decreases the chances of losing volatile analytes, and minimizes thermal degradation. Small to microscale extracts can be concentrated simply by gently blowing off the solvent with clean N_2.

Aqueous extracts of semi- to nonvolatile compounds are readily concentrated by freeze-drying, which also has the advantage that it produces powdery or glassy material, whereas rotary evaporation often produces syrups or gums, which can be a nuisance to remove from the flask and handle. Furthermore, freeze-drying minimizes the possibility of microbial degradation, which is always a consideration with dilute aqueous extracts.

1.6. Preliminary Fractionation and Cleanup

1.6.1. Fractionation Principles

Extracts can be fractionated based on molecular size (dialysis; size exclusion chromatography, chapter 2), volatility (distillation; gas chromatography, chapter 3), polarity (solvent partitioning, acid-base properties, chromatography), or reactivity (derivatization). Each of these is discussed below or in subsequent chapters. Several points should be borne in mind. First, preliminary fractionation steps should be chosen either to cut the bulk of the extract rapidly to move on to high-resolution, low-capacity chromatography methods quickly, or to selectively remove groups or classes of compounds (e.g., alkaloids, or volatiles). Second, the number of fractions to acquire merits some consideration. A small number of fractions is easier to bioassay, particularly if recombinations are required to track down synergistic compounds, and easier to reproduce. On the other hand, taking many fractions cuts the bulk more quickly, and may reduce the number of sequential fractionation and bioassay steps, but large numbers of fractions may be difficult to handle, particularly if recombinations are required. The number of fractions to take thus depends on how compounds separate and partition in a given fractionation step, mediated by the number of fractions that can be handily bioassayed. Some thought should also be given to additive bioassays, in which compounds are sequentially recombined, versus subtractive bioassays, in which fractions are sequentially removed from total recombinations (Byers 1992). The latter method may be much more efficient at identifying additive or synergistic compounds from multiple fractions.

1.6.2. Solvent Partitioning

A crude extract can be roughly fractionated on the basis of solubility classes by partitioning between an aqueous or partially aqueous phase and one or a series of less polar, immiscible organic phases. Partitioning of a solute A between two immiscible solvents X and Y can be expressed in terms of the partition coefficient K_D, where

$$K_D = \frac{[A]_X}{[A]_Y}$$

Solvent partitioning works best when values of K_D are very large or very small, so that solute A has a much higher affinity for one solvent than the other. When

K_D is large, several sequential extractions with solvent X will effectively remove A from solvent Y completely. This principle underlies countercurrent distribution methods of separation.

In practice, solvent partitioning can be used to efficiently divide extracts into smaller more manageable portions prior to chromatography, but some thought should be given to the solvents chosen and the desired fractionation. For example, the portion of a hexane extract that will partition from hexane into water will probably be small, whereas partitioning with 90% aqueous MeOH or acetonitrile may work better. In analogous fashion, partitioning of a water extract with hexane will probably remove little material, whereas more polar solvents may enable substantial fractionation. In general, solvent partitioning may be most effective for fractionating extracts made with intermediate polarity solvents, which contain both polar and nonpolar compounds. For example, extracts made with methanol can be diluted with water and extracted with a sequential series of solvents of increasing polarity (e.g., hexane, $CHCl_3$, ethyl acetate, and finally n-butanol). Extracts made with solvents such as EtOH, 70% EtOH in water or acetone can be concentrated and then partitioned between water and $CHCl_3$, for example, or between water and a series of solvents, as described above. Extracts made with water-immiscible solvents of intermediate polarity, such as $CHCl_3$ or ethyl acetate, can be partitioned with water to separate the most polar constituents of the extract.

In a general protocol developed by Kupchan et al. (1975), the sample is extracted with MeOH, the crude extract is evaporated, and the residue is partitioned between $CHCl_3$ and water. The $CHCl_3$ extract is partitioned between 10% aqueous MeOH and petroleum ether (= hexane). The 10% aqueous MeOH extract is adjusted to 20% aqueous MeOH and partitioned with CCl_4. The 20% aqueous MeOH is adjusted to 40% aqueous MeOH and extracted with $CHCl_3$. This provides a petroleum ether extract, a CCl_4 extract, a $CHCl_3$ extract, a 40% aqueous MeOH extract, and a water extract, each containing compounds of different polarity. Each fraction is monitored by thin-layer chromatography, bioassay, or some other analytical method. Additional steps can be used with this method. For example, VanWagenen et al. (1993) added an extra step by evaporating the MeOH from the 40% aqueous MeOH solution then extracting the aqueous suspension with ethyl acetate. Others have continued by extracting the aqueous suspension with n-butanol.

There are any number of variations on this general theme. For example, Harborne (1984) suggests extraction of homogenized fresh plant material with 4:1 MeOH:water (10× the weight of sample), followed by extraction of the residue with ethyl acetate (nonpolar lipids fraction). The MeOH:water layer is concentrated to 1/10 the volume, acidified, and extracted with $CHCl_3$ (terpenoids and phenolics fraction). The aqueous layer is made basic with NH_4OH, and back extracted with $CHCl_3$:MeOH (3:1, twice) and $CHCl_3$, to yield an alkaloids fraction, with the aqueous residue containing the most polar compounds.

At various stages in solvent partitions, interstitial layers may form that are not soluble in either phase. These should be filtered and treated separately. In a similar fashion, precipitates or crystals may form in stored extracts, particularly if they are cooled. These should be filtered off and tested for homogeneity and biological activity.

1.6.3 Acid-Base Separations

Acids, bases, phenols, and neutral compounds can be separated roughly as classes by partitioning extracts in organic solvents with aqueous solutions of different pH. For example, back extraction of a solvent extract with 0.5-M HCl will remove amine bases such as alkaloids as their hydrochloride salts (along with other water-soluble compounds). A second extraction of the remaining crude solvent extract with dilute $NaHCO_3$ will remove carboxylic acids as their salts. A final extraction with 0.5-M NaOH will remove phenols, which are ionizable at high pH. The organic solution remaining contains nonionizable neutral compounds. The basic, acidic, and phenolic compounds are recovered from each of the aqueous extracts respectively by adjusting the pH of each to the point where the compounds will be present in their uncharged forms (alkaloids, pH ~ 11; acids, pH ~ 3; phenols, pH ~ 7). The solutions may turn milky due to the non-water-soluble neutral forms coming out of solution, which are then back extracted into the original solvent used to make the extract. This whole process is shown schematically in Figure 1.4. Note that the order of the steps can be changed.

The success of this process as a fractionation step obviously depends on the relative amount of each class of compounds present, and which particular class

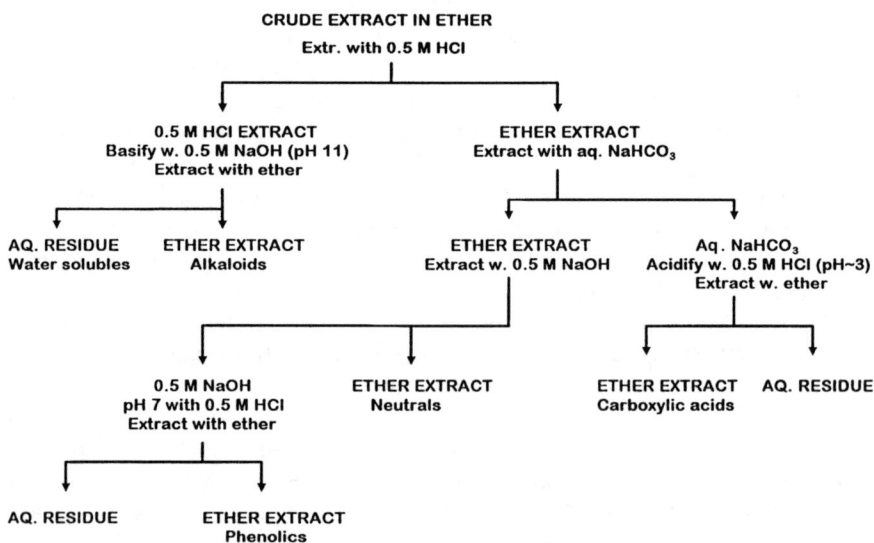

Figure 1.4. Acid-base partitioning scheme for fractionation of crude extracts.

of compounds is of interest. For example, this is a standard method for separation of the alkaloids fraction, and will probably cut the bulk of material substantially if the alkaloids fraction is of interest. On the other hand, the neutrals fraction will probably still comprise the majority of the material; even though not much mass has been cut from this fraction, it may still be useful to determine that a biologically active material is a neutral compound. Bioassays should be performed to ensure that the activity has not been destroyed by the acidic and basic conditions used. Furthermore, neutral compounds, which are significantly soluble in water, may partition into the aqueous extracts to some extent.

1.6.4. Distillation

Some extracts, such as those of plant tissues containing significant quantities of terpenoids, can be roughly fractionated by distillation. The fractionation can work either way; if we are interested in the volatiles, distillation provides a fraction free from nonvolatiles that can be used directly for gas chromatography with no further cleanup. On the other hand, if the nonvolatiles are of interest, distillation may remove large amounts of interfering volatiles. Monoterpene hydrocarbons can be mostly removed by distillation to ~80°C at 15 mm, with sesquiterpenes removed at somewhat higher temperatures and vacuum (~80–120°C at 15–0.1 mm Hg; Millar et al. 1986). Flash or Kugelrohr distillation, in which a flask containing the concentrated extract is heated under vacuum while rotating to minimize bubbling, is fast and simple and minimizes the heating time. An apparatus is available (e.g., Aldrich Chemical Co.) that can handle several hundred grams at a time. Alternatively, the concentrated extract can be fractionally distilled to afford a finer degree of fractionation, although the total heating time will be increased substantially.

1.6.5. Removal of Tannins and Polyphenols

Plant extracts may be dominated by excessive tannins or other phenolic material, which can interfere with bioassays, leading to unreliable results. Wall et al. (1996) compared seven methods for removal of tannins from a large number of plant extracts, including chromatography on polyamide, silica gel or Sephadex® LH-20, precipitation of tannins with polyvinylpyrrolidine or collagen, and two methods of solvent partitioning. The authors concluded that the simplest and most broadly applicable method was to prepare plant extracts in methanol, defat the extracts with hexane, and concentrate the remaining methanol extract. The residue was partitioned between chloroform and water, with the last traces of tannin being removed from the chloroform layers by backwashing with 1% aqueous NaCl. Alternatively, if the tannins were of interest, a tannins fraction could be isolated by chromatography on Sephadex® LH-20 (Wall et al. 1996).

1.6.6. Solid-Phase Extraction

Solid-phase extraction (SPE) is a term used to describe the quick cleanup and/or rough fractionation of (usually small) extracts with short beds of chromatographic packings. It has also been used to concentrate analytes from dilute solution. The devices consist of a small bed of packing, either in a cartridge that fits onto a syringe or retained in the bottom of a polypropylene syringe barrel. The amount of packing is usually 50–500 mg (beds up to 10 g are available, e.g., Mega BondElut® columns, Analytichem International), in all of the standard liquid chromatographic packings. The size of bed to use depends on the total mass of analytes in the sample volume; loadings should be less than 5% of the weight of the packing. SPE devices have been heavily marketed, and they are available from most chromatographic suppliers. In practical terms for infrequent use, it can be just as convenient and cheaper to make up your own SPE columns from a disposable pipette and a small amount of silica or reverse phase packing.

SPE is particularly useful for rapid cleanup of multiple samples of the same type, such as extracts from multiple replicates. By appropriate choices of packings and solvents, the devices can be used to retain undesired compounds while compounds of interest are eluted, or to retain analytes while undesired compounds are washed through. Retained analytes are then eluted as a group with a single strong solvent, or roughly fractionated by sequential elution with aliquots of solvent of increasing strength. Characteristics of chromatographic packings and solvent strengths are discussed in detail in the next chapter. Briefly, so-called normal phases (silica, alumina, Florisil®) strongly or irreversibly retain undesired polar compounds, while reverse-phase (octadecylsilyl; ODS) packings strongly retain lipophilic compounds. Unlike silica, virtually all analytes can be recovered from reverse-phase packings by appropriate manipulation of solvents.

SPE devices use slight pressure or centrifugation to push or slight vacuum to pull extracts and solvents through the bed at a rate of ~2 ml/min for smaller cartridges (50–100 mg packing). Multiple port vacuum manifolds allow simultaneous processing of many samples. Operation is simple. Silica (normal phase) cartridges are rinsed with several volumes of nonpolar solvent to wash off impurities and remove air bubbles from the adsorbent bed. The extract is applied and pulled or pushed into the bed. The bed is then rinsed/eluted with one or more aliquots of the loading solvent (~1 ml for a 100-mg bed), followed by aliquots of solvent of increasing polarity, to provide fractions containing analytes of increasing polarity. For unknown compounds, fractionation is arbitrary, while for known compounds, methods can easily be fine-tuned by adjusting the size and polarity of the aliquots of eluting solvent, to provide a fraction free from most contaminants.

Use of reverse-phase cartridges is analogous, except that the cartridge is first rinsed with MeOH or acetonitrile to wet the packing and remove impurities, followed by flushing with water. The sample is loaded as an aqueous solution, and the column is rinsed with the loading solvent and then eluted with a stepwise gradient of aliquots of solvent of increasing strength (aqueous MeOH or acetonitrile, then the pure solvent, followed by THF and CH_2Cl_2 if necessary).

The method has been extended to large-volume aqueous samples by the use of large disks containing thin beds of packing retained in inert matrices (e.g., 3M Empore™ disks, J.T. Baker Speedisk™). The solution is pulled through the disk in a filter apparatus, the disk is rinsed, and retained analytes are eluted with a strong solvent.

Detailed technical brochures and extensive applications bibliographies are available from most suppliers of SPE cartridges or disks, and the devices are usually supplied with detailed instructions for use.

1.7. Acknowledgments

The authors thank J. Steven McElfresh for preparing the figures and Dr. Horace G. Cutler for reviewing the manuscript.

1.8. References

Aldrich, J.R., J.E. Oliver, W.R. Lusby, J.P. Kochansky, & M. Borges. 1994. Identification of male-specific volatiles from nearctic and neotropical stink bugs (Heteroptera: Pentatomidae). J. Chem. Ecol. **20:**1103–1111.

Attygalle, A.B. & E.D. Morgan. 1988. Pheromones in nanogram quantities: structure determination by combined microchemical and gas chromatographic methods. Angew. Chem. Int. Ed. Engl. **27:**460–478.

Baker, T.C., L.K. Gaston, M.M. Pope, L.P.S. Kuenen & R.S. Vetter. 1981. A high-efficiency collection device for quantifying sex pheromone volatilized from synthetic sources. J. Chem. Ecol. **7:**961–968.

Bartelt, R.J. 1997. Calibration of a commercial solid-phase microextraction device for measuring headspace concentrations of organic volatiles. Anal. Chem. **69:**364–372.

Bestmann, H.J., J. Erler & O. Vostrowsky. 1988. Determination of diel periodicity of sex pheromone release in three species of Lepidoptera by "closed-loop-stripping." Experientia **44:**797–799.

Biedermann, M. & K. Grob. 1991. GC "ghost" peaks caused by "fingerprints." J. High Res. Chromatog. **14:**558–559.

Bjostad, L.B., L.K. Gaston & H.H. Shorey. 1980. Temporal pattern of sex pheromone release by female *Trichoplusia ni*. J. Insect Physiol. **26:**493–498.

Blight, M.M. 1990. Techniques for isolation and characterization of volatile semiochemicals of phytophagous insects. *In:* Chromatography and Isolation of Insect Hormones and Pheromones, eds. A.R. McCaffery & I.D. Wilson, pp. 281–287, Plenum Press, New York.

Boland, W., P. Ney, L. Jaenicke & G. Gassmann. 1984. A "closed-loop-stripping" technique as a versatile tool for metabolic studies of volatiles. *In:* Analysis of Volatiles, ed. P. Schreier, pp. 371–373, Walter de Gruyter, Berlin, New York.

Bone, L.W. & K.P. Bottjer. 1984. Characterization of and male adaptation to pheromone of female *Trichostrongylus colubriformis* (Nematoda). J. Chem. Ecol. **10:**1749–1758.

Borg-Karlson, A.-K. & R. Mozuraitis. Solid phase micro extraction technique used for collecting semiochemicals. Identification of volatiles released by individual signalling *Phyllonorycter sylvella* moths. Zeit. fur Naturforsch. **51c:**599–602.

Brooks, J.S., E.H. Williams, & P. Feeny. 1996. Quantification of contact oviposition stimulants for black swallowtail butterfly, *Papilio polyxenes,* on the leaf surfaces of wild carrot, *Daucus carota.* J. Chem. Ecol. **22:**2341–2357.

Browne, L.E., D.L. Wood, W.D. Bedard, R.M. Silverstein & J.R. West. 1979. Quantitative estimates of the western pine beetle attractive pheromone components, *exo*-brevicomin, frontalin, and myrcene in nature. J. Chem. Ecol. **5:**397–414.

Burger, B.V., Z.M. Munro & J.H. Visser. 1988. Determination of plant volatiles 1: Analysis of the insect-attracting allomone of the parasitic plant *Hydnora africana* using Grob-Habich activated charcoal traps. J. High Res. Chromatog. **11:**496–499.

Byers, J.A. 1992. Optimal fractionation and bioassay plans for isolation of synergistic chemicals: the subtractive-combination method. J. Chem. Ecol. **18:**1603–1621.

Byrne, K.J., W.E. Gore, G.T. Pearce & R.M. Silverstein. 1975. Porapak-Q collection of airborne organic compounds serving as models for insect pheromones. J. Chem. Ecol. **1:**1–7.

Calogirou, A., B.R. Larsen, C. Brussol, M. Duane & D. Kotzias. 1996. Decomposition of terpenes by ozone during sampling on Tenax. Anal. Chem. **68:**1499–1506.

Campos, M., A. Peña & A.J. Sánchez Raya. 1994. Release of ethylene from pruned olive logs: influence of attack by bark beetles (Coleoptera: Scolytidae). J. Chem. Ecol. **20:**2513–2521.

Certic, M., P. Andrasi & J. Sajbidor. 1996. Effect of extraction methods on lipid yield and fatty acid composition of lipid classes containing gamma-linoleic acid extracted from fungi. J. Am. Oil Chem. Soc. **73:**357–365.

Chang, J.F., J.H. Benedict, T.L. Payne, B.J. Camp & S.B. Vinson. 1989. Collection of pheromone from atmosphere surrounding boll weevils, *Anthonomus grandis.* J. Chem. Ecol. **15:**767–777.

Charlton, R.E. & R.T. Carde. 1982. Rate and diel periodicity of pheromone emission from female gypsy moths (*Lymantria dispar*) determined with a glass-adsorption collection system. J. Insect Physiol. **28:**423–430.

Chauret, D.C., C.B. Brenard, J.T. Arnason & T. Durst. 1996. Insecticidal neolignans from *Piper decurrens.* J. Nat. Prod. **59:**152–155.

Charron, C.S., D.J. Cantliffe, R.M. Wheeler, A. Manukian, and R.R. Heath. 1996. Photosynthetic photon flux, photoperiod, and temperature effects on emissions of (Z)-3-hexenal, (Z)-3-hexenol, and (Z)-3-hexenyl acetate from lettuces. J. Am. Soc. Hort. Sci. **121:**488–494.

Cole, R.A. 1980. The use of porous polymers for the collection of plant volatiles. J. Sci. Food Agric. **31:**1242–1249.

Colegate, S.M. & R.J. Molyneux, eds. 1993. Bioactive Natural Products: Detection, Isolation, and Structural Determination. CRC Press, Boca Raton, FL.

Cossé, A.A., J.J. Endris, J.G. Millar & T.C. Baker. 1994. Identification of volatile compounds from fungus-infected date fruit that stimulate upwind flight of female *Ectomyelois ceratoniae*. Entomol. Exp. et Appl. **72:**233–238.

Cossé, A.A., J.L. Todd, J.G. Millar, L.A. Martinez & T.C. Baker. 1995. Electroantennographic and coupled gas chromatographic electroantennographic responses of the Mediterranean fruit fly to male-produced volatiles and mango odor. J. Chem. Ecol. **21:**1823–1836.

Deml, R. & K. Dettner. 1994. *Attacus atlas* caterpillars (Lepidoptera: Saturniidae) spray an irritant secretion from defensive glands. J. Chem. Ecol. **20:**2127–2138.

Dobson, H.E.M. 1991. Analysis of flower and pollen volatiles. *In:* Essential Oils and Waxes, eds. H.F. Linskens and J.F. Jackson, pp. 231–251, Springer-Verlag, New York.

Dobson, H.E.M., J. Bergstrom, G. Bergstrom & I. Groth. 1987. Pollen and flower volatiles in two *Rosa* species. Phytochemistry **26:**3171–3173.

Doolittle, R.E., A. Brabham, & J.H. Tumlinson. 1990. Sex pheromone of *Manduca sexta* (L). Stereoselective synthesis of (10E, 12E, 14Z)-10,12,14-hexadecatrienal and isomers. J. Chem. Ecol. **16:**1131–1153.

Dressler, M. 1979. Extraction of trace amounts of organic compounds from water with porous organic polymers. J. Chromatog. **165:**167–206.

Ebeler, S.E., R.M. Pangborn & W.G. Jennings. 1988. Influence of dispersion medium on aroma intensity and headspace concentration of menthone and isoamyl acetate. J. Agric. Food Chem. **36:**791–796.

Golub, M.A. & I. Weatherston. 1984. Techniques for extracting and collecting sex pheromones from live insects and artificial sources. In: Techniques in Pheromone Research, eds. H.E. Hummel and T.A. Miller, pp. 223–285, Springer-Verlag, New York.

Grenacher, S. & P. M. Guerin. 1994. Inadvertent introduction of squalene, cholesterol, and other skin products into a sample. J. Chem. Ecol. **20:**3017–3025.

Gries, G., K.N. Slessor, R. Gries, G. Khaskin, P.D.C. Wimalaratne, T.G. Gray, G.G. Grant, A.S. Tracey & M. Hulme. 1997. Z6, E8-Heneicosadien-11-one: synergistic sex pheromone component of Douglas-fir tussock moth, *Orgia pseudotsugata* (McDunnough) (Lepidoptera: Lymantridae). J. Chem. Ecol. **23:**19–34.

Grob, K. & F. Zurcher. 1976. Stripping of trace organic substances from water: equipment and procedure. J. Chromatogr. **117:**285–294.

Hamilton, J.G.C., G.W. Dawson & J.A. Pickett. 1996. 9-Methylgermacrene-B; proposed structure for novel homosesquiterpene from the sex pheromone glands of *Lutzomyia longipalpis* (Diptera: Psychodidae) from Lapinha, Brazil. J. Chem. Ecol. **22:**1477–1491.

Hamilton-Kemp, T.R., J.G. Rodriquez, D.D. Archbold, R.A. Andersen, J.H. Loughrin, C.G. Patterson & S.R. Lowry. 1989. Strawberry resistance to *Tetranychus urticae* Koch: effects of flower, fruit, and foliage removal—comparisons of air- vs. nitrogen-entrained volatile compounds. J. Chem. Ecol. **15:**1465–1473.

Harborne, J.B. 1984. Phytochemical Methods, 2nd ed. Chapman & Hall, New York.

Haynes, K.F. & R.E. Hunt. 1990. Interpopulational variation in emitted pheromone blend of cabbage looper moth, *Trichoplusia ni*. J. Chem. Ecol. **16**:509–519.

Haynes, K.F., L.K. Gaston, M.M. Pope & T.C. Baker. 1983. Rate and periodicity of pheromone release from individual female artichoke plume moths, *Platyptilia carduidactyla* (Lepidoptera: Pterophoridae). Environ. Entomol. **12**:1597–1600.

Heath, R.R. & A. Manukian. 1992. Development and evaluation of systems to collect volatile semiochemicals from insects and plants using a charcoal-infused medium for air purification. J. Chem. Ecol. **18**:1209–1226.

Heath, R.R. & A. Manukian. 1994. An automated system for use in collecting volatile chemicals released from plants. J. Chem. Ecol. **20**:593–608.

Heath, R.R., J.R. McLaughlin, F. Proshold & P.E.A. Teal. 1991. Periodicity of female sex pheromone titer and release in *Heliothis subflexa* and *H. virescens* (Lepidoptera: Noctuidae). Ann. Entomol. Soc. Am. **84**:182–189.

Heath, R.R., P.J. Landolt, B. Dueben & B. Lenczewski. 1992. Identification of floral compounds of night-blooming jessamine attractive to cabbage looper moth. Environ. Entomol. **21**:854–859.

Heitz, J.R. & K.R. Downum. 1995. Light-activated pest control. ACS Symposium Series no. 616, American Chemical Society, Washington, DC.

Helmig, D. & L. Vierling. 1995. Water adsorption capacity of the solid adsorbents Tenax TA, Tenax GR, Carbotrap, Carbotrap C, Carbosieve SIII, and Carboxen 569 and water management techniques for the atmospheric sampling of volatile organic trace gases. Anal. Chem. **67**:4380–4386.

Hemming, F.W. & J.N. Hawthorne. 1996. Lipid Analysis. BIOS Scientific, Herndon, VA.

Hibbard, B.E. & L.B. Bjostad. 1988. Behavioral responses of western corn rootworm larvae to volatile semiochemicals from corn seedlings. J. Chem. Ecol. **14**:1523–1527.

Hibbard, B.E. & L.B. Bjostad. 1989. Corn semiochemicals and their effects on insecticide efficacy and insecticide repellency toward western corn rootworm larvae (Coleoptera: Chrysomelidae). J. Econ. Entomol. **82**:773–781.

Hollingdale-Smith, P.A. 1975. A simple, versatile, vapour sampling tube. Chem. Ind. pp. 226.

Honeycutt, M.E., V.A. McFarland & D.D. McCant. 1995. Comparison of three extraction methods for fish. Bull. Environ. Contam. Toxicol. **55**:469–472.

Kendall, D.M. & L.B. Bjostad. 1990. Phytohormone ecology. Herbivory by *Thrips tabaci* induces greater ethylene production in intact onions than mechanical damage alone. J. Chem. Ecol. **16**:981–991.

Krumperman, P.H. 1972. Erroneous peaks from Porapak-Q traps. J. Agr. Food Chem. **20**:909.

Kupchan, S.M., R.W. Britton, J.A. Lacadie, M.F. Ziegler & C.W. Sigel. 1975. The isolation and structural elucidation of bruceantin and bruceantinol, new potent antileukemic quassinoids from *Brucea antidysenterica*. J. Org. Chem. **40**:648–654.

Langenfeld, J.J., S.B. Hawthorne & D.J. Miller. 1996. Optimizing split/splitless injection port parameters for solid-phase microextraction. J. Chromatog. A **740**:139–145.

Laus, G. & D. Kepplinger. 1994. Separation of stereoisomeric oxindole alkaloids from *Uncaria tomentosa* by high performance liquid chromatography. J. Chromatog. A **662**:243–249.

Leal, W.S., Y. Ueda & M. Ono. 1996. Attractant pheromone for male rice bug *Leptocorisa chinensis:* semiochemicals produced by both males and females. J. Chem. Ecol. **22**:1429–1438.

Lewis, M.J. & A.A. Williams. 1980. Potential artifacts from porous polymers for collecting aroma components. J. Sci. Food Agric. **31**:1017–1026.

Lewis, J.A., C.J. Moore, M.T. Fletcher, R.A. Drew & W. Kitching. 1988. Volatile compounds from the flowers of *Spathiphyllum cannaefolium.* Phytochemistry **27**:2755–2757.

Loomis, W.D., J.D. Lile, R.P. Sandstrom & A.J. Burbott. 1979. Absorbent polystyrene as an aid in plant enzyme analysis. Phytochemistry **18**:1049–1054.

Lorbeer, E., M. Mayr, B. Hausmann & K. Kratzl. 1984. Zur Identifizierung fluchtiger Substanzen aus biologischem Material mit Hilfe der CLSA (Closed Loop Stripping Apparatus). Monatshefte fur Chemie. **115**:1107–1112.

Loughrin, J.H., A. Manukian, R.R. Heath, T.J. Turlings & J.H. Tumlinson. 1994. Diurnal cycle of emission of induced volatile terpenoids by herbivore-injured cotton plants. Proc. Natl. Acad. Sci. USA **91**:11836–11840.

MacLeod, G. & J.M. Ames. 1986. Comparative assessment of the artefact background on thermal desorption of Tenax GC and Tenax TA. J. Chromatog. **355**:393–398.

Maier, I. & M. Fieber. 1988. Retention characteristics of volatile compounds on Tenax TA. J. High Res. Chromatog. **11**:566–576.

Maier, I., G. Pohnert, S. Pantke-Bocker & W. Boland. 1996. Solid-phase microextraction and determination of the absolute configuration of the *Laminaria digitata* (Laminariales, Phaeophyceae) spermatozoid-releasing pheromone. Naturwissenschaften **83**:378–379.

Malosse, C., P. Ramirez-Lucas, D. Rochat & J.-P. Morin. 1995. Solid-phase microextraction, an alternative method for the study of airborne insect pheromones (*Metamasius hemipterus,* Coleoptera, Curculionidae). J. High Res. Chromatog. **18**:669–670.

Manura, J.J. 1994. Adsorbent Resins—Part I: Calculation and Use of Breakthrough Volume Data. The Mass Spec Source **7**:3–11; Scientific Instrument Services, Ringoes, NJ.

Martinson, T.A. & F.G. Plumley. 1995. One-step extraction and concentration of pigments and acyl lipids by sec-butanol from *in vivo* samples. Analyt. Biochem. **228**:123–130.

Matich, A.J., D.D. Rowan & N.H. Banks. 1996. Solid phase microextraction for quantitative headspace sampling of apple volatiles. Anal. Chem. **68**:4114–4118.

Matile, P. & R. Altenburger. 1988. Rhythms of fragrance emission in flowers. Planta **174**:242–247.

Meade, T., J.D. Hare, S.L. Midland, J.G. Millar, & J.J. Sims. 1994. Phthalide-based hostplant resistance to *Spodoptera exigua* and *Trichoplusia ni* in *Apium graveolens.* J. Chem. Ecol. **20**:709–726.

Middleditch, B.S. 1989. Analytical Artifacts. Elsevier, New York.

Millar, J.G., C.-H. Zhao, G.N. Lanier, D.P. O'Callaghan, M. Griggs, J.R. West & R.M.

Silverstein. 1986. Components of moribund American elm trees as attractants to elm bark beetles, *Hylurgopinus rufipes* and *Scolytus multistriatus.* J. Chem. Ecol. 12:583–608.

Mozuraitis, R., A.-K. Borg-Karlson, A. Eiras, P. Witzgall, A. Kovaleski, E.F. Vilela & C.L. Unelius. 1996. Solid phase microextraction technique used for collecting volatiles released by individual signalling *Bonagota cranaodes* moths. pp. 193, Abstracts of the Int. Soc. Chem. Ecol. 13th Annual Meeting, Aug. 18–22, Prague.

Nakanishi, K. 1989. Natural products chemistry—past and future. *In:* Natural Products of Woody Plants I, ed. J.W. Rowe, pp. 13–25, Springer-Verlag, Berlin.

Patt, J.M., D.F. Rhoades & J.A. Corkill. 1988. Analysis of the floral fragrance of *Platanthera stricta.* Phytochemistry 27:91–95.

Pelizzari, E.D. & K.J. Krost. 1984. Chemical transformations during ambient air sampling for organic vapors. Anal. Chem. 56:1813–1819.

Peters, R.J.B., J.A.D.V. Renesse, V. Duivenbode, J.H. Duyzer, & H.L.M. Verhagen. 1994. The determination of terpenes in forest air. Atmos. Environ. 28:2413–2419.

Pham-Delegue, M.H., P. Etievant, E. Guichard & C. Masson. 1989. Sunflower volatiles involved in honeybee discrimination among genotypes and flowering stages. J. Chem. Ecol. 15:329–343.

Phelan, P.L. & H. Lin. 1991. Chemical characterization of fruit and fungal volatiles attractive to dried-fruit beetle, *Carpophilus hemipterus* (L.) (Coleoptera: Nitidulidae). J. Chem. Ecol. 17:1253–1272.

Queiroz, E.F., F. Roblot, A. Cave, M. de Q. Paulo & A. Fournet. 1996. Pessoine and spinosine, two catecholic berberines from *Annona spinescens.* J. Nat. Prod. 59:438–440.

Rittschof, D., R. Shepherd, & L.G. Williams. 1984. Concentration and preliminary characterization of a chemical attractant of the oyster drill, *Urosalpinx cinera.* J. Chem. Ecol. 10:63–79.

Robacker, D.C. & R.J. Bartelt. 1996. Solid-phase microextraction analysis of static-air emissions of ammonia, methylamine, and putrescine from a lure for the Mexican fruit fly (*Anastrepha ludens*). J. Agric. Food Chem. 44:3554–3559.

Robacker, D.C. & R.A. Flath. 1995. Attractants from *Staphylococcus aureus* cultures for Mexican fruit fly, *Anastrepha ludens.* J. Chem. Ecol. 21:1861–1874.

Rothweiler, H., P.A. Wager & C. Schlatter. 1991. Comparison of Texas TA and Carbotrap for sampling and analysis of volatile organic compounds in air. Atmos. Environ. 25B:231–235.

Schaefer, J. 1981. Comparison of adsorbents in head space sampling. *In:* Flavour '81, pp. 301–313, Walter de Gruyter & Co, New York.

Schäfer, B., P. Hennig & W. Engewald. 1995. Analysis of monoterpenes from conifer needles using solid phase microextraction. J. High Res. Chromatog. 18:587–592.

Shani, A. 1990. Calling behavior of almond moth (*Ephestia cautella*) females kept in glass cages and airborne pheromone deposited on glass surfaces by airstream. J. Chem. Ecol. 16:959.

Stammiti, L., S. Derridj & J.P. Garrec. 1996. Leaf epicuticular lipids of *Prunus laurocerasus*—importance of extraction methods. Phytochemistry 43:45–48.

Sturaro, A., G. Parvoli & L. Doretti. 1992. Artifacts produced by Porapak Q sorbent tubes on solvent desorption. Chromatographia **33:**53–57.

Swedenborg, P.D. & R.L. Jones. 1992. (Z)-4-Tridecenal, a pheromonally active air oxidation product from a series of (Z,Z)-9,13-dienes in *Macrocentrus grandii* (Goidanich) (Hymenoptera: Braconiidae). J. Chem. Ecol. **18:**1913–1931.

Takeoka, G.R., R.A. Flath, M. Guntert & W. Jennings. 1988. Nectarine volatiles: vacuum steam distillation versus headspace sampling. J. Agric. Food Chem. **36:**553–560.

Teranishi, R., R.G. Buttery & H. Sugisawa. 1993. Bioactive Volatile Compounds from Plants. ACS Symposium Series 525. American Chemical Society, Washington, DC.

Tollsten, L. & G. Bergstrom. 1988. Headspace volatiles of whole plants and macerated plant parts of *Brassica* and *Sinapis. Phytochemistry* **27:**4013–4018.

Tumlinson, J.H., R.R. Heath & P.E.A. Teal. 1982. Analysis of chemical communication systems of Lepidoptera. *In:* Insect Pheromone Technology: Chemistry and Applications, eds. B.A. Leonhardt & M. Beroza, pp. 1–25. ACS Symposium Series 190, American Chemical Society, Washington, DC.

VanWagenen, B.C., R. Larsen, J.H. Cardellina II, D. Randazzo, Z.C. Lidert & C. Swithenback. 1993. Ulosantoin, a potent insecticide from the sponge *Ulosa ruetzerli.* J. Org. Chem. **58:**335–337.

Wall, M.E., M.C. Wani, D.M. Brown, F. Fullas, J.B. Olwald, F.F. Josephson, N.M. Thornton, J.M. Pezzuto, C.W.W. Beecher, N.R. Farnsworth, G.A. Cordell & A.D. Kinghorn. 1996. Effect of tannins on screening of plant extracts for enzyme inhibitory activity and techniques for their removal. Phytomedicine **3:**281–285.

Williams, A.A., H.V. May & O.G. Tucknott. 1978. Observations on the use of porous polymers for collecting volatiles from synthetic mixtures reminiscent of fermented ciders. J. Sci. Food Agric. **29:**1041–1054.

Witzgall, P. & B. Frérot. 1989. Pheromone emission by individual females of carnation tortrix, *Cacoecimorpha pronubana.* J. Chem. Ecol. **15:**707–717.

Witzgall, P., M. Bengtsson, G. Karg, A.-C. Backman, L. Streinz, P.A. Kirsch, Z. Blum & J. Lofqvist. 1996. Behavioral observations and measurements of aerial pheromone in a mating disruption trial against pea moth *Cydia nigricana* F. (Lepidoptera, Tortricidae). J. Chem. Ecol. **22:**191–205.

Yang, X. & T. Peppard. 1995. Solid-phase microextraction of flavor compounds—a comparison of two fiber coatings and a discussion of the rules of thumb for adsorption. L.C.-G.C. **13:**882–886.

2

Liquid Chromatography

Jocelyn G. Millar

2.1. Introduction

Liquid chromatography (LC) forms one of the cornerstones of modern separation methodology. There is an extensive body of literature available, from the primary literature (e.g., *Journal of Chromatography, Chromatographia, Journal of High Resolution Chromatography*) through reference manuals (e.g., *CRC Handbooks of Chromatography*) and texts. Applications bulletins and bibliographies are also available from many manufacturers of chromatography products. However, the bulk of this literature focuses on the separation of known compounds, and may not be useful for developing a separation scheme for an unknown.

There are three basic characteristics to liquid chromatography. First, it is carried out with two immiscible phases, usually a liquid (mobile) and a solid (stationary) phase. Second, separation occurs because the analytes are distributed between the two phases with unique coefficients of distribution. Third, efficient chromatography requires a high ratio of stationary phase to sample, which limits the amount that can be loaded on a column. Consequently, for large crude extracts, every attempt should be made to reduce the bulk of the extract, for example by solvent partitioning or acid-base fractionation (chapter 1), before beginning chromatography.

The sections below provide a general description of the major LC methods, with descriptions of apparatuses, operating parameters, and methods of detection. Advantages and limitations of each method are described. I have focused on those methods that are of general applicability and that are most useful for fractionation of extracts and isolation of unknown compounds. Specific methods for isolation and/or analysis of known compounds are easily found in, for example, *Chemical Abstracts*.

2.1.1. Liquid Chromatography Phases

In most LC methods, separation occurs by differential distribution of analytes between a moving liquid phase and a solid stationary phase. The mechanism of separation varies with the type of stationary phase. So-called normal phases (silica gel, Florisil®; alumina) work by reversible adsorption of compounds by polar functional groups on the surface of the stationary phase. Reverse phases work slightly differently; analytes are partitioned between the eluting solvent and the hydrophobic layer on the stationary phase surface. In general terms, choice of a stationary phase depends on the polarity and solubility of the analytes.

There are also a wide variety of specialized phases with properties intermediate between these two endpoints, but these phases would not normally be used in fractionation of extracts of unknowns, but rather for analyses of specific classes of known compounds. The only specialty phase method we discuss is argentation chromatography, in which differential complexation of alkenes by silver ions in the stationary or mobile phase is used to effect separation.

For "normal phase" LC packings, separations can be manipulated by varying the chemical structure and activity level of the packing materials. Silica gel, a silicon-oxygen matrix with free silanol groups on the surface of particles, is the most commonly used packing, and it is also one of the most retentive materials for polar compounds. Retention is further increased by oven drying silica gel (~125°C overnight) to remove adsorbed water. Silica gel is slightly acidic, and acid-labile compounds may be degraded on silica. This can be avoided to some extent by including a small amount of triethylamine (1%) in the eluting solvent. Alumina (Al_2O_3) comes in acidic, basic, and neutral forms, and in different activity grades (based on the water content), with activity I having the lowest water content and being the most retentive. Florisil®, a magnesium silicate, is less retentive than silica gel. Choice of a packing is initially somewhat arbitrary, with silica gel being the most widely used material. However, if problems are encountered with silica, such as degradation, strong retention making it difficult to recover compounds, or poor separation, then one of the other less-retentive materials can be tried.

Because normal phase separations are based on differences of the adsorption of analytes onto the polar surface of the packing material, separations are sensitive to the spatial configuration of molecules. Molecules whose polar functional groups are sterically hindered will be less strongly adsorbed, and so normal phase chromatography is often the method of choice for separation of diastereomers, many of which can be completely separated in multigram quantities, even using low-resolution methods such as flash chromatography.

Normal phase is usually used with nonaqueous solvents varying from hydrocarbons to alcohols. However, particularly for methods where the stationary phase is used only once, such as thin-layer or low-pressure LC, a small amount of water may be used in solvent systems for elution of polar compounds. Elution order roughly follows the series alkanes; alkenes; aromatic hydrocarbons and halocarbons; ethers; esters, aldehydes, ketones; alcohols; amides; carboxylic acids. Obviously, more than one polar functional group per molecule results in greater retention. For the separation of acids or amines (alkaloids), acetic acid and triethylamine, respectively, are added to the mobile phase to minimize tailing.

With reverse-phase LC, the basis of separation is relative hydrophobicity, so that reverse phase is good for separation of homologs, for example, which may not separate well by normal phase. Reverse-phase packings come in two main types. The most common consists of beads with hydrophobic coatings bonded to a silica core. These materials are characterized by their surface areas, the efficiency of the coating, and the "carbon load," or the percentage of the bead that consists of hydrophobic coating; the higher the carbon load, the more retentive the material. Carbon load is determined primarily by the length of the carbon chains (C_1 to C_6, C_8, C_{18}, C_{30} or phenyl) bonded to the bead surface, with C_{18} (octadecylsilyl, or ODS) coatings being the most common. Retention increases exponentially with chain length, and thus can be manipulated by choice of packing material; C_1–C_8 may work well with hydrophobic analytes such as nonpolar lipids, while more polar and water soluble compounds may be separated best by C_{18} packings. Very polar compounds with limited hydrophilic character will be retained and separated best by packings with higher carbon loads and/or longer alkyl chains bonded to the beads. The phenyl reverse-phase packings may be particularly good for aromatic and electron deficient compounds due to interactions between the aromatic groups in the analytes and the packing material. Reverse-phase packings with silica cores are often "endcapped" or otherwise deactivated; endcapping consists of blocking residual silanol groups with methyl groups to minimize tailing of ionizable and other polar compounds such as amines. Reverse phase columns are usually eluted with aqueous-organic solvent mixtures, such as water and methanol or acetonitrile, and often including a buffer. In general, they cannot be used with strongly acidic or basic solvent systems, which dissolve silica-based materials.

A second type of reverse-phase packing consists of beads of hydrophobic organic polymers (e.g., styrene-divinylbenzene polymers) or even graphitized carbon. These materials fill a useful niche because they are inert to acidic or basic solvents. They are used with aqueous-organic solvents, particularly buffered aqueous acetonitrile. Aqueous methanol is less useful because it does not wet the bead surface as well.

A number of factors affect how much material can be separated with an LC method and the efficiency of the separation. As a general rule, smaller, more uniform packings (spherical rather than irregularly shaped beads) provide the highest inherent resolution, but are the most expensive, are the most sensitive to overloading, have the highest backpressures (limiting column lengths) due to the resistance to flow, and have shorter lifetimes because they clog up faster. Preparative separations are carried out with columns packed with larger, irregularly shaped, and cheaper packings, with higher surface areas per gram of packing for higher capacity. Even then, preparative runs usually use much larger sample loads than would be required for maximal chromatographic efficiency.

Longer columns provide greater separating power, but chromatographic theory dictates that resolution is proportional to the square of column length; to double resolution, column length must be increased by a factor of 4. Increasing length also translates into higher backpressure. These two factors limit increases in separation efficiency obtainable by increasing column length. In general, much greater gains in separation can be made by changing the solvent composition, or the stationary phase, particularly for thin-layer chromatography (TLC) and high-pressure, or high-performance, liquid chromatography (HPLC), for which there are many stationary phases available. Switching stationary phases is obviously limited by the number of columns or phases on hand, or the size of one's budget!

Normal- and reverse-phase liquid chromatography complement each other extremely well, and separations in sequential fractionation steps can be optimized by switching from one to the other technique. Each method has limitations. For example, irreversible adsorption or degradation of compounds tends to be more of a problem with normal-phase chromatography on silica, and low-pressure columns may be used once and discarded rather than reused, because silica is relatively cheap. Irreversible adsorption is less of a problem with more expensive reverse-phase packings, in large part because analytes are essentially dissolved in rather than adsorbed by the surface layer. Consequently, reverse-phase materials can be cleaned and reused by stripping a column with the strong solvent component (e.g., methanol or acetonitrile), followed by reequilibration with the elution solvent.

Reverse phase has distinct advantages in the separation of certain classes of compounds. In particular, reverse phase is more sensitive to lipophilicity and size than is normal phase, with the result that homologs and other closely related compounds with similar functional groups but different lipophilicity, which may separate poorly if at all on silica, are resolved on reverse phase. However, for preparative HPLC of hydrophobic materials, loading samples onto the column can be an intractable problem due to the limited solubility of the materials in the eluting solvent. This is discussed in more detail in section 2.4.

Ion exchange stationary phases bind ionized compounds to charged moieties bonded to an inert matrix. Separation occurs due to differences in the pKa's (or isoelectric points for amphoteric compounds) of the analytes, with compounds being sequentially eluted by changes in pH or ionic strength, or both, of the mobile phase. Size exclusion packings separate compounds on the basis of effective molecular size in solution, although in some cases, adsorption effects can be significant or even predominate. The mechanisms of these more specialized chromatographic phases are covered in sections 2.6 and 2.7, respectively.

2.1.1.1. Chromatography with Ion-Pairing Reagents

Ion pairing is used in the chromatography of ionizable compounds (e.g., carboxylic acids and amines) and compounds permanently ionized at normal chromatographic pH (e.g., sulfates and quaternary amines). In ion-pair chromatography, the buffered mobile phase contains a lipophilic ion with the opposite charge to the analytes. The pairing reagents form neutral complexes with the analytes, which then are retained well on reverse phase columns. Alternatively, the analyte ions may form transitory ion pairs with ion-pairing reagents adsorbed or partioned onto the stationary phase. For anionic analytes, lipophilic cations such as tetralkylammonium salts are used, while for cations, C_4–C_{12} alkyl sulfonic acid salts are used. The size, type, and concentration (usually 5–10 mM) of ion-pairing reagent, optimal pH, and solvent blend are determined by experiment. Premixed ion-pairing reagent-buffer cocktails, needing only dilution with water, are available (e.g., Alltech).

2.1.2. Solvent Strength and Selectivity

Manipulation of the solvent system provides the simplest method of optimizing most chromatographic separations because selectivity (i.e., the separation between peaks) is linearly related to resolution, whereas the relationship between chromatographic efficiency and resolution is quadratic. Solvents are characterized by the sum of their Van der Waals, dipolar, hydrogen-bonding, and dielectric interactions with solutes, which relates to their polarity and solvent strength. Increasing or decreasing retention volumes without changing the order of elution of compounds is accomplished by manipulation of the solvent strength, that is, changing the ratio of solvents, while keeping the solvent components the same (e.g., 80:20 to 75:25 hexane-ethyl acetate). Generally, a weak solvent (hexane for normal phase, water for reverse phase) is blended with a strong solvent to obtain a continuously variable blend from which the best blend can be selected (eluting all analytes within 2 to 10 column volumes). As a rough rule of thumb, the retention is halved by every 10% increase in the percentage of the strong solvent. For normal phase, solvent strengths increase in the order hydrocarbons < $CHCl_3$ < CH_2Cl_2 < ether < EtOAc < THF < acetonitrile < MeOH < H_2O (Snyder & Kirkland 1979). For reverse phase, the order is approximately reversed: H_2O < MeOH < acetonitrile < THF < CH_2Cl_2 < hydrocarbons.

However, compounds may not separate well with a given solvent blend, no matter how much the strength is changed. Changing the relative separation between compounds, and often their elution order, is accomplished by changing the types of interactions between the solvent and solutes, that is, by changing the solvent composition while keeping the solvent strength approximately the same (e.g., changing from aqueous methanol to aqueous acetonitrile or tetrahydrofuran in reverse phase). Changing the solvent components has the most dramatic effect with compounds with different functional groups, rather than for homologs in which the functional groups are the same. Tables of "equieluotropic" solvent mixtures, i.e., mixtures of different solvents of similar strengths, are often given in the reference pages of chromatography catalogs (e.g., EM Separations).

To aid the optimization of solvent blends, Snyder (1974) classified solvents into eight groups, based on their selectivity characteristics (Table 2.1). Thus, a series of binary solvent blends consisting of a weak solvent with a strong solvent from different solvent groups can be rapidly screened by TLC to find a likely blend. For normal phase, blends of hexane with ether, methylene chloride, and ethyl acetate provide a good starting point, while for reverse phase, blends of

Table 2.1.   Common Solvents Used in Liquid Chromatography, Classified by Selectivity Characteristics[a]

Solvent group	Solvent
I	aliphatic ethers; trialkylamines
II	aliphatic alcohols
III	amides, glycol ethers, sulfoxides, THF
IV	glycols, acetic acid, formamide
V	methylene chloride
VI	aliphatic ketones and esters, dioxane, acetonitrile
VII	aromatic hydrocarbons
VIII	chloroform, water

[a]Snyder (1974).

water with MeOH, acetonitrile, and tetrahydrofuran (THF) can be tried. For difficult cases, ternary and even quaternary solvent blends can be tried.

2.1.3.   Choosing a Chromatographic Method

The quantity, complexity, and matrix of a sample will usually dictate the initial chromatographic step. If submilligram amounts of active compounds are available in a few microliters of extract, then at most, a single cleanup step should be used before taking the material directly to a microscale column or HPLC. For a few milligrams to a few hundred milligrams, one or more fractionations with low-pressure or thin-layer chromatography may be appropriate before HPLC, and compounds are often purified by these methods alone. For extracts comprising hundreds of milligrams to multigram amounts, low-pressure chromatography will be required to cut the bulk of material down before going on to medium pressure or HPLC.

The complexity of the sample will determine whether isocratic (fixed solvent blend) or gradient elution should be used. Samples containing compounds with widely different properties will require gradient elution, while fractions containing compounds with similar properties may be separated either way. The properties of the analytes, such as polarity, solubility, and stability will determine whether normal- or reverse-phase methods should be used. Aqueous extracts are not generally run on normal phase materials, except in TLC, but in most cases, either normal or reverse phase will suffice for a first chromatographic step. A representative decision tree, giving a rough idea where to start, is shown in Figure 2.1. For separation of complex mixtures, alternating between normal- and reverse-phase (or other) chromatographic methods in sequential fractionation steps often works best.

Finally, the choice of LC method depends on whether an analyte only needs to be detected and/or quantified, or whether preparative isolation of the analyte is required, for example, for bioassay or spectroscopy. These requirements will

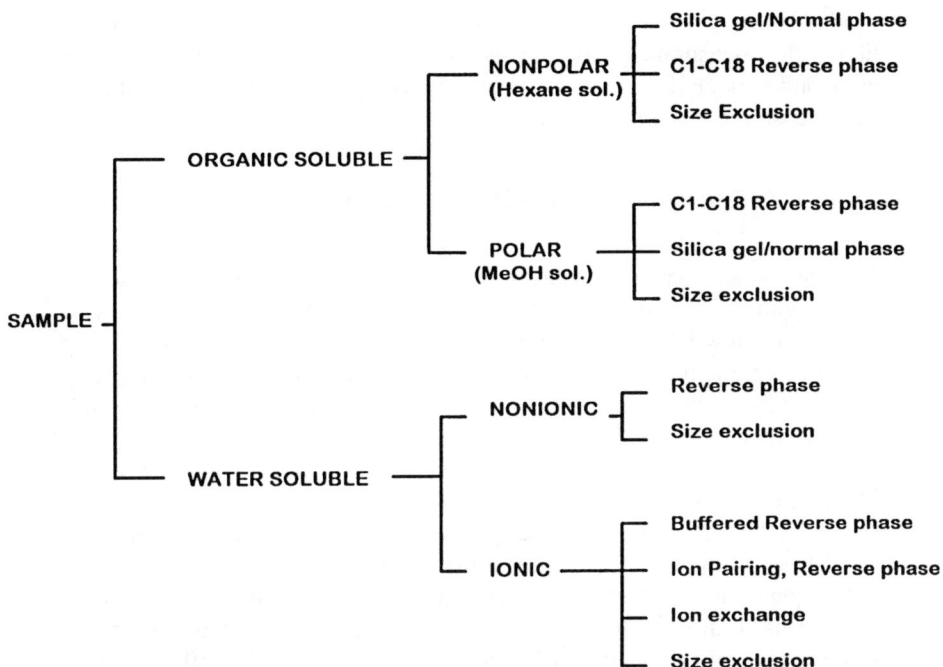

Figure 2.1. Stationary-phase selection guidelines. The probable method of first choice is listed at the top of each category.

determine the scale of the separation, the efficiency of the columns required, and whether destructive or nondestructive detection methods can be used.

2.2. Thin-Layer Chromatography

Thin-layer chromatography is widely used for qualitative and quantitative screening of extracts, for quick cleanup of extracts, and for smaller-scale, low- to medium-resolution preparative separations (e.g., Fell 1996). A minimum of expensive equipment and technical know-how is required, and multiple samples can be run simultaneously, giving results in minutes to a few hours. Unlike column LC, where compounds retained on the column are not seen, the total sample can be visualized. Thus TLC is a very useful way for checking for irreversible adsorption of compounds onto expensive HPLC columns. There are also a large number of selective detection methods that selectively visualize only specific classes of compounds. Furthermore, standards can be run simultaneously with samples of the same TLC plate, whereas shifting retention times can be a problem in HPLC. Because TLC plates are used only once, very crude samples, which might clog or damage HPLC columns, can be used.

TLC is usually used qualitatively, but it can be used quantitatively in conjunction with a densitometer to read the size and intensity of developed spots. In preparative mode, however, complete recovery of samples can be difficult, depending on the resolution of sample bands.

2.2.1. TLC Equipment, and Development of Plates

TLC plates are available from a number of manufacturers, in various sizes, coated with various stationary phases on various backings, and with varying bed thicknesses; few labs coat their own plates anymore because of the high quality and moderate expense of commercial plates. The layers are held together and onto the plate with organic or inorganic binding agents, and fluorescent indicators are often included in the layer (section 2.2.2). Plates also come with a preadsorbent strip, which refocuses applied samples as a sharp band, for maximum resolution. The properties of normal and reverse phase stationary phases were discussed in section 2.1.2. Detailed applications bibliographies for these and the more specialized phases are available from most manufacturers.

For qualitative analyses, plates with stationary phase thicknesses of 0.1–0.25 mm are used, in sizes from 4×8 cm to 20×20 cm, enabling ~3–15 samples to be run simultaneously. Analytical samples (1–2 µl of 1–10% solution) are spotted about 1 cm from the bottom of plates with 5–10 µl glass capillaries (Fisher Scientific).

For preparative work, thicker layers (1–2 mm) on 20×20 cm plates are used. The solvent system is first optimized on analytical plates, and the sample load per plate is estimated. For Rf (Relative to solvent Front) differences of >0.25, 10–20 mg/plate can be used; for Rf differences of <0.15, smaller loads of 2–5 mg/plate must be used. The sample is applied with a "streaker," or manually by repeated spotting of the sample as a horizontal line of small overlapping spots ~1 cm up the plate. To minimize edge effects, the vertical edges of the plate should be cleaned of silica, and spots/streaks should start 1–1.5 cm in from the edge of the plate.

TLC plates tend to adsorb contaminants from packaging materials and the lab atmosphere. These contaminants frequently show up as a band at the top of the plate after development and visualization. When doing preparative work or sensitive analyses, it is essential to predevelop the plates in a stronger solvent than the one to be used, possibly even several times, to drive all the adsorbed impurities to the top of the plate as a narrow band where they are unlikely to contaminate the sample. The predeveloped plate is then oven dried at ~80°C for 1 h before use.

Plates come with glass, aluminum, or plastic backings. Glass and aluminum backings are best for those applications requiring heating during development of the spots. Aluminum and plastic backed plates are convenient because they can be cut to any desired size with scissors; cleanly scoring and breaking glass-backed plates is more tricky.

Many sources recommend activation of silica gel plates by heating at ~110°C for an hour or more before use. However, in our experience, activation is rarely required for routine applications such as screening samples, and it can be a nuisance because plates begin adsorbing moisture as soon as they are removed from the oven, with the amount adsorbed varying with time and ambient humidity. However, activation can be useful when separating very nonpolar compounds; the higher activity of the desiccated silica provides better resolution than with nonactivated plates.

Beakers or jars with tight-fitting lids can be used as TLC "tanks" for developing narrow analytical plates. Rectangular glass tanks are used for preparative plates. In both cases, the tank wall should be partially lined with a thick sheet of filter paper to saturate the tank with solvent vapor, and the tank should be tightly closed so that it stays saturated during development. The tank is filled to a depth of ~5 mm with solvent, soaking the saturating paper as well, and making sure that the sample spots or streaks on the plate are well above the solvent level so that the sample is not leached off. Plates are developed until the solvent front is about 1 cm from the top, then removed and air-dried thoroughly, so that the solvent does not interfere with detection of analytes. The analytes are then visualized, using selective or universal, destructive or nondestructive methods (described below), depending on requirements.

Plates can be developed a number of times with the same or different solvents to increase resolution, visualizing the spots at each stage (nondestructively!) or after the final elution. Sequential elutions may be particularly useful with samples containing compounds varying widely in polarity; for example, a relatively nonpolar solvent system (hexane:toluene) is used for a first separation of nonpolar compounds on silica, followed by elution with solvent(s) of increasing polarity (e.g., hexane-ethyl acetate blends) for secondary separation of compounds of higher polarity.

Two-dimensional development has been used extensively for separation of complex mixtures of compounds of specific classes (e.g., amino acids, carbohydrates, plant pigments). With this technique, a single spot is placed in one corner of the plate, and the plate is developed in the first solvent mixture. After drying, the plate is rotated 90°, so that spots from development in the first direction now form a line along the new bottom of the plate. Development with a different solvent system now helps separate those compounds which coeluted in the first dimension. This technique is used primarily for analytical work because of the small sample loading (i.e., a single spot).

Solvents commonly used with normal-phase TLC include mixtures of aromatic and aliphatic hydrocarbons, ether, chloroform and methylene chloride, ethyl acetate, acetone, and alcohols, acetic acid, and even water. For reverse phase, aqueous MeOH, acetonitrile, or THF are often used. Extensive tables of solvent mixtures, phases, and detection methods for specific classes of compounds are available (e.g., Harborne 1984; Sherma & Fried 1991; Stahl 1969; Fried & Sherma 1994; Wagner et al. 1996; Kirchner 1990; Fried & Sherma 1996; *CRC Handbooks of Chromatography* for various classes of compounds).

2.2.2. Detection and Visualization Methods

TLC spots and bands may be visualized by spraying the developed plates with various reagents to produce colored or ultraviolet (UV)-active products, often after heating. Several essentially universal reagents produce colored spots with almost any organic compounds, or if a specific class of compounds are sought, selective spray reagents are available. Lists, recipes, and uses of several hundred of these specialized reagents are tabulated in Stahl (1969), Zweig and Sherma (1972a, 1972b), and Jork et al. (1990). The latter reference is particularly good because it includes detailed descriptions of reagent preparation, the color-producing reaction, and the classes of compounds visualized. It must be emphasized that many of the reagent cocktails are highly toxic, corrosive, or unstable, so they should be prepared and handled with due caution. The reagent is usually sprayed evenly onto the plate as an aerosol mist, using a TLC sprayer (available from most chromatography suppliers) in an enclosed cabinet; a cabinet for occasional use is readily prepared by lining a cardboard box with polyethylene film. The sprayed plate is usually heated to hasten the color development, using either a hot plate or, for quick development, a "hairdryer" heat gun. As the plate is heated, the developing spots often pass through a series of colors characteristic of the compounds, i.e., the color development is a useful piece of information for characterization of a compound.

Several general-purpose detection methods are described below. Nondestructive methods include

1. Absorption of UV light; substances with a UV chromophore (e.g., a conjugated double bond system) can be detected by irradiation of the plate in a UV light box (usually 254 and 360 nm). On plates with a fluorescent indicator incorporated into the layer, UV-absorbing compounds produce dark spots on a yellow-green background. However, it must be emphasized that this detection method is not universal, and works poorly if at all with many non-UV-absorbing compounds. The method is useful with all TLC phases.

2. Use of lipophilic fluorescent dyes, which when sprayed on a normal phase plate, are concentrated in the analytes, and show up as darker bands against a lighter background when the plates are irradiated with long-wavelength UV (360 nm). Common dyes include 2′, 7′-dichloro-fluorescein (0.2% in 95% EtOH) and rhodamine B (0.05–0.5% in EtOH). These dyes are strongly adsorbed to the silica, so that the

compounds of interest can be eluted from the silica without contamination from the dyes. These dyes work for most organic compounds, particularly larger, lipophilic molecules. However, they are not useful with lipophilic reverse phases.

3. Use of iodine vapor, which is the best nondestructive universal detection agent. The dry, developed plate is placed in a closed chamber with a gram of iodine crystals. The subliming iodine vapor is reversibly adsorbed by the analytes, resulting in brown spots against a lighter background after a few minutes. For faster development, the plate can be sprayed with an iodine solution (1% in $CHCl_3$ or MeOH). In either case, the iodine rapidly evaporates on leaving the plate in clean air for a hour or so, so that compounds can be recovered cleanly. However, spots can be made permanent by spraying the plate with 10% starch solution, resulting in deep blue spots. Occasionally, analytes may react irreversibly with iodine (Jork et al. 1990). The method is useful with normal and reverse phases.

Universal but destructive visualization methods include

1. Spraying with sulfuric acid solutions followed by heating to >100°C to char organic compounds. The reagent is prepared by cautiously adding 5–10 ml of concentrated sulfuric acid to 85 ml chilled water, EtOH, or MeOH. More concentrated solutions have been used but are probably not necessary because the acid is concentrated during heating of the plate. Characteristic colors evolve as the plate is heated. Anisaldehyde-sulfuric acid has also been used widely (8 ml conc. sulfuric acid and 0.5 ml anisaldehyde added cautiously to 85 ml chilled water or alcohol + 10 ml AcOH). The reagent is stable for several weeks at 4°C. Both reagents are suitable for normal or reverse phase plates.

2. Spraying with 5% phosphomolybdic acid in EtOH. Some compounds produce blue spots on a yellow background at room temperature, while heating to 110°C is required for others. The method works for most organic compounds, with either normal or reverse phase plates. Fresh solution should be made every few weeks.

2.2.3. Recovery of Compounds

Having visualized and marked a spot or band with a nondestructive method, the band is then scraped off the plate with a razor blade, placed in a Pasteur pipette with a small glass wool plug or a small sintered glass funnel, and the analyte is eluted with several aliquots of the strongest solvent used to develop the plate (e.g., EtOAc, if the plate was developed with EtOAc-hexane mixtures).

2.3. Paper Chromatography

Paper chromatography (PC), chromatography on paper sheets, is in many ways analogous to TLC, and often similar equipment and visualization techniques are used. The stationary phase can be the cellulose matrix of the paper itself, or the paper may be impregnated or saturated with a liquid or solid modifier (materials as diverse as silica gel to paraffin or silicone oil). Analytes are spotted or streaked ~2 cm from one edge of a square or rectangular sheet of paper (or in a thin arc around a central point with a wick for radial development). The paper edges are clipped together to form a cylinder, which is then placed in a tightly closed developing tank with a few millimeters depth of the developing solvent, so that the solvent migrates up the paper by capillary action (so-called ascending mode). Alternatively, particularly for large sheets, the top edge of the sheet is placed in a raised trough of solvent, and the solvent migrates down the paper (descending mode). Spots or bands on the developed paper are then visualized by dipping or spraying with reagents, which form diagnostic colored derivatives with different classes of compounds, often after heating. Several hundred of these reagents are listed in Zweig and Sherma (1972b). Many of these reagents are similar or the same as those used with TLC. Papers are often developed in two dimensions, with a different solvent system in each dimension, to increase the effectiveness of the separation. PC can be used as an analytical method, or in preparative mode with thick sheets (1–2 mm; Whatman) designed for this purpose. Separated compounds are easily recovered by first cutting off a strip along one edge of the sheet and visualizing the compounds, then using the resulting pattern of bands to cut the rest of the sheet into strips, followed by elution of the compounds from each strip.

Paper chromatography has the advantage that it is inexpensive and easy to use (requiring a minimum of expensive equipment), and in the past it was widely used for all manner of separations. However, separations may be complicated by poor spot or band shape, and by the dependence of Rf values on concentration, small changes in solvent composition, and the nature of the paper used, among many other factors (Zweig and Sherma 1972b). Consequently, standards should always be used, on the same sheet with one-dimensional development, or by developing a set of standards on an identical sheet under identical conditions for two-dimensional runs.

Paper chromatography now has been largely superseded by TLC on cellulose or other layers, or HPLC, reflected in the declining number of PC references and texts published in the last two decades. However, PC can still be useful for qualitative or semiquantitative analyses, particularly with pigments and medium- to high-polarity compounds such as small organic acids, amino acids, and sugars. For further information, see the detailed descriptions of PC methods and the tables of solvents and spray reagents in Zweig and Sherma (1972a, 1972b).

A number of applications of PC to phytochemical analysis are described in Harborne (1984).

2.4. Low-Pressure Column Chromatography

2.4.1. Column Flash Chromatography

Low-pressure LC uses columns varying in size from a capillary tube to a column 10 cm or more in diameter. Gravity flow chromatography has been largely superseded by flash chromatography, in which the column is either slightly pressurized or the solvent is pulled through the column from the bottom by vacuum. Either variation greatly increases the flow rate (and speed of the separation) and the resolution, and because the entire separation can be achieved in an hour or so (as opposed to many hours to days for the gravity flow technique), problems with on-column degradation of unstable compounds are minimized; generally, sample recovery is excellent. Resolution is moderate to good, and in numerous instances, natural products have been isolated using flash chromatography alone. The equipment required is inexpensive and readily available (e.g., Aldrich), consisting of a heavy-walled glass column about 60 cm in length with a bottom stopcock and a gas inlet/pressure release valve on top (Fig. 2.2), and a source of clean compressed air or nitrogen. A screw-on top, rather than a clamped ball-joint top, is preferred to minimize leakage. Flash chromatography, with either normal-phase (silica, alumina, Florisil®) or reverse-phase packings is the method of choice for preliminary fractionation of several tens of milligrams to 10 g or more of extracts.

The method was first outlined in detail by Still et al. (1978), from which the following points can be summarized. A separation is first worked out with TLC, finding a solvent system that separates the compounds of interest by at least 0.1 Rf unit, with the compound of interest having an Rf value of ~0.25 to 0.35; compounds with higher Rf values will elute very quickly and be poorly resolved. Silica gel with a particle size of 230–400 mesh (40–63 µm, E. Merck #9385 or equivalent) is optimal, with a normal bed length of ~15 cm, but with bed lengths of 25 cm or more for more difficult separations or larger sample sizes. However, it should be remembered from basic chromatographic theory that column length must be increased by a factor of 4 to double resolution. The packing is loaded into the column dry (with a small plug of glass wool in the throat of the column), the column is tapped gently with the stopcock open to settle the bed, and then filled with solvent, which is then forced through the bed with a pressure of no more than 20 psi of N_2 until the bed is evenly packed with no trace of retained air (about one to two column volumes). During packing and operation, the bed must not run dry, and is refilled as necessary. The packing solvent can be reused. Once packed, the solvent is allowed to run down until the top of the bed is just dry and ready for the sample to be loaded. The original report (Still et al. 1978)

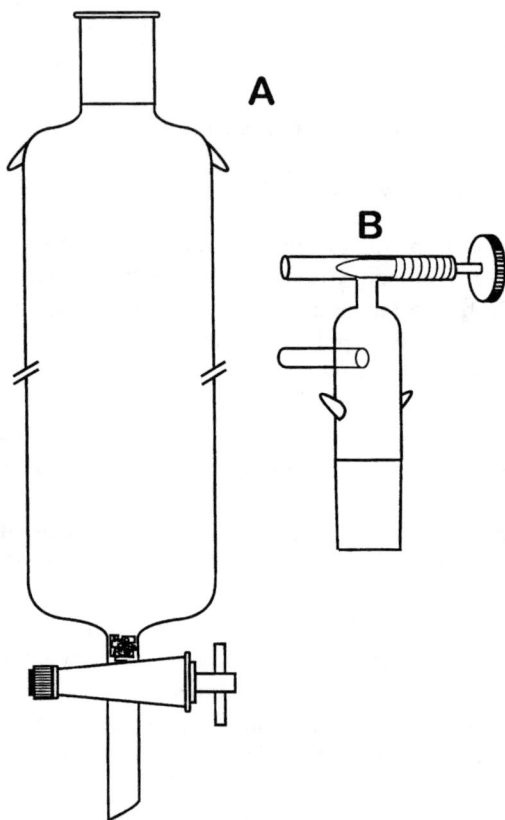

Figure 2.2. Heavy-wall column (*A*) and flow controller (*B*) used with flash chromatography. The throat of the column is fitted with a plug of glass wool to retain the packing material.

suggested using a fine layer of sand at the top and bottom of the column; in our experience, this is not necessary, as long as care is taken to avoid disturbing the bed (an uneven bed causes uneven solvent flow and poorer resolution). An uneven bed can be leveled by gentle tapping of the column with 1–2 mm of solvent above the bed.

We have also found that better and faster packing results from loading the column with a slurry of the packing in the elution solvent, followed by packing the bed under pressure, as described above. This avoids problems from localized heating of the packing due to exothermic interactions of the packing with polar solvents, which can cause gas pockets and a fragmented bed. Slurry packing is also particularly useful when using a solvent system with a small percentage of a polar solvent in the mixture, with which it may be difficult to equilibrate the column properly by the original dry packing method.

The sample is carefully loaded as a solution (~20%) in either the weakest of the solvents in the elution mixture or in the eluting solvent itself, using a long pipette (a 5 to 25-ml volumetric pipette works well) to apply it evenly dropwise onto the bed, taking care not to disturb the bed. The sample must not be too concentrated because viscous fingering will occur, causing uneven application to the bed, nor should the sample volume be too large, which will cause increased bandwidths. The applied sample is run down into the bed, and then rinsed onto the bed twice by carefully applying ~1 cm of solvent, rinsing down the walls of the column. Solvent is then applied by pipette to a depth of about 2 cm, after which, with care, solvent can be poured down the walls to fill up the column. The top is then screwed on, the bottom stopcock is opened, and the column is carefully pressurized. All these operations must be done without disturbing the bed. If the bed is disturbed during loading or rinsing, it can be leveled again as described above. For the novice, a little practice loading the column using only the elution solvent may be time well spent.

The loaded column is eluted by applying N_2 pressure at such a rate that the linear flow of low-viscosity solvents (e.g., mixtures of hexane, ethyl acetate, and acetone) down the column is 5 cm/min. This results in very rapid flows of 100 ml/min or more for large columns, so fractions are often collected by hand (one fraction every few seconds!) rather than with an automated fraction collector to minimize spillage as tubes are changed. For higher viscosity solvents, slower flows are needed for optimal separations (Still et al. 1978). In all cases, care must be taken not to overpressurize the glass columns.

When the solvent level is down to ~2–4 cm above the top of the bed, the pressure is slowly released with the bleed valve (fast release may fragment the bed!), and the column is carefully reloaded with solvent. If the column does accidentally run dry, it can usually be refilled and the chromatography continued without significant loss of resolution. Refilling and elution is repeated as often as necessary to elute the desired material off the column.

A 20-mm i.d. column can be loaded with ~160 mg of material for a more difficult separation (RF difference $\leq$ 0.1), and 400 mg or more for an easy separation (Rf $\geq$ 0.2). Sample loadings for columns of different sizes are adjusted in accordance with their cross-sectional area. For a 20-mm i.d. column, 10-ml fractions are collected initially, with fraction size increasing as isocratic elution progresses (bands become wider as elution progresses). Fraction sizes with larger or smaller diameter columns are adjusted accordingly based on cross-sectional area. Smaller fractions can be taken with difficult separations to minimize overlap of compounds. For difficult separations, a recent report has suggested that very small amounts of a polar modifier (0.1–0.3% isopropanol or methanol) added to the solvent mixture increases resolution noticeably with silica gel chromatography by decreasing peak tailing (Khlebnikov 1996).

Fractions are screened versus the crude material by TLC, spotting every second or third fraction on a 20 × 5 cm plate (i.e., a regular analytical plate turned sideways), which can be developed in a few minutes. Having located the group of fractions of interest, each fraction at each end of that group can then be rechecked by TLC, or by a complementary analytical method (e.g., gas chromatography [GC]). A complementary method is recommended, because it is common for compounds to coelute, particularly with complex samples.

Flash-chromatography-grade silica is not cheap (cheaper in bulk!), and silica can be reused, by first stripping with a polar solvent (or Soxhlet extraction of the packing with a polar solvent), then reequilibrating with the desired solvent system. However, either method takes time, uses a lot of solvent, and may not completely clean the silica, so the relative costs and benefits should be weighed, particularly for chemical ecology work where trace contamination can be important. On the other hand, expensive reverse-phase packings have fewer problems with irreversible adsorption, and can routinely be stripped with strong solvents (e.g., methanol, then THF or CH_2Cl_2) and reused.

Flash chromatography was originally developed for organic synthesis (Still et al. 1978) where the objective was to separate one or a few compounds from unwanted side products. The objective in the fractionation of extracts of biologically active materials is somewhat different because the compounds in the extract usually cover a broad range of polarities, and any one of the compounds in the extract, from the least to the most polar, may be the desired one. Three modifications to the basic flash chromatographic method may be required to accommodate these differences. First, to both maximize the separation of all compounds and ensure that the active material is recovered, the column is eluted with stepwise gradients (one to two column volumes per step) of increasing amounts of a stronger solvent (e.g., ether, ethyl acetate, or acetone for normal phase) in a weaker solvent (e.g., hexane, for normal phase), finishing with a really strong solvent (e.g., ethanol for normal phase) to ensure all materials are recovered. Because the solvent strength is being continually increased, fraction size is held constant instead of being increased, as suggested above for isocratic runs. With extracts of higher polarity, solvent mixtures of increased strength are used (e.g., chloroform/methanol for normal phase). The column packing should never be thrown away until one is sure that the active material has been recovered, or in the worst possible case, destroyed by interaction with the packing.

Second, the extract may not be completely soluble in the weak solvent used to start the elution. In this case, for normal phase, the extract can be coated onto a portion of packing, or finely powdered anhydrous sodium sulphate (extract:powder, 1:5 to 1:10; Dawson 1985) or neutral alumina by concentration of a slurry with the extract in an organic solvent. If this is done on a rotary evaporator, the neck of the flask must be plugged (e.g, with a wad of glass wool) so that the powder does not migrate up into the innards of the rotary evaporator. Alternatively, an adapter with an internally mounted glass frit is available (Aldrich Z10, 747-6). The resulting coated powder is carefully added to the top of the prepacked bed, and the adsorbed extract is then rinsed into the bed by addition of a few aliquots (3 × 3 cm) of the starting solvent. The resulting layer of extracted powder also serves to protect the bed surface during subsequent additions of solvent.

Third, because the extracts contain compounds of widely differing polarity that will be sequentially eluted with a solvent gradient, larger sample loads can be used than would be used in an isocratic elution. Column loadings as high as 1 part extract to 10 parts silica gel (wt/wt) have been used, and loading ratios of 1:25 to 1:50 are common.

For reverse-phase flash chromatography (e.g. Huang et al. 1993, Blunt et al. 1987), the extract often will have limited solubility in the weak starting solvent (e.g., water, or water-methanol mixtures). In this case, the extract is coated onto a portion of the reverse-phase packing material, as described above, and the coated packing is loaded onto the prepacked column (packed as described above), either as a dry powder or an aqueous slurry (Blunt et al. 1987), and the column is eluted with a stepwise gradient of solvents of increasing strength (water, water-MeOH, MeOH, MeOH-CH_2Cl_2). Expensive reverse-phase packings are usually stripped of all applied analytes by this solvent series, and consequently can be reused a number of times, after reequilibration. Because the packings are reused, poorly soluble samples should be adsorbed on packing only for loading (see above), not sodium sulfate, as the latter will contaminate the packing.

Flash chromatography cartridge columns are now commercially available in several sizes, with either normal or reverse phase packings (Biotage, Charlottesville, VA). However, these columns require concomitant purchase of appropriate fittings and hardware.

2.4.2. Vacuum Flash Chromatography

This variation uses a vacuum to pull solvent through a bed of packing instead of pushing it through with pressure. The equipment required and procedures are even simpler than those of column flash chromatography. All that is required is a sintered glass funnel, an adapter with a vacuum takeoff, and a vacuum source. The first reported uses of the method were apparently from the laboratory of J.C. Coll, and the method has been very nicely described in several references (Harwood 1985; Coll & Bowden 1986; Pelletier et al. 1986; Yau & Coward 1988).

The procedure is as follows (Fig. 2.3). Flash chromatography packing is poured into a sintered glass funnel to a depth of about 5 cm or 5 mm below the lip. A vacuum is applied while leveling the top of the bed, with particular care that there are no gaps at the edges. A firm settled bed should result, with some space above it to add solvent. Carefully fill the funnel to the lip with the elution solvent, and suck it through the bed with the vacuum, adding more solvent in the top until solvent starts coming through at the bottom. Then suck the bed dry. With the vacuum off, load the sample in the eluting solvent (or even better, in the

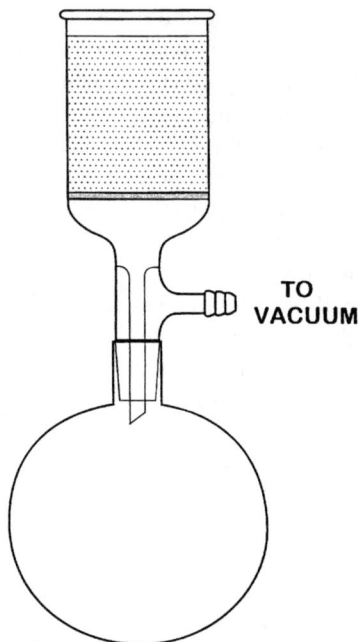

Figure 2.3. One-piece sintered glass funnel-vacuum adapter for use with vacuum flash chromatography. Fractions are collected in individual round-bottom flasks.

weak solvent in the blend) dropwise onto the top of the dry bed in an even layer; the sample drains into the dry bed readily. Turn on the vacuum briefly to pull the sample into the bed, and rinse on with several aliquots of solvent. Then place a well-perforated piece of filter paper on top of the bed so that solvent can be added quickly without disturbing the bed. With the vacuum off, carefully start adding the first aliquot of eluting solvent, then turn on the vacuum and pull through until the column is dry. Repeat with the next aliquot of solvent, etc., collecting each aliquot in a different flask. Recommended sample sizes are 0.5–3 g for a 4-cm i.d. × 5-cm-high bed, taking fractions of 15–30 ml (Harwood 1985).

For samples of limited solubility in the starting solvent, the sample can be adsorbed onto a portion of packing or inert material as described in section 2.4.1. The dry powder is then evenly spread over the surface of the preequilibrated bed, a vacuum is briefly applied to settle it, and the first aliquot of solvent is applied.

This method can be used with stepwise gradients of solvent mixtures of increasing strength. In this and most other respects, it is analogous to flash chromatography in columns, and because of its flexibility and ease of use, it has found extensive use in natural products isolations.

2.4.3. Medium-Pressure Chromatography

The term "medium-pressure chromatography" is used to describe chromatography systems operated at pressures less than about 6–7 bar, usually with some type of pumping system and longer, reusable columns. These systems are essentially permanent, preparative scale units and are intermediate in cost between flash chromatography and HPLC. System design and components have been described (Meyers et al. 1979; Bailey & Patt 1992), and components or complete turnkey systems are available from several manufacturers (e.g., Isco, Lincoln, NE; Fluid Metering Inc., Oyster Bay, NY; EM Separations, Gibbstown, NJ). Because of longer column lengths, better resolution can be obtained than with flash chromatography (i.e., separation of compounds with a TLC Rf difference of ~0.05). However, because of the expense involved in purchasing and operating medium-pressure systems, they are usually found only in labs doing extensive preparative scale separations. In many instances, these systems have been superseded by taking fractions from flash chromatography directly to HPLC, particularly for more difficult separations.

2.5. High-Pressure or High-Performance Liquid Chromatography

HPLC is primarily a separation method for analytical and smaller scale high-resolution preparative separations, and as such, it is usually used in the later stages of a separation scheme once the mass of material has been reduced to a gram or less. Even though a column can be overloaded considerably, it must be remembered that an HPLC column holds only a few grams of packing, so sample capacity is inherently limited (i.e., for good resolution, a high ratio of stationary phase to sample is required). Provided that there is some separation between the compounds of interest, this loading limitation may be partially overcome by eluting with a large number of column volumes of weak solvent instead of eluting the incompletely resolved compounds more rapidly with a stronger solvent. The critical factor is the difference in the reciprocals of the TLC Rf values, which translates into the difference in the numbers of column volumes (ΔCV) of solvent required to elute two compounds. For example, with a strong solvent with TLC Rf values of 0.3 and 0.4 for two compounds, ΔCV is only $3.3 - 2.5 = 0.8$ column volumes. For a weaker solvent, giving Rf values of 0.1 and 0.05, for example, ΔCV is now 10 column volumes, and a better separation may result. This strategy is easy to use with HPLC, where the column volume is small in comparison to the size of the solvent reservoir; it is less convenient to use with flash chromatography because of the necessity of having to refill the column numerous times and because of the large volume of solvent consumed.

HPLC systems are permanent systems consisting of a number of integrated parts, each of which is considered in detail below. Overall, HPLC has the best resolution of the LC methods described in this chapter, and it is also the most versatile because separations can be manipulated and optimized by changing many characteristics of both the stationary phase and the mobile phase. Furthermore, it is usually used with on-line, real-time detection, so that precise fractions can be cut, minimizing the amount of overlap between compounds and the volume of the fraction containing the compound of interest. Our discussion focuses on the use of HPLC for the fractionation of extracts of unknowns. Methods for the isolation or analysis of compounds of known structure are readily available in reference sources or the primary literature.

2.5.1. Connections

Microbore tubing and zero-dead-volume fittings are essential to minimize peak broadening. Tubing and fittings used to be primarily stainless steel, but over the last decade, polymeric fittings and tubing (e.g., Upchurch Scientific, and licensed distributors) have become popular for a number of reasons, particularly for connections that are changed frequently (e.g., injector to column). First, polymeric fittings eliminate the possibility of damaging the threads of metal column endfittings. Second, many polymeric fittings are "universal," or come with "universal" ferrules, which self-adjust to different fittings, so that they can be used with columns from any manufacturer. Because there are at least eight nut sizes used by different column manufacturers (three in common use in the United States), this is a major advantage. Many chromatographers have cursed column manufacturers for not using a standard nut size! Furthermore, because polymeric fittings distort with tightening, they can often be used to make good seals to damaged or distorted metal fittings. Third, polymeric fittings are made for finger-tightening, so no wrenches are required. Fourth, polymeric fittings are reusable, so that they can be used for many connections, whereas stainless fittings become permanently swaged onto the tubing. Polymeric tubing is also useful because it is flexible, inexpensive, and easy to cut to the exact length required. Flexible, thin-walled stainless tubing may come in preset lengths because the thickened end portions, where the ferrules seal, are welded on. Thicker walled stainless tubing can be cut to any length, but it is much less flexible. In the rare instance that a polymeric fitting breaks off in a metal seat due to overtightening, the broken part is easily removed by melting a slot with a heated screwdriver and unscrewing it (Dolan & Upchurch 1988).

2.5.2. Solvents

HPLC uses high-purity solvents, which can be simultaneously filtered and degassed by vacuum filtration directly into the solvent reservoirs through 0.2-μm filters. Further degassing is done by sparging with clean helium, and solvents are maintained under a blanket of helium during operation. Filtration removes particulates, which contribute to pump wear and column clogging. In-line filters also are used between the pump and the column. Degassing minimizes the formation of gas bubbles, particularly during solvent mixing. Gas bubbles upset flows and cause baseline disturbances as they go through the detector, and dissolved oxygen may contribute to background when using a UV detector. Although in theory most solvents could be used for HPLC, in practice only a handful are commonly used (Table 2.2), for a number of reasons. First, only solvents with a very low or no UV absorbance in the range above 210 nm can normally be used with UV detectors. Second, solvent viscosity must be considered because high viscosity solvents are difficult to force through columns at satisfactory flow rates. Solvents such as chloroform should be used with prudence because traces of HCl and phosgene can cause damage. Ether, THF, and dioxane used in HPLC normally do not have preservatives, so care must be taken to prevent dangerous peroxide formation. Furthermore, silica-based packing materials are sensitive to acids and bases, so that aqueous solvents should be maintained at roughly neutral pH (pH 5–8). Polymeric reverse phase packings will withstand acidic or basic solvents, provided these corrosive solvents do not damage the pump and lines.

Solvents commonly used with silica columns include hexane, diethyl ether, THF, ethyl acetate, acetone, and isopropanol (Table 2.2). Care should be taken to minimize water contamination in hygroscopic solvents such as THF and acetone because water will affect column performance. As with flash chromatog-

Table 2.2. Properties of Solvents Commonly Used in either Normal or Reverse Phase HPLC, or Both, in Order of Increasing Polarity

Solvent	Miscible with water	UV Cutoff	Refractive index	Viscosity (cP)	Boiling point (°C)	Solvent group[a]
hexane	no	190	1.375	0.31	69	—
CH_2Cl_2	no	233	1.424	0.44	40	VII
diethyl ether	no	218	1.352	0.23	34	I
methyl *t*-butyl ether	no	210	1.369	0.27	55	I
tetrahydrofuran	yes	212	1.408	0.55	66	III
ethyl acetate	slightly	256	1.372	0.45	77	VI
acetone	yes	330	1.359	0.30	56	VI
dioxane	yes	215	1.422	1.32	101	VIa
isopropanol	yes	190	1.378	2.30	82	II
acetonitrile	yes	190	1.344	0.37	82	VIb
acetic acid	yes	—	1.372	1.30	118	IV
methanol	yes	190	1.329	0.60	65	II
water	—	190	1.333	1.00	100	VIII

[a]Solvent group designations as determined by Snyder (1974), based on solvent properties such as hydrogen bonding abilities, polarities, and dipole moments.

raphy, small quantities of acetic acid or triethylamine (1% or less) can be added to mobile phases to minimize tailing of acidic and basic compounds respectively.

With reverse phase, common solvents include aqueous methanol or acetonitrile (Table 2.2), of which acetonitrile is both less viscous and a better solvent. In general, mixtures of aqueous methanol, acetonitrile, or THF of equal eluting strength give quite different separations, even to the extent of reversing peak elution orders. This can be very useful when optimizing separations. If stronger solvents are required, methylene chloride or even hexane can be used, bearing in mind that these solvents are immiscible with water, so an intermediate solvent will be required.

Buffers are routinely used in reverse phase or other modes of chromatography of ionizable compounds (such as carboxylic acids and amines). Because many ionizable compounds have profoundly different retention characteristics (and UV spectra) depending on whether or not they are ionized, for good chromatography, it is crucial to ensure that the analytes are present in either the neutral form or the fully ionized form. In practice, this means working at a pH at least 2 pH units higher or lower than the pK_a of the analytes. Extremes of pH may cause damage to silica-based packings, and for those applications where one must work at pH higher or lower than 3–8, alternate methods should be used (e.g., polymer-based columns, saturator columns between the pump and the injector to saturate the solvent with silica, or ion-pairing reagents, section 2.1.2.1; Poole & Poole 1991).

Several other points must be considered. First, buffer salts must not precipitate out when mixed with the organic portion of the solvent blend; the resulting particulates can damage the pump and clog lines and columns. This limits buffer concentrations of organic or inorganic salts, but buffer concentrations of 0.1 M or less are usually entirely adequate. Furthermore, the system should always be purged of buffer with water before shutdown. Second, for use with a UV detector, buffers must have minimal UV absorbance. Third, apparent pH will change as the buffer is mixed with the organic solvent, so that the pH may need some adjustment. Some of the most commonly used buffers and their characteristics are listed in Table 2.3. Note that acetic acid and triethylamine, unlike the salt buffers, are volatile, and thus can be removed easily from fractions or purified compounds. Furthermore, because they are nonaqueous, they can be used with normal phase chromatography.

Pure water for HPLC can be purchased or generated with HPLC or other high-grade water purification systems (e.g., Hewlett-Packard, or MilliQ® system from Waters Scientific). Even highly purified water is susceptible to microbial contamination and should be refrigerated when not in use and changed frequently.

A final word of caution: HPLC solvents from different manufacturers contain different stabilizers and additives, and even in small amounts these can have a very significant effect on chromatography. For example, in a normal phase separation of linear furanocoumarins, eluting with 81:19 hexane:THF, markedly better

Table 2.3.   Characteristics of HPLC Buffers

Buffer	pKa	Buffering range	UV cutoff (nm)
phosphate (Na or K, 0.01 M)			
pK$_a$ 1	2.1	1.1–3.1	190
pK$_a$ 2	7.2	6.2–8.2	
pK$_a$ 3	12.3	11.3–13.3	
sodium citrate			
pK$_a$ 1	3.1	2–6	225
pK$_a$ 2	4.7	2–6	
pK$_a$ 3	5.4	2–6	
sodium acetate (0.01 M)	4.7	4.2–5.4	205
tris (0.02 M)	8.3	7–9	204
acetic acid, 1%	4.7		230
triethylamine, 1%	10		235

separation was obtained with THF from Aldrich than from Fisher Scientific (Trumble et al. 1992).

2.5.3. Injectors

Injectors enable the introduction of a sample aliquot into the high-pressure flow stream going to the column. They consist of a variable volume sample loop (<5 μl to >2 ml) which is filled off-line, then switched into the solvent flow path. Several points are important. First, if a sample volume equal to the loop volume is loaded into the loop, 15–25% of the injected sample may be lost out of the end of the loop because injected sample flows through the center of the loop at about twice the velocity that it flows along the walls, so that some sample reaches the end of the loop when only about half of the loop volume has been loaded. Thus, for preparative work, the volume injected should be no more than half the loop volume. For quantitative work with no internal standard, and where there is ample material, the problem can be overcome by injecting an excess of sample (two to five times the loop volume) to ensure that the loop is completely filled.

The single most intractable problem when using HPLC in preparative separations, particularly with gradient elution, is loading. Ideally, one would like to load the sample in the eluting solvent or better, a weaker solvent than the eluting solvent so that the injected sample is refocused as a narrow band at the head of the column. In practice, the sample may be poorly soluble or even insoluble in the initial solvent, particularly with reverse-phase analysis of extracts containing nonpolar compounds, in which an aqueous-organic solvent systems is used. This problem is solved somewhat unsatisfactorily by injecting the sample in a stronger solvent (e.g., methanol, when using aqueous methanol). The volume of strong solvent must be kept to a minimum, and even then, the slug of strong solvent tends to carry the analytes down the column, degrading resolution and distorting peak shapes. Even worse, after injection, as the sample is mixed and diluted with the eluting solvent, the sample can oil or precipitate out, potentially clogging the column or tubing. If this occurs, it may be possible to clear the tubing blockage by disconnecting the column and using a slow flow of the strong solvent. If the blockage is in the column or frits, it may be possible to clear the blockage by reversing the column and flushing with a slow flow of the strong solvent. Overall, poor solubility of analytes represents a significant problem in preparative HPLC fractionations.

2.5.4. Pumps

HPLC pumps use pistons or reciprocating metal diaphragms to pump solvent, with the flow regulated by altering the piston stroke volume or speed. HPLC differs from GC in that published separations may not be easily reproduced due to subtle differences in column packings, pump designs, or solvent compositions. This is particularly true when one is trying to reproduce gradient elutions. Thus, for HPLC, there is no equivalent of the Kovats system of retention indices of compounds, which has proven to be so useful and reproducible in GC (chapter 3).

In their simplest form, HPLC pumps deliver a single solvent mixture at a controlled rate. More sophisticated designs use gradient elution, which immeasurably increases the power of the method because it allows the separation of mixtures of solutes with a wide range of polarities. For example, elution of a silica column with a gradient of 100% hexane to 100% THF will separate many tens and possibly more than one hundred compounds in a complex extract. In practical terms, gradient elution as opposed to isocratic elution allows for improved resolution at the start of the run, while decreasing overall run time by elution of more strongly retained compounds by the increasing solvent strength. With the appropriate system, gradients can be created from up to four solvents, either sequentially or simultaneously. However, gradients, or isocratic blends mixed by the pump, are created in two ways, and the differences between them have important effects on the chromatography.

First, in systems with low-pressure mixing, a metering pump blends the desired portions of each solvent at low pressure, and the solvent mixture is then delivered with a single high-pressure pump. Good degassing is critical because solvents often release dissolved air on mixing (e.g., water:MeOH). The resulting gas bubbles can both ruin the prime on the high-pressure pump and cause detector noise. Low-pressure mixing is done in a mixing chamber, which adds to the total precolumn volume; the effects of this are discussed below.

Pumps with high-pressure mixing use two or more high-pressure pumps to deliver individual solvents in the programmed ratio, with mixing occurring downstream of the pumps. While being less sensitive to outgassing problems because the solvents are mixed under high pressure, another potential problem arises; that is, when two or more solvents are mixed, the resulting volume is often different from the sum of each individual volume. Thus, the flow that is delivered to the column may be different from the flow delivered by a pump with low-pressure mixing.

Both types of pumps also may have difficulty in reproducibly supplying blends or gradients created with small proportions of one or more constituents, particularly at low flow rates. Instead of having the pump mix and deliver a gradient of, for example, 0–5% ether in hexane by drawing solvent from reservoirs containing pure hexane and pure ether, more reproducible results may be obtained by filling reservoir B with 5% ether in hexane, and programming the gradient from 100% A (hexane) to 100% B (5% ether in hexane). Furthermore, premixing the two solvents may reduce problems with outgassing.

In gradient elutions, different pumps and systems have different precolumn or "dwell volumes" (i.e., the volumes between the point where the solvents are mixed and the top of the column). The dwell volume includes the volume of mixing chambers, injectors, filters, tubing, and for low-pressure mixing, the pump volume. When the dwell volume is increased, the gradient arrives at the head of the column at a later time. Furthermore, the larger the dwell volume, the more the gradient differs from the programmed gradient, due to mixing and diffusion of the solvent within the dwell volume. Dwell volume effects can be very significant; measured dwell volumes for several common HPLCs varied from 0.5 to >6 ml (Snyder & Dolan 1990). Obviously, with flow rates of 1–2 ml/min, and the void volume of a typical analytical HPLC column (4.6 × 150 mm) being a couple of milliliters, differences in dwell volumes can have profound effects on gradient elutions. In the best cases, although retention times may be shifted, the overall separation and resolution may remain similar. However, in the worst cases, peaks resolved on one HPLC can coelute or even reverse in elution order on a second instrument (Snyder & Dolan 1990). Effects can be manifested even on the same instrument if the dwell volume is changed, for example, by addition of an autosampler. Dwell volumes can be estimated from manufacturers specifications for system components or determined by introducing a UV absorbing compound into solvent B, with the column removed and the line from the injector going directly to the detector (Snyder & Dolan 1990).

Dwell volume effects are particularly important for early eluting compounds, which tend to be better separated under more isocratic conditions (e.g., large dwell volume). Overall, larger dwell volumes will cause increased retention times. Furthermore, it is crucial to remember that for systems with large dwell volumes upstream of the injector, sufficient time must be allowed for the column to be reequilibrated when doing multiple runs (i.e., for the isocratic, preinjection solvent to be flowing through the injector when the injection is made). Otherwise, the new injection will be made into the tail end of a solvent mix of decreasing strength rather than the isocratic initial blend.

In practical terms, one should not be unduly alarmed if the conditions reported for a gradient HPLC method do not result in a good separation on one's own equipment. The differences can be due to any one or a combination of pump and dwell volume effects, slight differences in solvent, or differences in the (nominally similar) column packings (see below). Instead, the reported conditions should be used as a starting point, and fine-tuning will usually result in a robust method for your own equipment.

2.5.5. Columns

HPLC columns are always used with short guard columns, packed if possible with the same material. The disposable guard column protects the analytical or preparative column from damage due to particulates, irreversibly adsorbed components, or other damaging materials.

The choice of column packing and dimensions is determined by several parameters, including the resolution required, the sample characteristics and solubility, and the scale (analytical or preparative). With samples of a few micrograms or less, the best resolution is obtained with long columns (~25 cm) packed with small (3 or 5 μm), spherical particles, using flow rates of a milliliter or more per minute with the typical 4.6-mm i.d. columns. Columns longer than 25 cm are not practical because of backpressure limitations. If the compounds are well resolved, shorter columns (5 or 10 cm) can be used for faster analyses and reduced solvent consumption. If greater sensitivity is required, narrower bore (2 mm) columns can be used; with these low-volume columns, analytes are eluted in a smaller volume, so detectability is increased. To maintain resolution, however, smaller sample volumes must be used. A few additional words of caution are appropriate. Columns with smaller particle size, while giving increased resolution, clog up faster and are more sensitive to overloading, so they are best suited for analytical rather than preparative purposes.

For preparative applications, columns of 1–2.5 cm i.d. are usually used, using larger, irregularly shaped particles (10 μm or more) with high surface areas for a large loading capacity. These columns can be used with flow rates of 10 ml/ min or more. For difficult separations, the effective column length can be increased manyfold by recycling the column effluent back through the column a number of times with a valving system (Kubo et al. 1986; Kubo & Nakatsu 1990; Hadacek et al. 1994; Smith & Kubo 1995).

The basic characteristics of normal and reverse phase packings were discussed in section 2.1.1. There are hundreds of HPLC packings available, particularly for reverse-phase HPLC, so at first glance the choice of packing material may seem daunting. However, many packings are made for specialty applications with known compounds and so may not be well suited to fractionations of mixtures of unknowns. For most general-purpose fractionations or separations, a C_{18} reverse-phase and/or a normal-phase silica column is used. Intermediate phase columns (e.g., cyano or amino), which tend to be less robust, may be most useful in later stages of the separation once some of the properties of the compound of interest are known. Furthermore, because columns are expensive, the purchase of large numbers of columns to try out separations is not feasible, nor is it normally required. Careful optimization of solvent blends and use of normal and reverse phases in sequential fractionation steps will usually allow isolation of the desired compound.

Unfortunately, due to minor differences in surface area and chemistry, carbon load, and the efficiency of bonding the reverse phase coating to the base material, columns with apparently similar specifications from different manufacturers often behave quite differently. In fact, even batch-to-batch variations from the same manufacturer are not uncommon. Rather than (often vainly) trying to reproduce a reported method, it may be easier to use the reported conditions as a starting point from which to optimize the method on your own system. This is particularly true in light of the subtle differences in solvent blends and gradients produced by different systems (section 2.5.4).

2.5.5.1. Column Regeneration

The chromatographic performance of HPLC columns inevitably decreases with use, and the backpressure increases as strongly retained materials and particulates accumulate. Columns can often be regenerated (see Supelco *HPLC Trouble-shooting Guide*) by reversing and flushing them with a series of solvents of increasing strength (~10 column volumes of each). For normal-phase columns, the series hexane, methylene chloride, isopropanol, anhydrous MeOH, and methylene chloride, can be used, followed by reequilibration with the mobile phase. For reverse-phase columns, the flushing solvents used depend on the likely contaminants. For water soluble compounds, the recommended sequence is hot water, methanol, acetonitrile, THF, methanol, and mobile phase, while for compounds poorly soluble in water, the series isopropanol, THF, methylene chloride, hexane, isopropanol, and mobile phase can be used. Protein contaminants are flushed with the sequence water, 0.1% trifluoroacetic acid and water, isopropanol, acetonitrile, water, and mobile phase. Other combinations of solvents are also suitable (Rabel & Palmer 1992; Alltech 1995).

2.5.6. Detectors

Sensitivity, dynamic range, and selectivity/universality are three important parameters to consider when choosing an HPLC detector for chemical ecology work. Of the many HPLC detectors available (e.g., Scott 1996), until recently only two, refractive index (RI) and UV detectors, were generally useful in the isolation of unknown compounds. Both are nondestructive, but both have limitations that must be considered, as discussed below. Other LC detectors, such as fluorescence or electrochemical detectors, while exquisitely sensitive, are of limited use because they are destructive and/or they detect limited classes of compounds. Standard references (e.g., *CRC Handbooks of Chromatography*) or the primary literature should be consulted for optimal methods of detecting specific classes of compounds.

Within the past few years, a third method of detection of LC analytes, coupled HPLC-mass spectrometry (MS), has become much more accessible for routine analyses due to decreases in cost and the development of more user-friendly systems suitable for use by nonspecialists. At the time of writing, the routine use of LC-MS is increasing rapidly, with a number of turnkey systems available from several manufacturers (e.g., Finnigan, Hewlett-Packard, Micromass). It is anticipated that LC-MS soon will become as standard as GC-MS in chemical ecology laboratories.

In LC-MS, the HPLC effluent is directed into the mass spectrometer, where it is stripped of solvent and ionized (usually by electrospray), and the mass spectra of all analytes are recorded. There are several significant advantages over other types of detection. First, the mass spectrometer is closer to a universal detector than UV or other types of HPLC detectors, with essentially all nonvolatile and semivolatile compounds being detected. Second, the sensitivity is good, extending into the low nanogram range or below. Third, analogous to GC-MS, a mass spectrum of each analyte is obtained, providing much larger amounts of information from each analyte than would be available from other types of detector.

There are some limitations. First, mass spectrometry is destructive so that the sample cannot be recovered. Thus, LC-MS is not useful for preparative separation unless the effluent is split. In fact, the method is only appropriate for analytical scale runs anyway, with a maximum flow of 1–2 ml/min, so that if larger columns or higher flows are required for preparative separations, the effluent must be split. Second, detection of volatile compounds (e.g., monoterpenes) may be problematic, because they will be stripped off with the solvent. However, GC rather than HPLC would probably to be method of choice for analyzing volatile compounds, so this is a minor limitation.

The operational details of LC-MS are described in chapter 4.

2.5.6.1. Refractive Index Detectors

RI detectors detect changes in the refractive index of the eluent caused by the presence of analytes. They are the least sensitive of HPLC detectors, which for fractionations and preparative separations may actually be an asset because the detector is less likely to saturate. Furthermore, models are usually available with either preparative or analytical scale flow cells. Although in theory they are universal detectors, in practice peak sizes depend on both the analyte concentration and the difference in RI between the analyte and the solvent. Thus, compounds with RIs very close to that of the solvent will give smaller peaks than compounds with RIs widely different from the solvent, for the same mass of material. Relative peak sizes will also change with solvent, as the ΔRI between analytes and solvent changes. Peaks can be either positive or negative, depending on whether the RI of the analyte is greater or less than that of the solvent. Nevertheless, RI detectors do have some advantages. Within the above limitations, they can detect any compound, whereas UV detectors may fail to detect compounds with no UV chromophore, including several large classes of compounds such as sugars and hydrocarbons. RI detectors can also be used with virtually any solvent, whereas UV detectors are restricted to non-UV-absorbing solvents. However, they cannot be used easily with solvent gradients, one of the more powerful features of HPLC, because of the change in RI. RI detectors are sensitive to changes in flow (and pressure) and temperature, requiring stabilization and continued corrections for drift under ambient lab conditions.

2.5.6.2. UV detectors

UV detectors are the most common HPLC detectors. There are three types: fixed wavelength, variable wavelength, or diode array. In practical terms, for fractionations, short wavelengths are usually used to detect as many compounds as possible in the sample, so all UV detectors suffer from the limitation that their use is restricted to non-UV-absorbing solvents and buffers (Tables 2.2 and 2.3). When fractionating unknowns, it must be borne in mind that these are not universal detectors, and they will not detect compounds without UV chromophores such as conjugated double bonds, aromatic rings, etc., except possibly very weakly at high concentrations where the bulk UV properties of the solvent are changed. Even for UV-absorbing compounds, because peak size is proportional to the magnitude of the absorption, which varies over several orders of magnitude, peak sizes may bear little relationship to the relative amounts of analytes. The largest peaks in a chromatogram may be due to strongly absorbing trace components, while the compounds of interest go virtually undetected. This can be both daunting and confusing to the unwary. Furthermore, peak sizes will be different at different wavelengths, so changing wavelengths can drastically alter the look of the chromatogram. Overall, the sensitivity of UV detectors is very poor for some compounds, and conversely, the dynamic range may be limited with higher concentrations of strongly UV-absorbing compounds due to saturation of the detector. In the latter case, detector saturation is avoided by monitoring at a wavelength some distance from the absorption maxima. For analytical purposes, with compounds of known structure, the selectivity of the detector can be put to good use by monitoring at a wavelength where the compounds of interest give a good response, and where (ideally) potential interferences are UV transparent. This can also be used to search for specific functional groups with characteristic absorptions (e.g., conjugated trienes versus conjugated dienes, 267 versus 230 nm; Tumlinson et al. 1989).

Fixed wavelength are the cheapest and least useful of the UV detectors, allowing detection at only one wavelength (e.g., 214, 225, or most commonly 254 nm, by appropriate combinations of lamps and filters). Obviously, these detectors will fail to detect many compounds. Variable wavelength detectors are more useful, covering 190–350 nm (deuterium lamp) or 350–800 nm (tungsten lamp). This greater flexibility in choice of wavelength is useful, but the detector is still limited because only a single wavelength can be monitored at any time. With unknowns, it is usually best to monitor at short wavelengths (e.g., 210 nm) where organic compounds have the greatest chance of having at least some absorbance. With newer models, it is also possible to obtain UV spectra of peaks by temporarily stopping the flow, and scanning over the UV-visible range.

Diode arrays are the most useful UV detectors because they monitor all wavelengths simultaneously. This has several results. First, chromatograms at a number of wavelengths can be taken and stored simultaneously, any one of which can be monitored in real time, and all of which can be accessed after the run. Thus, only a single run is needed to determine the optimum wavelength to monitor. Second, UV spectra can be taken and stored, based either on time (i.e., one spectrum every few seconds) or on peak detection. The former method is more reliable, because if the wrong wavelength is chosen to monitor for a particular compound and no peak is detected, no spectrum will be taken. In any event, the spectrum of each UV-absorbing compound in the extract can be obtained from a single run. Third, diode array detectors have peak purity algorithms, which work by comparing the ratio of absorbences at two or more wavelengths across a peak. If the peak is pure, this ratio will be constant from the front to the back of the peak. If the peak contains two or more compounds with different UV spectra, then the ratio will change as the peak is crossed. The catch is that the peak purity check will not detect non-UV-absorbing impurities, and it may not detect impurities with very similar UV spectra, for example, coeluting compounds in the same structural class.

The limitations of HPLC detectors must be borne in mind at all times when isolating unknowns. It is all too easy to collect large, apparently clean peaks from an HPLC run and be seduced into thinking that you have isolated significant quantities of pure material, when in reality, you have (a) isolated trace amounts of a strongly UV-absorbing (or high-refractive-index) compound (b) isolated a readily detected compound heavily contaminated with coeluting compounds which are transparent to the particular detector, or conversely (c) isolated a detector-transparent active compound heavily contaminated with coeluting UV-absorbing material. If possible, the purity of isolated materials should always be checked by a complementary method (e.g., GC or HPLC under different conditions).

When doing quantitative analysis, choice of internal standards may not be trivial. Both the chromatographic properties and the detectability of internal standards need careful consideration. For example, for an HPLC method with UV detection, an ideal internal standard should have polarity and solubility properties similar to the analytes of interest to ensure that it is carried through extraction and cleanup steps in similar fashion, and it should also have a similar UV absorption. Furthermore, the internal standard must obviously not be present in the extract, which often excludes the use of naturally occurring analogs of the analytes as internal standards. One possible solution is to use simple, stable derivatives of the analytes or analogs (e.g., Trumble et al. 1992), particularly ones that are not natural products. This technique has the further advantage that the properties of the internal standard can be fine-tuned by preparation of several different derivatives for testing.

2.5.7. Troubleshooting HPLC Systems and Separations

With so many integrated parts operating at high pressure, and solvent flowing through very narrow bore tubing and frits, problems are inevitable. In addition to the help provided by the troubleshooting section of the instrument manual, most suppliers of chromatography columns and hardware publish useful trouble-shooting bulletins and maintain technical help lines. Some of the more common HPLC problems, causes, and solutions are summarized in Table 2.4.

To a large extent, problems can be minimized by good operating practices. As a matter of routine, all solvents and buffers should be of high quality and properly filtered and degassed. After using buffers or corrosive aqueous solvents, the system should always be flushed with water. Guard columns must be used, preferably in combination with in-line filters before the injector. If possible, samples should be subjected to at least one cleanup step, and filtered or centrifuged to remove particulates. Low-dead-volume syringe filters for HPLC are readily available.

2.6. Size Exclusion Chromatography

Size exclusion chromatography (SEC, = gel filtration or gel permeation chroma-tography) uses porous gels to effect separations on the basis of molecular size. In theory, separation depends only on the degree of penetration of analytes into gel pores; in practice, adsorption and partition effects may play minor to dominant roles in the separation as well (see below). When operated in strictly SEC mode, molecules larger than the pores are completely excluded and elute with the void volume V_0 as an unresolved mixture, and smaller molecules then elute in descending order of size within a volume equal to the total pore volume V_p. Thus, the useable separation volume of an SEC column is only V_p, which limits the number of separable peaks to about 10. Under optimal conditions, compounds differing in molecular weight by about 10% may be separated. In isolations of bioactive compounds, SEC is most useful as an intermediate fractionation step, in which a group of molecules of similar molecular weights are separated from larger and smaller compounds. It is particularly useful, for example, as a cleanup step for GC (Tumlinson & Heath 1976; Coeffelt et al. 1979) or HPLC (Persoons

Table 2.4.   Common HPLC Problems, Probable Causes, and Solutions

Problem	Possible cause	Solutions
Poor peak shape	Overloading	Reduce sample size.
	Injection volume too large	Concentrate; inject in weak solvent so that peak is refocused at head of column.
	Dead volume in system	Check connectors match fittings; fill void at top of column by tightening piston; manually fill void with packing.
	Poor efficiency	Use smaller, more uniform packing.
	Blocked frit	Backflush or replace frit.
	Coelution of two compounds	Adjust solvent selectivity.
	Injection solvent too strong	Reduce injected solvent strength.
	Ionizable solutes	Use and/or adjust pH of buffers.
Ghost peaks	Contamination from previous run	Use longer gradient, elution time.
	Contaminated solvent	Replace solvent.
Baseline problems, spikes	Gas bubbles in solvent	Degas solvent; install backpressure regulator at outlet.
	Difference between injection solvent and mobile phase	Use mobile phase as injection solvent.
Baseline problems, drift	Contaminated mobile phase	Prepare fresh mobile phase.
	Temperature fluctuations	Use column oven; cover RI detector.
Baseline problems, noise	Lamp/detector problem	Replace lamp, clean flow cell.
	Pump problems	Check pump operation.
	Contamination	Flush periodically with strong solvent.
Lack of sensitivity	Inappropriate detector/wavelength	Change detector/wavelength.
	Sample too small	Increase sample size.
	Irreversible adsorption	Use stronger solvent or different stat. phase.
Changing retention times	Leaks in connections	Check/tighten all connections; replace damaged fittings; try a polymeric connector in distorted fitting instead of stainless steel.
	Pump leaks/pump seal or check valve failure	Replace pump seals and/or check valves.
	Injector leaks	Replace injector rotor.
	Insufficient buffering capacity	Increase buffer concentration.
	Contamination	Flush column periodically with strong solvent.
	Solvent mixing problems	Check pump mixing and delivery.
	Insufficient equilibration time	Increase equilibration time.
	Evaporation of volatile solvent component	Sparge less vigorously; replace solvent.
	Damage to packing material	Use less corrosive solvent, more neutral pH.
	Adsorbed water on normal phase column	Regenerate column.

Continued on next page

Table 2.4.   Continued

Problem	Possible cause	Solutions
Fluctuating pressure	Leaks	See leak correction procedures above.
	Air bubble in pump	Degas solvents and purge/reprime pump.
High backpressure	Flow rate too fast for solvent viscosity	Decrease flow and/or solvent viscosity.
	Lines, frits, column blocked	Clean or replace lines and frits; backflush and regenerate column.
	Salt precipitation	Decrease buffer concentration; use nonsalt buffer.

et al. 1979), separating large, nonvolatile and/or lipophilic molecules from smaller molecules.

SEC gels are characterized by pore size, which determines the approximate size of molecules that will be completely excluded (the exclusion limit), and the distribution of pore sizes, which affects the range of molecular sizes that can be separated. This is most easily visualized in a plot of log molecular weight versus elution volume (Fig. 2.4). Gel A with a narrow pore size distribution has the flattest slope over the separation range (V_p), providing the best separation over a comparatively short molecular weight interval, while gel B with a broader range of pore sizes has a steeper slope, indicating that it is not as effective at separating molecules of similar size, but will separate a broader size range of molecules. Note also that for both gels (Fig. 2.4), molecules larger than the exclusion limits elute together at the void volume V_0, and all molecules have eluted from the column at the volume $V_0 + V_p$. In practice, the key parameter for separation is the effective molecular size, rather than the actual molecular size, with compact, polycyclic compounds having relatively smaller effective sizes than linear molecules. Consequently, when being used for estimation of molecular weights, the column should be calibrated with compounds of similar structure to the analytes.

SEC gels are made from a variety of materials, including rigid glass and silica, or softer dextrans, polystyrene-divinyl benzene, and acrylamide polymers. For the polymeric materials, the degree of cross-linking determines both the pore size and rigidity of the gels. This has two results. First, some polymeric gels swell in solvent (as much as a factor of 5 or more!), and the degree of swelling and the consequent exclusion limit of the gel depend on the affinity of the gel for the particular solvent. If the solvent is changed, the gel may shrink or expand, so some care needs to be exercised when using these materials. Second, the swollen gels, particularly those with larger pore sizes, can collapse if overpressurized, resulting in loss of resolution and reduced flow. Thus, softer gels must be used at low pressures and slow flow rates.

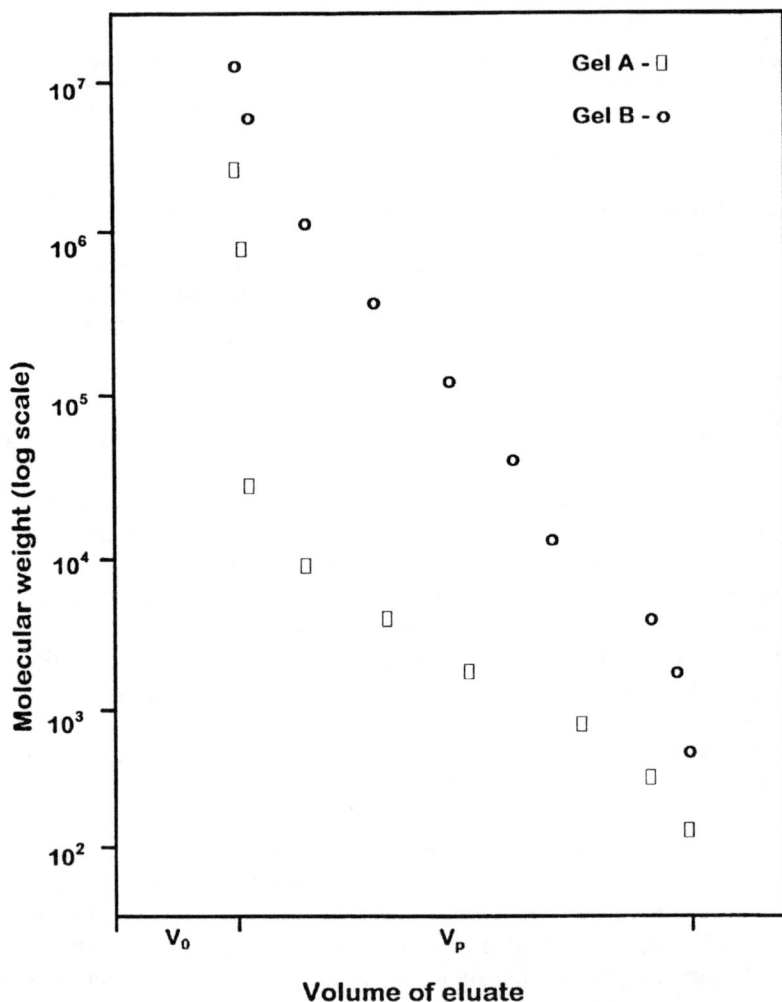

Figure 2.4. Calibration graphs for size exclusion chromatography gels. Gel A exhibits good resolution over a relatively narrow molecular weight range (~10^2 to 10^4), while gel B has lower resolving power, but a much larger operational range (~10^3 to 10^6).

Most SEC gels are designed for separation of water-soluble macromolecules, but there are several types, such as Sephadex® LH-20 (a hydroxypropylated dextran, Pharmacia), and the BioBeads® S-X and TSK-H® series (polystyrene-divinylbenzene, BioRad and Tosohaas, respectively) specifically made for use with organic solvents (e.g., THF, CH_2Cl_2, $CHCl_3$, and toluene). Some care still needs to be exercised when changing solvents because of differential swelling problems. The technical literature for each gel contains details of solvent compatibilities, and good pamphlets on the characteristics and use of SEC materials are available (e.g., Pharmacia 1980, 1991a; Phenomenex 1990a).

Fortunately, for most applications in chemical ecology, the target molecules are relatively small (molecular weight <5000), appropriate for the small pore size, most rigid gels. These gels will usually tolerate some degree of pressurization, so that reasonably fast flow rates can be used to speed up analyses (see below). SEC gels designed for HPLC are the most rigid, and can be used at typical HPLC flow rates; band broadening due to slow diffusion and mass transfer is minimal with small molecules, so there is no reason to run at the slow rates required with macromolecules, in either low-pressure or HPLC applications.

2.6.1. Practical Consideration in SEC Chromatography

SEC parameters are determined by the resolution required and the sample characteristics. The gel is chosen based on the solvent to be used, and so that the molecules of interest elute within the fractionation range of the gel (i.e., within the shallowest part of the slope in Fig. 2.4). The solvent should be chosen to minimize adsorption effects; for example, polar solvents should be used with polar silica or dextran based materials, and less polar solvents with polystyrene-divinylbenzene gels. Factors affecting resolution include gel bead size and uniformity, uniformity of packing, and column length. Column length is dictated by the resolution required, and column diameter by the sample volume. Longer columns provide better resolution, but in practice, column length is usually restricted to 1 m or less. For better resolution, several columns can be connected in series. In all cases, dead volumes in tubing and the like must be minimized to maintain resolution.

Sample size determines column diameter, because resolution depends on a low ratio of sample volume to column volume. For optimal resolution, sample size should be no more than 1–5% of the total column volume to maintain a narrow starting zone. Concentration of the applied sample volume is important only to the extent that it affects viscosity; viscous samples move down the column unevenly, and the viscosity of the applied sample should be less than twice the solvent viscosity. Thus, small organic molecules can often be used in concentrations of 5–25%, whereas molecules which form high-viscosity solutions (e.g., proteins) must be applied as dilute samples.

For group separations, where molecular weights differ by an order of magnitude of more, short columns of coarser beads can be used with high loadings. For example, in desalting proteins, sample size can be >25% of the column volume!

2.6.2. Loading and Operation of Low-Pressure SEC Columns

The dry gel is swollen overnight in the desired solvent, remembering that the gel will take up about two to five times its volume of solvent (depending on solvent). A thick slurry of the swollen gel (~75% gel, 25% solvent) is then poured down the wall of the column in one continuous motion, the remainder of the column is filled with solvent, and flow is started immediately so that the bed packs evenly. Several column volumes of solvent are passed through the column to finish the packing process. The quality of the packing process can be checked by running a colored, nonretained standard (e.g., blue dextrans), which should move down the column as a uniform band. The column is then calibrated with a set of appropriate standards. Kits of standards are available, but it may be more useful to use a personal set of standards tailored to one's specific application.

If using SEC frequently, purchase of proper columns with adjustable, low-dead-volume end fittings for containing the bed and sample introduction adapters is worthwhile. Kits are available (e.g., Pharmacia, BioRad). For occasional use, standard glass LC columns can be used, if care is taken to minimize dead volumes. The sample is introduced by running the solvent down to the top of the bed, carefully adding the sample solution and running it into the bed, rinsing it in with a few milliliters of solvent, and then carefully filling the column with solvent. The column can be eluted with gravity flow, using a simple siphoning system to maintain a constant head pressure. With a heavy-wall flash chromatography column, the column can be pressurized slightly to increase flow, or a pump can be used (e.g., an HPLC pump is readily adapted), but make sure that the column will not be overpressurized, because of the potential to burst the column and because highly swollen gels can be compressed, ruining their operation. HPLC pumps usually have adjustable maximum pressure safety switches, to stop the flow if the set pressure is exceeded. Separations can be monitored in real time with a refractive index or UV detector, or checked, for example, by TLC of fractions.

2.6.3. Partition and Adsorption Chromatography with SEC Gels

In practice, partitioning and adsorption effects may dominate chromatography with some SEC gels. These effects are documented best for Sephadex® LH-20, a hydroxypropylated dextrans gel with both hydrophilic and lipophilic properties. When used with mixed polar and nonpolar solvents (e.g., hydrocarbon-CHCl₃ mixtures), this gel preferentially takes up the polar solvent. Analytes are then partitioned between the polar "stationary phase" (i.e., the gel swollen with the polar solvent) and the eluting solvent. In fact, in most applications in which Sephadex® LH-20 has been used with a mixture of solvents, partitioning effects rather than molecular size may dominate the separation (Pharmacia 1980).

For example, aromatic compounds are strongly adsorbed by Sephadex® LH-20, particularly when used with alcohol solvents. This effect has been put to good use in the separation of tannins (Strumeyer & Malin 1975; Navon et. al 1993; also see section 1.6.5). The effect is minimized by using $CHCl_3$ instead of alcohols as eluting solvent. However, compounds with hydroxyl or carboxyl functions may be retarded when using $CHCl_3$ due to H-bonding interactions with the hydroxyl functions of the gel matrix. In short, a judicious choice of solvent enables considerable manipulation of compound separations, and as a result, Sephadex® LH-20 chromatography is widely used in both size exclusion mode and partition/adsorption mode with mixed solvent systems (Henke 1995).

2.7. Ion-Exchange Chromatography

In ion-exchange chromatography, charged analytes in aqueous solutions are reversibly bound to oppositely charged moieties bonded to an inert matrix. Compounds are eluted by decreasing the affinity of the analytes for the ion exchanger, by shifting the pH of the eluent toward the value at which the analytes become neutral, or by increasing the ionic strength, which increases the competition for charged sites. Either stepwise or linear elution gradients can be used. Separations occur due to differences in pK_a (isoelectric points for amphoteric compounds like amino acids) between the analytes. The method has good resolving power, high capacity, and considerable flexibility in both stationary and mobile phases. In its simplest form, ion exchange is used for crude fractionation of extracts into cationic, anionic, and neutral fractions. With a little more care, high-resolution separations are readily achieved.

Packings are characterized by the base matrix (silica, polystyrene-divinylbenzene, cellulose, dextrans), the fixed ion bound to the matrix, and the capacity, expressed as milliequivalents per gram of packing. Capacity, rigidity, and flow properties all vary considerably with the matrix, so some care must be exercised in choosing a packing. The designations "weak" and "strong" refer not to the strength of interactions with analytes, but instead to the type of fixed ion; strong cation (sulfonic acids) or anion (quaternary amines) exchangers are fully ionized at working pH, while the degree of ionization of weak cation (carboxylic acids) or anion exchangers (1°, 2°, or 3° amines) depends on the pH of the mobile phase. Nonrigid packings may change volume with solvent, pH, or ionic strength, particularly with weak ion exchangers as they change from charged to neutral form with changing pH, so some degree of care must be exercised. Instructions supplied with the packings/columns provide details of solvent compatibility. As usual, small, uniform beads, packed uniformly in long columns provide the best resolution.

Samples are loaded onto the column at a pH at which the analytes are fully charged (at least 1 pH unit away from neutral point) so that they bind tightly to the packing. Nonretained compounds are then flushed off with the loading buffer, followed by progressive elution of analytes by shifting the pH toward the point where analytes become neutral, by increasing the ionic strength, or both. As an estimate, substrates start to dissociate from the fixed ions at about 0.5 pH units away from the neutral point (Pharmacia 1991b).

The choice of whether to use a strong or weak exchanger is primarily dependent on the working pH range; if high or low pH ranges are needed, strong exchangers should be used. With strong ion exchangers, the capacity and separation mechanism also do not change with pH, while with weak exchangers, retention characteristics change markedly as the fixed ions become neutral with changing pH. The working range of each packing is clearly specified.

Sample size depends on several factors. For gradient elutions, where the sample loaded is strongly retained, short columns (length about four to five times diameter) loaded to 10–20% of their capacity give good resolution. However, if the column is eluted with the loading buffer, then the sample volume should not exceed 1–5% of the column volume.

The choice of initial conditions can be somewhat arbitrarily selected based on a rough knowledge of the types of compounds to be separated, if gradient elutions are used. The choice of initial conditions is easily determined more accurately by simple experiments. To determine the starting pH, ~0.1 g of packing is placed in each of 10 test tubes. The gel in each tube is equilibrated by washing with 10 aliquots of 0.5-M buffer, with the pH of the buffer used with each tube varying by 0.5 pH units, to provide a series spanning 5 pH units. Normal ranges to test are pH 5–9.5 for anion exchangers, and 3.5–8 for cation exchangers. The packing in each tube is then equilibrated by washing five times with 10 ml of buffer of the same pH but lower ionic strength. Aliquots of buffer are then added to each tube, along with aliquots of sample, the tubes are mixed for 20 min, and the supernatant is assayed for analyte. The pH to use is the first one at which there is no analyte in the supernatant (i.e., where the analytes are just bound). To determine the initial ionic strength, the above experiment is repeated, keeping the pH fixed, and varying the ionic strength from 0.05 to 0.5 M. The initial ionic strength to use is the last one in which there is no analyte in the supernatant.

The choice of buffer depends on the pH range required, and buffering ions with the same charge as the fixed ions are usually used so that they do not compete with the analytes for the fixed ions. The initial buffer concentration will depend somewhat on the nature of the buffer used. Some common buffers are listed in Table 2.5. Volatile buffers are convenient for preparative purposes because they are easily removed under vacuum. More extensive tables of buffers are found in manufacturers' brochures (e.g., Pharmacia 1991b; Phenomenex 1990b).

Ion exchangers usually come with one of several different counterions. Dry gels must be swollen at the pH to be used by heating an unstirred slurry of the gel in buffer on a boiling water bath for 2 h, replacing the buffer several times. Columns are loaded as described for SEC (section 2.6.2). The gel should contain the same counterion as in the initial solvent. If not, it can be exchanged by suspending the gel in a 1-M solution of a salt of the new counterion for 1 h. The supernatant is then decanted off, and the process is repeated several times with the starting buffer.

Factors affecting retention, besides pH and ionic strength, include the nature

Table 2.5. Common Buffers for Ion-Exchange Chromatography (from Pharmacia 1991b).

Substance	pK$_a$	Working pH	Concentration (mM)	Counterion
cation exchange				
citric acid	3.1	2.6–3.6	20	Na$^+$
formic acid	3.8	3.8–4.3	50	Na$^+$, Li$^+$
acetic acid	4.8	4.8–5.2	50	Na$^+$, Li$^+$
malonic acid	5.7	5.0–6.0	50	Na$^+$, Li$^+$
phosphate	7.2	6.7–7.6	50	Na$^+$
anion exchange				
piperazine	5.7	5.0–6.0	20	Cl$^-$, formate
bis-tris propane	6.8	6.4–7.3	20	Cl$^-$
triethanolamine	7.8	7.3–8.0	20	Cl$^-$
tris	8.1	7.6–8.0	20	Cl$^-$
N-methyl diethanolamine	8.5	8.0–8.5	50	Cl$^-$, acetate
1,3-diaminopropane	8.6, 10.5	8.5–9.0, 9.8–10.3	20	Cl$^-$
ethanolamine	9.5	9.0–9.5	20	Cl$^-$
volatile buffers				
pyridine/acetic acid		3–6		
trimethylamine/HCl		6.8–8.8		
trimethylamine/CO$_2$		7–12		

of the competing ion, temperature, and flow rate. Increased temperatures and slow flow rates both contribute to larger elution volumes due to increased interaction of analytes with the fixed ions on the packing. It may be necessary to change counter-ions in order to elute firmly-bound analytes. Ion affinities for strong cation exchangers roughly follow the series $K^+ > NH_4^+ > Na^+ > H^+ > Li^+$, while strong anion exchanger affinities follow the series citrate $> Cl^- > HCO_3^- > H_2PO_4^- >$ acetate $>$ formate $> F^- > OH^-$.

2.8. Argentation Chromatography

LC with silver ions either adsorbed on the stationary phase or dissolved in the mobile phase separates compounds on the basis of the number, geometry, and substitution patterns of carbon-carbon double bonds. Alkynes are also strongly retained but aromatic compounds are not. The strength of the complexation between Ag^+ ions and alkenes is strongly affected by steric hindrance, and so both cyclic and acyclic compounds differing only in the position of the double bond can be separated (Phelan & Miller 1981, Friedel & Matusch 1987). Argentation chromatography is used with TLC, and low- and high-pressure LC, using normal or reverse phases, and silver-loaded ion exchangers. It has been widely used in the separation of fatty acid esters and other lipids (Juaneda et al. 1994), terpenoids (Friedel & Matusch 1987), and insect pheromones (Phelan & Miller

1981). It has been used on a small scale in disposable pipette columns (Blomquist et al. 1980) to a large, preparative scale, using ratios of stationary phase to sample of about 100:1 or less, particularly for reverse phase. It is the most useful method of separating olefin isomers, particularly on a preparative scale. An applications bibliography is available from EM Separations (Gibbstown, NJ).

A number of silver-loaded chromatography phases are commercially available, but it is also simple to make one's own. However, a few words of caution. First, silver salts are light sensitive, so all silver-loaded media must be protected from light during and after their preparation. Second, AgNO$_3$ is toxic, and it rapidly stains skin black, so adequate precautions should be taken during handling. Third, AgNO$_3$ is corrosive, so all metal equipment, particularly HPLC equipment, should be thoroughly flushed with water after use. Furthermore, silver may be leached off packings and must be removed from fractions before long-term storage by thorough washing of a solution of the purified compound with dilute aqueous NaCl (Heath et al. 1975), or by passing a solution of the compound through a plug of powdered NaCl (Arn et al. 1979).

TLC plates coated with AgNO$_3$-impregnated silica gel are available with 5–20% AgNO$_3$ (Analtech, Alltech). Alternatively, standard silica (or other normal phase plates) with a binding agent can be dipped in aqueous AgNO$_3$ (5–25%) for 20 min, air-dried, then oven-dried overnight at 80°C. Plates can also be prepared from scratch by simply incorporating the required amount of AgNO$_3$ into the aqueous silica slurry used to coat the plates. For relatively nonpolar compounds, plates are developed with mixtures of hexane with ether, toluene, or 1-hexene, with stronger solvent being used as required for more polar analytes. Developed plates can be visualized with sulfuric or phosphomolybdic acid and heat (section 2.2.2).

For low-pressure LC, silver-impregnated silica and alumina are available from Aldrich or are readily prepared by slurrying silica in an acetonitrile solution of the appropriate amount of AgNO$_3$ (Evershed et al. 1982). The solvent is removed by rotary evaporation, followed by warming under vacuum. Water can be used in place of acetonitrile, but it is more difficult to remove. Before use, packings are activated by heating overnight at 110°C. A fritted adapter (Aldrich #Z10,747-6) or a cotton or glass wool plug should be used in the throat of the rotary evaporator to prevent migration of abrasive packing into the innards of the apparatus and the pump. The material is then packed as usual (section 2.4), and eluted using the solvents described above. Column sizes can be as small as a packed pipette (Blomquist et al. 1980).

Low-pressure LC has also been used with a silver-loaded macroporous strong cation exchange resin (Lewatit SP 1080; Houx et al. 1974). The water-swollen gel was flushed with 0.4-M aqueous AgNO$_3$ until silver ions were detected in the eluate, then rinsed thoroughly with water and equilibrated with MeOH. Long columns (1.5 and 2.1 m) with small, uniform particle sizes were used to separate monoenes (up to ~500 mg/run), eluting with MeOH. This resin is no longer available, but analogous macroporous strong cation exchange resins (e.g., Toyo-pearl SP 650 resins; TOSOHAAS Montgomeryville, PA) may work equally well.

Argentation HPLC has been carried out with normal and reverse phases, and with ion exchangers; a normal phase column can be purchased (ChromSpher® Lipids, Chrompack, Middelburg, The Netherlands). Silver-loaded normal phase columns are easily prepared from any standard column (Heath & Sonnet 1980). The column is mounted vertically, with a stainless steel reservoir connected on top (e.g., an empty preparative HPLC column). The reservoir is filled with an acetonitrile solution of AgNO$_3$ (0.35 g in 15 ml for a 4.6 mm × 25 cm column), then topped up with hexane. The system is connected to the pump, and hexane is pumped in, forcing the AgNO$_3$ solution through the silica bed. The coated column is then equilibrated with the mobile phase to be used (e.g., hexane with ether or benzene). This method has the advantages that minimal quantities of expensive AgNO$_3$ are required, and the injector and pump are not exposed to corrosion.

Argentation chromatography is also useful in C$_{18}$ reverse-phase HPLC (e.g., Vonach & Schomburg 1978). In an excellent study, Phelan and Miller (1981) demonstrated complete separation of positional and geometric isomers of mono-ene alcohols, acetates, and aldehydes, all four isomers of some nonconjugated diene acetates, and farnesene and farnesol isomers, in both analytical and prepara-tive scale. Using commercial C$_{18}$ columns, monoene isomers were separated best with mobile phases of 0.1-M AgNO$_3$ in 20–30% water in MeOH (depending on chain length), while lower concentrations (0.05 and 0.012 M) were used to separate dienes and polyenes respectively. Working in reverse phase has the advantages that the method is simplest to use, there are a variety of different packing chemistries that can be used to manipulate the separations, and the silver ion concentration can be manipulated at will to optimize separations. Disadvantages include the high cost of AgNO$_3$ and the corrosive nature of the mobile phase.

Cation exchange HPLC columns have also been loaded with silver ions by flushing with ~20 column volumes of 1-M AgNO$_3$ and then water, followed by a series of organic solvents of decreasing polarity (e.g., 10 micron Partisil® SCX, Merritt & Bronson 1977; 5 micron RSilCat®, Powell 1981). Analytes are eluted with mixtures of dioxane, methanol, and CHCl$_3$, with a small percentage acetoni-trile, which apparently competes for active sites. A simpler method of loading has recently been described, in which a 5 micron 4.6 mm × 25 cm Nucleosil® SA column was flushed sequentially with 60 ml 1% NH$_4$NO$_3$ and water, followed by injection of 10–100 µl aliquots of 20% AgNO$_3$ over 10 min, while eluting with water (Juaneda et al. 1994). The column was rinsed with water and MeOH, then equilibrated with mobile phase (CH$_2$Cl$_2$-MeOH), and used to separate all eight isomers of linolenic acid phenacyl esters.

2.9. Countercurrent Chromatography

Countercurrent chromatography (CCC) is undergoing rapid development, and the technique has now been used in the isolation of numerous natural products. With the development of commercial instruments by several companies (e.g., Sanki Engineering, Kyoto, Japan; P.C. Inc., Potomac, MD; Pharma-Tech Research, Baltimore, MD), CCC is rapidly becoming more accessible to natural products chemists. Its main advantages are the total recovery of all applied sample in one or the other of the two liquid phases, its versatility and good resolution, and its reproducibility. However, each run may take from hours to days, which may render method development tedious. It is particularly useful for the isolation of milligram to multigram quantities of large, polar, and/or labile compounds, which may be irreversibly adsorbed or degraded by solid LC stationary phases.

CCC is analogous to normal liquid chromatography with the exception that both the mobile and stationary phases are immiscible liquids, one of which is usually aqueous based. Either phase can be the mobile phase, or in some cases, both phases are mobile, and are passed through each other. In analogy to other chromatographic methods, analytes are partitioned between the two immiscible liquids, with separation occurring as a result of differences in the partition coefficients of analytes between the two phases. Any two immiscible liquids can be used, allowing for virtually infinite manipulation of a given separation by adjustments to the composition of one or both phases. Analytes are limited only by the requirement that they have at least limited solubility in both phases.

In its original form, droplet CCC, droplets of the mobile phase are passed through the immiscible stationary phase contained in several hundred vertical tubes connected in series, top to bottom, by capillary tubing. The sample is loaded into the first tube, and then the mobile phase is introduced as droplets (from the bottom of the first tube, if mobile phase is lighter than the stationary phase, from the top if the reverse). The droplets rise to the top of the first tube, then pass through the connecting tube to the bottom of the second tube, and so on. The analytes partition between the mobile phase droplets and the stationary phase, with those analytes with the highest affinity for the mobile phase eluting first. In the original apparatus, the speed of the droplets of mobile phase was determined by the relative density of the two phases, and separations routinely took hours to days. However, it was rapidly realized that the speed of the mobile phase could be increased by rotating the apparatus to provide centrifugal driving forces much larger than gravity. This principle forms the basis of centrifugal partition chromatography, and it is used in commercial instruments from Sanki Engineering (Kyoto, Japan). The apparatus and its applications have been described in a recent monograph (Foucault 1995).

In a more recent innovation, one or more long helical coils filled with mixed mobile and stationary phases are rotated about a central axis. At the correct combination of coil diameter, solvents, and rotational speed, the stationary phase is maintained in the coil by centrifugal force, while the mobile phase is pumped through the coil, eluting the analytes. There are many modifications on the basic design. The mechanistic details are complex and have not yet been completely modeled. Space does not permit even a cursory description of the major basic designs, and the interested reader is referred to several excellent mongraphs (Mondava & Ito 1988; Conway 1990; Ito & Conway 1996) for thorough discussions of both the method and its expanding applications in natural products research.

2.10. Acknowledgments

I thank J. Steven McElfresh for preparation of the figures and Alan Renwick for reviewing the manuscript.

2.11. References

Alltech. 1995. Three ways to extend HPLC column life. Alltech Bulletin #326:8–9. Alltech Assoc., Deerfield IL.

Arn, H, S. Rauscher & A. Schmid. 1979. Sex attractant formulations and traps for the grape moth *Eupoecilia ambiguella* Hb. Bull. Soc. Entomol. Suisse **52**:49–55.

Bailey, P.F. & K.C. Patt. 1992. The LPLC Handbook: A Guide to the Selection, Installation and Operation of Low Pressure Liquid Chromatography Instruments. Isco Inc., Lincoln NE.

Blomquist, G.J., R.W. Howard, C.A. McDaniel, S. Remaley, L.A. Dwyer & D.R. Nelson. 1980. Application of methoxymercuration-demercuration followed by mass spectrometry as a convenient microanalytical technique for double-bond location in insect-derived alkenes. J. Chem. Ecol. **6**:257–269.

Blunt, J.W., V.L. Calder, G.D. Fewnwick, R.J. Lake, J.D. McCombs, M.H.G. Munro & N.B. Perry. 1987. Reverse phase flash chromatography: a method for the rapid partitioning of natural product extracts. J. Nat. Prod. **50**:290–292.

Coffelt, J.A., K.W. Vick, P.E. Sonnet & R.E. Doolittle. 1979. Isolation, identification, and synthesis of a female sex pheromone of the navel orangeworm, *Amyelois transitella* (Lepidoptera: Pyralidae). J. Chem. Ecol. **5**:955–966.

Coll, J.C. & B.F. Bowden. 1986. The application of vacuum liquid chromatography to the separation of terpene mixtures. J. Nat. Prod. **49**:934–936.

Conway, W.D. 1990. Countercurrent Chromatography: Apparatus, Theory, and Applications. VCH, New York.

Dawson, I. 1985. Lab notes. Aldrichimica Acta **18**:2.

Dolan, J.W. & P. Upchurch. 1988. Troubleshooting LC fittings, part II. L.C.-G.C. **6**:886–892.

Evershed, R.P., E.D. Morgan & L.D. Thompson. 1982. Preparative-scale separation of alkene geometric isomers by liquid chromatography. J. Chrom. **237**:350–354.

Fell, R.D. 1996. Thin-layer chromatography in the study of entomology. *In:* Practice of Thin Layer Chromatography: A Multidisciplinary Approach, eds. B. Fried & J. Sherman, pp. 71–104, CRC Press, Boca Raton, FL.

Finidori-Logli, V., A.-G. Bagneres, D. Erdmann, W. Francke & J.-L. Clement. 1996. Sex recognition in *Diglyphus isaea* Walker (Hymenoptera: Eulophidae): role of an uncommon family of behaviorally active compounds. J. Chem. Ecol. **22**:2063–2079.

Foucault, A.P. 1995. Centrifugal Partition Chromatography. Marcel Dekker. New York.

Fried, B. & J. Sherma. 1994. Thin Layer Chromatography. Techniques and Applications, 3rd ed. Marcel Dekker, NY.

Fried, B. & J. Sherma, eds. 1996. Practical Thin-Layer Chromatography. A Multidisciplinary Approach. CRC Press, Boca Raton, FL.

Friedel, H.D. & Matusch, R. 1987. Separation of non-polar sesquiterpene olefins from tolu balsam by high-performance liquid chromatography; silver perchlorate impregnation of a prepacked preparative silica gel column. J. Chrom. **407**:343–348.

Hadacek, F., C. Müller, A. Werner, H. Greger & P. Proksch. 1994. Analysis, isolation and insecticidal activity of linear furanocoumarins and other coumarin derivatives from *Peucedanum* (Apiaceae: Apioideae). J. Chem. Ecol. **20**:2035–2054.

Harborne, J. 1984. Phytochemical Methods, 2nd ed. Chapman & Hall, London.

Harwood, L.M. 1985. "Dry-column" flash chromatography. Aldrichimica Acta **18**:25.

Heath, R.R. & P.E. Sonnet. 1980. Technique for in situ coating of Ag$^+$ onto silica gel in HPLC columns for the separation of geometrical isomers. J. Liq. Chrom. **3**:1129–1135.

Heath, R.R., J.H. Tumlinson, R.E. Doolittle & A.T. Proveaux. 1975. Silver nitrate-high pressure liquid chromatography of geometrical isomers. J. Chrom. Sci. **13**:380–382.

Henke, H. 1995. Preparative Gel Chromatography on Sephadex LH-20. Translated by A.J. Rackstraw. Hüthig GmbH, Heidelberg.

Houx, N.W.H., S. Voerman, & W.M.F. Jongen. 1974. Purification and analysis of synthetic insect attractants by liquid chromatography on a silver-loaded resin. J. Chromatog. **96**:25–32.

Huang, X., J.A. Renwick & K. Sachdev-Gupta. 1993. A chemical basis for differential acceptance of *Erysimum cheiranthoides* by two *Pieris* species. J. Chem. Ecol. **19**:195–210.

Ito, Y. & Conway, W.D. 1996. High-speed Countercurrent Chromatography. John Wiley, New York.

Jork, H., W. Funk, W. Fischer & H. Wimmer. 1990. Thin Layer Chromatography Reagents and Detection Methods, Vol. 1a. VCH, Weinheim, Germany.

Juaneda, P., J.L. Sebedio & W.W. Christie. 1994. Complete separation of the geometrical isomers of linolenic acid by high performance liquid chromatography with a silver ion column. J. High Res. Chrom. **17**:321–322.

Khlebnikov, V. 1996. Lab notes. Aldrichimica Acta **29**:58.

Kirchner, J.G. 1990. Thin-Layer Chromatography: Techniques of Chemistry, Vol. XIV, 2nd ed. John Wiley & Sons, 1978, reprinted 1990 by Sigma Chemical Co., St. Louis, MO.

Kubo, I. & T. Nakatsu. 1990. Recent examples of preparative-scale recycling high performance liquid chromatography in natural products chemistry. L.C.-G.C. **8**:933–939.

Kubo, I., S. Komatsu, T. Iwagawa & D.L. Wood. 1986. Analytical and preparative separation of bark beetle pheromones by high-performance liquid chromatography. J. Chromatogr. **363**:309–314.

Mandava, N.B. & Y. Ito. 1988. Countercurrent Chromatography: Theory and Practice. Marcel Dekker, New York.

Merritt, M.V. & G.E. Bronson. 1977. High Performance liquid chromatography of p-nitrophenacyl esters of selected prostaglandins on silver-ion loaded microparticulate cation-exchange resins. Anal. Biochem. **80**:392–400.

Meyers, A.I., J. Slade, R.K. Smith & E.D. Mihelich. 1979. Separation of diastereomers using a low cost preparative medium-pressure liquid chromatograph. J. Org. Chem. **44**:2247–2249.

Navon, A., J.D. Hare, & B.A. Federici. 1993. Interactions among *Heliothis virescens* larvae, cotton condensed tannin, and the CryIA(c) δ-endotoxin of *Bacillus thuringiensis*. J. Chem. Ecology **19**:2485–2499.

Pelletier, S.W., H.P. Chokshi & H.K. Desai. 1986. Separation of diterpenoid alkaloid mixtures using vacuum liquid chromatography. J. Nat. Prod. **49**:892–900.

Persoons, C.J., P.E.J. Verwiel, E. Talman & F.J. Ritter. 1979. Sex pheromone of the American cockroach, *Periplaneta americana*. J. Chem. Ecol. **5**:221–236.

Pharmacia. 1980. Sephadex LH-20: chromatography in organic solvents. Pharmacia Fine Chemicals, Uppsala, Sweden.

Pharmacia. 1991a. Gel Filtration: Principles and Methods, 5th ed. Pharmacia LKB Biotech, Piscataway, NJ.

Pharmacia. 1991b. Ion Exchange Chromatography: Principles and Methods, 3rd ed. Pharmacia LKB Biotech, Piscataway, NJ.

Phelan, P.L. & J.R. Miller. 1981. Separation of isomeric insect pheromonal compounds using reversed phase HPLC with $AgNO_3$ in the mobile phase. J. Chrom. Sci. **19**:13–17.

Phenomenex. 1990a. A Users Guide to Gel Permeation Chromatography. Phenomenex, Torrance, CA.

Phenomenex. 1990b. A Users Guide to HPLC Trouble-shooting. Phenomenex, Torrance, CA.

Poole, C.F. & S.K. Poole. 1991. Chromatography Today. Elsevier, Amsterdam.

Powell, W.S. 1981. Separation of icosenoic acids, monohydroxyicosenoic acids, and prostaglandins by high pressure liquid chromatography on a silver ion-loaded cation exchange column. Analyt. Biochem. **115**:267–277.

Rabel, F. & K. Palmer. 1992. Regeneration of silica based HPLC columns. Am. Lab. **August**: pp. 65–66.

Scott, R.P.W. 1996. Chromatographic Detectors. Marcel Dekker, NY.

Sherma, J. & B. Fried. eds. 1991. Handbook of Thin-Layer Chromatography. Marcel Dekker, New York.

Smith, L.R. & I. Kubo. 1995. The racemization of crinitol. Resolution of crinitol and of 3-nonen-2-ol enantiomers by recycle-HPLC, via MTPA esters. J. Nat. Prod. **58**:1608–1613.

Snyder, L.R. 1974. Classification of the solvent properties of common liquids. J. Chromatogr. **92**:223–230.

Snyder, L.R. & J.W. Dolan. 1990. Reproducibility problems in gradient elution caused by differing equipment. L.C.-G.C. **8**:524–537.

Snyder, L.R. & J.J. Kirkland. 1979. Introduction to Modern Liquid Chromatography, 2nd ed. John Wiley & Sons, New York.

Stahl, E. ed. 1969. Thin-Layer Chromatography: A Laboratory Handbook. Springer-Verlag, Berlin.

Still, W.C., M. Kahn & A. Mitra. 1978. Rapid chromatography technique for preparative separations with moderate resolution. J. Org. Chem. **43**:2923–2925.

Strumeyer, D.H. & M.M. Malin. 1975. Condensed tannins in grain sorghum: isolation fractionation, and characterization. J. Agr. Food. Chem. **23**:909–914.

Supelco. 1986. HPLC Troubleshooting Guide. Technical bulletin #826, Supelco Inc., Bellefonte, PA.

Touchstone, J.C. & M.F. Dobbins. 1978. Practice of Thin Layer Chromatography. John Wiley & Sons, New York.

Trumble, J.T., J.G. Millar, D.E. Ott & W.C. Carson. 1992. Seasonal patterns and pesticidal effects on the phototoxic linear furanocoumarins in celery, *Apium graveolens*. J. Agr. Food Chem. **40**:1501–1506.

Tumlinson, J.H. & R.R. Heath. 1976. Structure elucidation of insect pheromones by microanalytical techniques. J. Chem. Ecol. **2**:87–99.

Tumlinson, J.H., M.M. Brennan, R.E. Doolittle, E.R. Mitchell, A. Brabham, B.E. Mazemenos, A.H. Baumhouer & D.M. Jackson. 1989. Identification of a pheromone blend attractive to *Manduca sexta* (L.) males in a wind tunnel. Arch. Insect Biochem. Physiol. **10**:255–271.

Vonach, B. & Schomburg G. 1978. High-performance liquid chromatography with Ag^+ complexation in the mobile phase. J. Chrom. **149**:417–430.

Wagner, H., S. Bladt & E.M. Zgainski. 1996. Plant Drug Analysis. A Thin Layer Chromatography Atlas, 2nd ed. Springer-Verlag, New York.

Yau, E.K. & J.K. Coward. 1988. Filtering-column chromatography—a fast, convenient chromatographic method. Aldrichimica Acta **21**:106–107.

Zweig, G. & J. Sherma. 1972a. CRC Handbook of Chromatography, Vol. I. CRC Press, Boca Raton, FL.

Zweig, G. & J. Sherma. 1972b. CRC Handbook of Chromatography, Vol. II. CRC Press, Boca Raton, FL.

3

Analytical and Preparative Gas Chromatography
Robert R. Heath and Barbara D. Dueben

3.1. Introduction

Gas chromatography (GC) is the standard method of analysis and often the method of choice for small-scale preparative separations of volatile and semivolatile compounds, for a number of reasons. First, capillary GC has very high resolution, so that many more than 100 compounds can be separated in one run. Second, GC is a simple, fast, flexible, robust, and relatively inexpensive method that can be used routinely with a minimum of specialized training. Third, the standard detector used with GC, the flame ionization detector, is both sensitive and essentially universal, so that only small amounts of sample are required (detection limit ~10–100 picograms) and all compounds in a sample are detected. Fourth, unlike most other chromatographic methods, relative retention times in GC (retention indices) are highly reproducible between instruments and between laboratories, so that retention indices form a valuable identification tool, particularly for compounds present in trace amounts where standard spectroscopic methods of identification cannot be used. Finally, GC equipment has relatively few moving parts and few things to wear out, requiring minimal maintenance, so that gas chromatographs often have a working lifetime of a decade or more.

Preparative GC can be carried out either with capillary (0.2–0.32 mm i.d.) or megabore (0.53 mm and larger) open tubular columns for separating subnanogram to microgram quantities of materials, or with larger capacity packed columns (~2–5 mm i.d.) for separating micrograms to milligrams per run, enough for almost all spectroscopic identification methods. In fact, preparative GC is really the only practical method of separation and recovery of submilligram quantities of volatile compounds, uncontaminated by solvent or other impurities. The standard flame ionization detector (FID) is destructive, but this limitation is easily circumvented for preparative separations by using a different detector (e.g., a thermal conductivity detector), or by splitting the sample so that only a small fraction goes to the detector, and the rest is collected (section 3.5).

In GC separations, analytes are partitioned between an inert carrier gas (mobile phase), which serves solely to sweep the analyte vapor through the column, and a (usually) liquid phase bonded or coated on the inner surface of the column, or on an inert support packed in the column. The retention of a particular analyte depends on its volatility and the interactions between its functional groups and those of the liquid phase. These interactions include Van der Waals nonpolar interactions, dipole-dipole interactions, hydrogen bonding, and interactions between aromatic rings. The driving force in GC is temperature; columns are generally heated within the range of 40–325°C, and as the temperature is increased, the analytes move into the vapor phase and move along the column. Alternatively, GC runs can be made isothermally provided the temperature is high enough to impart at least some volatility to the analytes. Higher temperatures (>400°C) can be used with special high-temperature columns, and temperatures <40°C can be used by installing cryogenic cooling in the GC oven (liquid CO_2

or N_2), which is available as an option on commercial instruments. Quick and dirty cooling of GC ovens to ~0°C can also be accomplished by putting 100 g or so of dry ice on the floor of the GC oven (well away from the column!).

There are two major requirements for GC samples. First, they must have at least some slight volatility at GC temperatures, and second, they must have sufficient thermal stability to survive analysis (Smith 1988). Thus, GC can be used to analyze most gases, many small and medium sized organic molecules (up to ~C_{30}, and much higher for some classes of compounds) and even some organometallic compounds. Furthermore, many other nonvolatile compounds such as sugars and amino acids can be easily converted to volatile derivatives amenable to GC analysis. The only compounds for which GC is entirely unsuitable are organic or inorganic salts, and large molecules such as synthetic or biological polymers (Poole & Poole 1991).

The basic design of the gas chromatograph has remained virtually unchanged since the early 1960s (Grob 1995). The essential components consist of a controlled flow of clean carrier gas, an injection port to introduce the sample, a temperature-controlled oven, an analytical column containing the stationary phase, and a detector with some sort of recorder. For preparative separations, this list is extended to include a fraction collector. Each of these components is discussed in detail below.

This chapter is not intended to supplant the operating manuals supplied with individual GC units, nor is it meant to provide a tutorial in the basics of gas chromatography. Instead, it is intended to provide practical information to enable the reader to obtain the best results in real-life analyses. For illustrative purposes we discuss techniques that are often used in chemical ecology.

The two components of the gas chromatograph that most affect the separation of the analyte or the resolution of compounds are the column and injector, and the bulk of the chapter concentrates on these. While other components and factors such as the column oven, temperature programming, data collection systems, detectors, and so forth are required for analyses, the contributions of these components generally have less effect on the separation than the column and injector. However, as a preamble, we begin with a brief discussion of factors crucial to optimizing separations.

3.2. Mechanisms of Separation and Useful Equations

Efficient use of gas chromatography requires some knowledge and familiarity of separation science technical terms. Whereas it is beyond the scope of this chapter to present a detailed summary of chromatographic theory and mathematics, a brief review of several of the basic equations is warranted to further a better understanding of the effects of chromatographic parameters on the separations of analytes.

The degree of separation of two compounds is a function of two parameters, the separation factor α and the column efficiency, usually defined by historical analogy with distillation theory as the number of theoretical plates N (or the height-equivalent theoretical plate [HETP] obtained by dividing the number of theoretical plates by the column length). The separation factor α depends on the interactions of the analytes with the stationary phase, which determines how long specific compounds are retained on the column. The efficiency of the column N is related to how broad or narrow peaks are and the uniformity of peak shapes.

The separation factor α is a measure of the relative amount of time each compound spends interacting with the stationary phase or support, excluding the time spent in the mobile phase. This is defined as the adjusted retention t_r' where

$$t_r' = t_r - t_m,$$

where t_r is the measured retention of the compound and t_m is the retention time of an unretained compound, i.e., the time taken for the carrier gas to pass through the column. The separation factor $\alpha_{a/b}$ for two compounds is then determined as

$$\alpha = \frac{t_r' \text{ of compound } b}{t_r' \text{ of compound } a}$$

By convention, the compound with the longer retention time is designated compound b so α is always greater than 1. The stationary phase is the primary factor in GC that determines α.

Calculation of column effective efficiency (or effective theoretical plates) requires the measurement of the t_r' and the peak width for a given peak. The efficiency of a column can be calculated in two ways.

$$N = 16 \times \frac{t_r' \text{ compound } a}{\text{width of peak } a \text{ at the base}} \text{ or}$$

$$N = 5.54 \times \frac{t_r' \text{ compound } a}{\text{width of peak } a \text{ at half height}}$$

where the peak width at the base (or width at half height) is measured in time units, and N is a measure of the resolving power of the column and is a function of column characteristics (column diameter, length, thickness of stationary phase, and uniformity of coating/packing). The efficiency of the separation, but not the inherent efficiency of the column, is also affected by the carrier gas flow rate.

A combination of efficiency N and separation factor α determine whether two compounds are resolved on a particular stationary phase, and either one or both can be manipulated to improve a separation. The combined effects of α and N on the separation of two compounds are expressed in a third parameter, the resolution R_s, calculated from the $t'_r a$ and $t'_r b$ and the idealized peak widths at the base (using triangulation):

$$R_s = \frac{2(t'_r b - t'_r a)}{(W_{base} a + W_{base} b)}$$

An additional factor affecting resolution (which is taken into account when physically measuring the parameters required in the resolution formula) is the ratio of a compound's adjusted retention time t'_r to the retention time of a nonretained compound t_m. This ratio, known as the capacity factor k, is defined as $k = t'_r/t_m$, and it is a measure of a column's ability to retain a compound. An expression for R_s can be written in a different form (for compounds eluting close to each other so that one can assume that $W_{base} a = W_{base} b$), which relates the three parameters number of plates N (i.e., efficiency), separation factor α, and capacity factor k directly to resolution R_s.

$$R_s = 1/4 \ N^{1/2} \times \frac{\alpha - 1}{\alpha} \times \frac{k}{k + 1}$$

Given that the goal of any separation is to achieve the best possible resolution, several useful conclusions can be drawn from this equation. First, resolution R_s is proportional to the square root of N, so small increases in efficiency will have negligible effect on resolution. Thus, switching from a low-efficiency packed column to a high-efficiency capillary column with an analogous stationary phase, where there is an order of magnitude or more increase in efficiency, may be worthwhile, whereas increasing efficiency by smaller increments (e.g., by increasing column length or decreasing column diameter), will be less effective. Second, using very long isothermal runs (retention times) to improve a difficult separation may not work because as k gets larger, the values of $k/(k + 1)$ approach unity, and further increases in k result in diminishingly small increases in R_s. Thus, values of k less than about 5 in isothermal runs will be most effective. For temperature programmed runs, however, larger values of k can be acceptable because later-eluting analytes may not start to move through the column at all until well into the run.

Third, resolution (proportional to $[\alpha - 1)/\alpha$) changes most dramatically with small changes in α when α is slightly >1, that is, for difficult separations. The implication from this is that switching to a stationary phase which gives even a slightly larger separation factor α will have a direct and dramatic effect on resolution. For example, if switching columns results in an increase in α from 1.01 to 1.05, the term $(\alpha - 1)/\alpha$ increases by a factor of almost 5, i.e., resolution is improved almost fivefold! In short, when attempting to improve a difficult separation, the first thing to try is switching to a column with a different stationary phase in the hope that the separation factor α will be increased.

3.3. GC System Components

3.3.1. Carrier and Makeup Gas

A regulated supply of clean carrier gas with stable and reproducible flow is crucial for effective GC operation. It should be remembered that analytes do not interact with the carrier gas, that is, the carrier gas is inert and serves solely to sweep the analytes through the column. Thus, unlike liquid chromatography, changing the mobile phase (carrier gas) has no effect on GC selectivity, but it may have major effects on chromatographic efficiency (see below).

The ideal carrier gas should be nonreactive, nontoxic, nonflammable, and cost effective. Three carrier gases are commonly used, hydrogen, helium, and nitrogen (argon-methane mixtures are also used with electron capture detectors), and each has advantages and disadvantages. The most important factor to consider is flow rate, and plots of chromatographic efficiency (as measured by minimizing the HETP) versus flow rates for several carrier gases (the famous Van Deemter equation) can be found in any GC reference book or catalog. The optimum carrier gas flow rate, as measured by linear velocity through the column, depends on molecular weight, with larger, more viscous gases such as nitrogen requiring low flow rates (Table 3.1). Linear velocity is easily measured by injection of a nonretained analyte, such as methane, propane, or butane (from a propane torch or butane cigarette lighter vented into a septum-capped tube; inject an aliquot of the gas). Nitrogen provides the best theoretical chromatographic efficiency, but only at a cost of very slow flow rates of ~10 cm/s and long retention

Table 3.1. *Effect of Carrier Gas Type on Capillary Column Efficiency[a] (Expressed as the Number of Effective Theoretical Plates/m.)*

Phase[b]	Gas	Linear flow (cm/s)	N_{eff}/m
low polarity	N_2	9.2	3500
	He	25.0	2000
	H_2	40.0	1470
high polarity	N_2	9.5	1790
	He	16.0	1560
	H_2	34.0	1250

[a]Data from Scott (1960).

[b]In the authors' experience, medium polarity phases have optimum flows between those of low- and high-polarity phases. It should be noted that the theoretical column efficiencies listed in this table assume 100% coating efficiency, which may not be the case with higher polarity phases.

times. Furthermore, efficiency with N_2 decreases sharply with either increasing or decreasing flow rate, and the slow flow rates translate into longer analyte residence times in hot injectors and columns, increasing the possibility of thermal degradation. These combined factors render nitrogen unsuitable as a general purpose carrier gas, despite its low cost.

Hydrogen is also inexpensive and has excellent properties as a carrier gas. The optimum flow linear rate is about 40 cm/s, and very importantly, chromatographic efficiency changes only slowly with increasing flow rate, so that at much higher flows (e.g., 80 cm/s), chromatographic efficiency is decreased by a factor of only ~1.5. Because hydrogen is also the smallest and least viscous gas, the overall result is short runs at the lowest possible temperatures, with a considerable tolerance for flow rates well above the theoretical optimum with little loss in efficiency. Thus, hydrogen is particularly good for large molecules with limited volatility, and for very small bore or long columns due to its low viscosity. The major disadvantage of hydrogen is its extreme flammability, so all connections must be checked frequently for leaks, and a built-in leak detector in the GC oven is advisable. Furthermore, the flow from the split and purge vents should be safely vented. In rare instances, hydrogen can also react with analytes (Liddle & Bossard 1985).

Helium represents a good compromise between nitrogen and hydrogen. It is more expensive, but there is no explosion hazard. Its chromatographic properties are very good, with optimum efficiencies at linear flow rates of about 20 cm/s, and like hydrogen, its chromatographic efficiency decreases slowly with increasing flow rate, so that helium can be used at flow rates much faster than the optimum with minimal loss of efficiency. These fast flow rates have two other benefits. First, fast flow transfers sample to the column more quickly, minimizing residence time in the hot injector, particularly for splitless injections. Second, if the GC is being operated in constant pressure mode, during temperature programming the flow through the column will decrease as the carrier gas expands. Thus, by setting the flow rate to 1.5–2 times the theoretical optimum, the efficiency will actually improve as the run progresses, because the decreasing linear flow rate will be moving back towards the optimum flow for maximum efficiency. Conversely setting the carrier gas flow rate very close to the optimum (at the initial temperature of a programmed run), may be detrimental, because efficiency decreases rapidly at flow rates below the optimum. Overall, helium is the carrier gas of choice, providing the best resolution when analytes elute over a wide temperature range. Because it is completely inert, there is also no chance of reaction with analytes. To minimize gas costs, late model GC systems frequently have flow controllers which adjust carrier gas flows to low levels when the GC is not in use. Alternatively, older model GC systems can be retrofitted (e.g., Gow-Mac Flowminder®; Supelco or Alltech).

In capillary GC, makeup gas is added to the postcolumn flow to sweep out the detector cell rapidly and maintain the narrow width of the eluting peaks.

When using helium as the carrier gas, it is most convenient to also use helium as the makeup gas, as it minimizes the plumbing and gas scrubbing devices required. However, helium is expensive. Thus, with either hydrogen or helium carrier gas, nitrogen can be used as the makeup gas to minimize explosion hazards and cut costs respectively. With FID detectors, N_2 makeup gas also results in slight increases in sensitivity.

For most applications, high-grade, oxygen- and moisture-free commercial gases are suitable for use as carrier gas. Trace levels of oxygen and/or moisture are particularly harmful to polar phases such as Carbowax®, and even nonpolar polysiloxane phases can be irreversibly damaged via oxidation at high temperatures, resulting in poor chromatographic performance and high levels of column bleed. Certain porous polymers used in gas analysis (Chromosorb®, Porapak®, Tenax®) may also be deactivated by moisture. Therefore, oxygen and moisture traps must be used, even with nominally high-purity carrier gases. A hydrocarbon trap is also usually used (hydrocarbons contribute to baseline noise), and these are built into some GC systems (e.g., Hewlett Packard). It should also be stressed that trying to use cheaper, less pure grades of gas is a false economy, because the expense and nuisance of frequent replacement of traps far outweighs the gas cost savings.

Traps that sequentially remove moisture and oxygen or a combination trap (e.g., Supelco high-capacity gas purifier) can be used, followed by an indicating oxygen trap to indicate when the high capacity trap is exhausted. If using two traps, the moisture trap should be installed first because moisture will deactivate the oxygen trap. Molecular sieve moisture traps (5A and 13X) remove water, some nonpermanent gases, and some of the lighter hydrocarbons. Both types change color when they are exhausted. If hydrogen is to be used as carrier gas, double check with the manufacturer to make sure the trap is compatible with hydrogen.

3.3.2. Connections

Plumbing of GC gas lines is usually done with refrigeration grade copper tubing, available from chromatographic suppliers, or much more cheaply from refrigeration suppliers. This grade of tubing is usually clean enough to use as is, and if it is solvent rinsed, make sure that all traces of solvent are removed before the tubing is connected to the GC system.

Column connections to GC fittings are usually made with graphite or Vespel ferrules, or composites of the two. Graphite ferrules withstand high temperatures and do not bind permanently to columns, but they are soft and easily deformed. Vespel ferrules are harder, but they often bind to the column, and so cannot be reused. Composite ferrules represent a good compromise, providing the increased mechanical strength of Vespel, without binding permanently to the column.

Besides the traditional nut and ferrule connections, two devices for making fast connections to fittings have been introduced. The Quadrex Quick-Connect® mounts on the GC fitting, and uses simple lever-action to compress the ferrule and form a tight seal. The J&W Connex® uses GC columns fitted with special ceramic ferrules, which seal to connector bodies mounted on the GC fittings. These connectors allow for columns to be changed easily, but the permanently mounted ceramic ferrules are a disadvantage if the column becomes contaminated. Furthermore, some users have apparently had problems making good seals in critical applications such as GC-mass spectrometry (GC-MS) connections (J&W Scientific, personal communication, 1997).

There are a variety of devices available for connecting two columns (e.g., a retention gap and an analytical column, see below). The simplest consist of glass or fused silica press-fit connectors (available from most GC suppliers). For these to seal properly, it is essential that the column ends be cut square, with no ragged edges, and rough handling once assembled can cause the connections to leak or fall apart. The connections can be made more durable by sealing them with polyimide (J&W) or high temperature epoxy resins (Bemgard & Ostman 1992). When the connections come apart, fragments of polyimide column coating are often left behind. However, with care, these fragments can be burnt out by treatment of the connector at 550°C for 3 h (Wesen & Mu 1992).

Various types of butt connectors, which join two columns by mounting them with nut-tightened ferrules to a connector body, are also available (e.g., Scientific Glass Engineering). However, these are less convenient to use, and the weight of the connector body is awkward.

There are also various fused silica press-fit Y's, T's, and crosses for connecting three or four columns, for example when splitting effluent flows. These connectors are expensive, but they can be reused after heat treatment as described above. Similar connections can also be made with butt connectors, by using two-hole ferrules at one or both ends of the connector, respectively.

3.3.3. Injectors and GC Injection Techniques

Successful GC separation depends on the introduction of the sample as a focused band at the head of the column. Creation of this narrow band is dependent on several factors, including volatility of the analytes, volatility of the solvent, concentration, column diameter, thickness of the stationary phase, and the initial column temperature. The more focused the sample band is at the beginning of the run, the sharper the peaks will be. Sample injection techniques must also minimize sample decomposition or absorptive losses. Injection techniques commonly used in capillary chromatography include split injection, splitless injection, direct injection, temperature-programmed vaporization, and on-column injection (often referred to as cool on-column injection). With the exception of on-column injection, all injection techniques utilize flash vaporization of the injected sample and solvent, which are then swept onto the column by the carrier gas.

The first potential problem with flash vaporization techniques is known as sample backflash (or flashback). Backflash occurs when the injected sample is too large and the vaporized sample expands beyond the capacity of the injector liner and into the body of the injector, resulting in a large, tailing solvent peak. Since the metal parts of the injector are not inert, analytes may be degraded when they come in contact with the metal surfaces. Furthermore, the vaporized sample may condense on contact with cool spots in the system, such as the septum or the gas inlet lines, resulting in carryover or "ghost peaks" during subsequent analyses.

Several practices can help to minimize backflash. First, and most obvious, injection volumes appropriate for the capacity of the liner should be used. Liners of various internal volumes are available, as discussed below. Second, septum purge flow should be set high enough so that traces of sample are purged from the septum before the next run. Third, injector temperatures should be optimized. If the injector temperature is too low, sample vaporization may be slow, resulting in ghost peaks or broad peaks. If the temperature is too high, sample backflash and component degradation will occur. For most routine samples, an injector temperature of 250–350°C is sufficient (Jennings 1987).

Finally, choice of injection solvent is critically important (Alltech 1997). The volume of vapor generated from injections of a given volume of different solvents is not the same, but is dependent on the molecular weight and density of the solvent. Furthermore, the volume of vapor generated by injection of a given volume of the same solvent is dependent on the column head pressure, which in turn depends on the column length and diameter, and the carrier gas used. These ideas are summarized in Table 3.2. Low-molecular-weight, dense solvents

Table 3.2. *Volumes of Solvent Vapor (in Microliters) Generated by Injection of 1 μl of Solvent into Columns of Various Diameters, Operated at Column Head Pressures Giving Optimal Flow Rates*[a,b]

Solvent	0.25 mm i.d.	0.32 mm i.d.	0.45 mm i.d.	0.53 mm i.d.
acetonitrile	575	680	788	831
CS_2	499	590	684	721
isooctane	182	216	250	264
methanol	744	879	1020	1079
CH_2Cl_2	469	555	644	679
heptane	206	243	282	297
hexane	230	272	316	333
pentane	261	309	358	377
toluene	283	334	388	409
water	1666	1970	2285	2409

[a]Values taken from Alltech (1996a), Bulletin 242.

[b]With 1-μl injection, injection port temperature at 325°C, and head pressure set for carrier gas flow at linear velocity, yielding optimum column efficiency.

such as water and methanol create the largest volumes of vapor. Other factors also need to be considered, such as the solvent compatibility with the detector (e.g., halogenated solvents cannot be used with electron capture detectors), the solvent boiling point, and the compatibility of the solvent with the stationary phase. The best solvent for a particular application is usually found by trial and error. For example, methylene chloride (CH_2Cl_2) is often used with a flame ionization detector, but would not be used at all with an electron capture detector and would only be used cautiously with a nitrogen-phosphorus detector (McMinn & Hill 1992).

3.3.3.1. *Split Injection*

Split injection is perhaps the most common of the injection techniques used with capillary GC. It is the easiest way to compensate for the relatively small solvent/sample capacity of most capillary columns because it places only a small fraction of the injected sample on the column. It is used when sample material is not limiting (i.e., concentrations of 0.1–10 mg/ml), either isothermally or with temperature programming. Furthermore, by minimizing the amount of material placed on the column, column degradation, e.g., by dirty samples, is minimized. Injection volumes of 1–2 µl are normally used, but volumes of up to 5 µl can be used without significant loss of resolution or efficiency.

In split injection, the injection port is fitted with two valves, one acting as a septum purge by allowing a small flow of carrier gas to purge the bottom of the septum, and the second directing carrier gas from the bottom of the injection port, near the column inlet, out of the split vent. After sample injection, a small portion of the carrier gas-sample vapor mixture enters the column, while the bulk is directed out the split vent. The fraction of the injected sample that enters the column is determined by the ratio of the carrier gas that exits through the split vent versus the amount directed onto the column (the split ratio). The split ratio is adjusted by increasing or decreasing the flow to the split vent with a needle valve; typically, split ratios of 10:1 to 500:1 are used. Lower split ratios are used when the loading capacity of the column is large (i.e., large i.d. columns, columns with a thick film of stationary phase) and the sample is dilute. High split ratios are used when the column capacity is low (small i.d. and/or thin-film columns) and the sample is concentrated (Grob 1986).

The split injector liner volume must be large enough to contain the vaporized sample without backflash problems but small enough to ensure that the sample is rapidly swept onto the column. Flow disruption within the liner to ensure thorough mixing of the sample is achieved with designs incorporating baffles or inverted cups, for example. Hourglass-shaped liners accomplish the same function and are easier to clean. Alternatively, packing an injector liner with a small piece of silanized glass wool (do not use regular laboratory grade glass wool), or even a small amount of prep GC packing material (i.e., 3% OV-1 on Chromosorb®

W HP) also works, with the additional benefit that the packing material protects the analytical column from contamination with nonvolatile residue.

The high efficiency of split injection is due to the rapid introduction of sample onto the column during the injection, but this also results in discrimination against less volatile compounds, which do not have sufficient time to vaporize before they are vented out of the split vent. However, these discrimination effects are reproducible and are readily compensated for by calibration with standards.

Split injection is easy to use, and very high resolution can be obtained. However, split injection cannot be used for trace analysis when maximum sensitivity is needed, as is often the case, for example, with analysis of insect pheromone samples containing a few nanograms or less. Instead, one of the injection techniques described below should be used.

3.3.3.2. Splitless Injection

Splitless injection is used for samples containing trace compounds in low concentrations, when it is necessary to transfer the maximum amount of sample onto the column to maximize sensitivity. The same injection port used for split injection can be used for splitless injection, with minor modifications. A narrower bore injector liner is used, along with a valve that closes the split vent line. Splitless injector liners are normally straight tubes without any flow disruption features, often with a restriction at the bottom to center the column and keep it away from the walls of the heated liner. For small injection volumes (<1 µl), a 2-mm i.d. liner (~250 µl volume) is used. To avoid backflash problems with larger injections (0.5–5 µl; Alltech 1996a), a 4-mm i.d. liner (~1000 µl volume) can be used, remembering that the increased liner volume also results in a longer time being required to transfer the sample to the column.

With splitless injection, the injection period is extended because the sample is introduced onto the column at a rate equal to the flow through the column, i.e., a much slower rate than with split injections. The long injection period would normally cause band broadening, so the vaporized sample has to be refocused as a sharp band on the top of the column. This can be achieved in two ways (Smith 1988). First, the column (or at least the head of the column) can be cooled to at least 100–150°C below the boiling points of the analytes of interest. These compounds are then trapped at the top of the column as they are injected, while the solvent is unretained. Cooling can be achieved simply by cooling the column oven, or by using a special cryogenic device to chill the head of the column (e.g., Scientific Glass Engineering, Austin, TX). The cryogenic device is also particularly useful for refocusing purge and trap or headspace volatiles. A simple way to achieve the same effect without elaborate equipment is to chill part of a loop at the front of the column in liquid nitrogen or powdered dry ice. With these cryogenic methods, the solvent will be condensed as well, acting as a temporary thick film of stationary phase providing an additional focusing effect, the solvent effect.

In more general terms, with the solvent effect or "Grob" method (Grob 1986), the sample is introduced onto the column at a temperature approximately 20°C below the boiling point of the solvent. When the vaporized solvent leaves the injector, it recondenses at the front of the column, forming a solvent film that traps and refocuses the sample. As with cold trapping, this condensed solvent acts as a thick film of stationary phase and provides a temporary increase in film thickness. The solvent and analyte are concentrated into a very narrow band at the front portion of the column. At the end of the injection period, the oven temperature is then raised rapidly to boil off the solvent, leaving the analytes in a narrow band at the head of the column.

Solvent selection is critical for obtaining a good solvent effect. Generally the solvent should boil about 10–30°C above the initial column temperature. Furthermore, the solvent should be compatible with the stationary phase, so that it dissolves in and wets the stationary phase, forming an even film to trap and focus analytes. If the solvent is not compatible with the phase, the condensed solvent may bead instead of forming a film, resulting in poor peak shape. This problem can be avoided by using a retention gap consisting of a piece of uncoated fused silica capillary (~1–10 m) attached to the front of the column. Furthermore, retention gaps allow slow injections of volumes >2 μl to be made. This, and other uses of retention gaps are described in greater detail in section 3.3.3.5.

After the sample transfer period (~20–60 s) the split valve is opened and the remaining solvent in the injector is purged from the system, so that the solvent peak drops sharply to the baseline. The period that the split valve remains closed depends on the solvent, the size of the injection and injector liner, and the injector temperature. Correct choice of injector temperature and sample transfer period results in virtually all of the sample being placed on the column, while avoiding an excessive solvent peak.

Splitless injection is used extensively for trace analyses, for the separation at the highest resolution of complex mixtures (such as plant oils and flavors), and for large volumes of gases such as environmental air (Ravindranath 1989). Inlet discrimination against higher boiling compounds is less severe for this injection method than for split injections. Poor peak shapes of early eluting compounds may cause problems, but these may be correctable by using a lower-boiling-point solvent to obtain better solvent or cold-trapping effects.

Optimum performance from splitless injection depends on several interacting factors. For example, injection port liners must have a large enough internal volume to avoid backflash. However, the larger the liner volume, the longer it takes to sweep the sample into the column, particularly with smaller bore columns with flow rates of 1 ml/min or less. Longer dwell time in the liner allows for both greater back diffusion, exacerbating the problem of emptying the liner quickly, and increases the time that the sample is in contact with the hot injector, which may be deleterious for thermally sensitive compounds. These problems can be reduced by a cumulative series of proactive steps. First, using a wider bore column (0.32 mm i.d.) allows a higher flow rate through the column, so the sample is transferred to the column more quickly, while still providing good resolution. Second, higher-molecular-weight, less dense solvents should be used to create the minimum volume of vapor (e.g., 1 μl of MeOH creates > four times as much vapor as 1 μl of isooctane, Table 3.2), which in turn allows the use of smaller i.d. injector liners for more rapid transfer of sample to the column. In fact, when using the solvent effect, the condensation of the solvent on the head of the column rapidly reduces the vapor volume, which aids in the rapid clearance of the liner, and this effect is most pronounced with narrower bore liners. Third, injection sizes should be kept as small as possible. When making manual injections, it must be remembered that it is the total solvent volume that counts. If the syringe contains a fraction of a microliter of the solvent used to clean the needle, as well as the sample drawn up for injection, the total volume injected will be increased. To eliminate this extraneous solvent, all traces of solvent should be removed from the syringe by suction with a piece of Teflon® spaghetti tubing connected to a vacuum source before the sample is drawn up. Fourth, to get the best solvent effect, samples should be injected at temperatures well below the boiling point of the solvent, using a solvent which will wet the stationary phase. Fifth, to eliminate possible problems with flooding with larger injections, a 1- to 10-m retention gap should be used. Sixth, the time interval before the split vent is opened should be optimized to provide optimal transfer of sample to the column without exposing the sample to the hot injector for excessively long periods. Seventh, a plug of silanized glass wool placed midway up the liner to wipe traces of solvent and sample from the syringe needle tip as it is withdrawn improves reproducibility. Finally, frequent cleaning of the injection port liner and, if necessary, periodically breaking off a turn or two of the column will help maintain good peak shapes and repeatability.

3.3.3.3. Direct Injection

Direct injection should not be confused with cool on-column injection, discussed in section 3.3.3.5. In many older GC instruments, packed column injectors have been converted to accept megabore capillary columns (0.45 mm i.d. and up). With these converted instruments with large bore capillaries, there is no flow splitting, and the entire injected sample is vaporized and swept onto the column. The gas flow through the injector is usually high (15 ml/min or higher), and so the sample is rapidly introduced onto the column. Optimum efficiency is attained by using initial column temperatures lower than the boiling point of the solvent. The injector is fitted with a fused silica liner, for which there are a variety of designs, including simple open tubes, or open tubes with either glass wool, baffles, deactivated beads, or other material coated with an appropriate stationary phase, analogous to injector liners used with split injections. The straight tube design is generally not recommended, as it will often produce a very broad and tailing solvent peak. Overall, the method is quite sensitive because the entire sample is directed onto the column. The injection port liner traps nonvolatilized compounds, extending the life of the column, and the liners are readily cleaned and reused. However, the injector liner has considerable dead volume, so that it can take some time to sweep the entire sample onto the column, leading to significant peak broadening and lower efficiencies.

3.3.3.4. Programmed Temperature Vaporization Injection

Programmed temperature vaporization (PTV) injection is a versatile technique which is gaining popularity, and several versions are available as options on commercial instruments. The method involves injection of the sample into a cold injector port, followed by rapid heating to vaporize the sample. This system can be operated hot or cold, split or splitless, in total sample introduction or solvent elimination modes.

For cold splitless injection, the sample is dissolved in an appropriate low-boiling-point solvent and injected into the vaporizer with the temperature at or below that of the boiling point of the solvent, with the split valve closed. After 2–3 s, the injector heater block is rapidly heated to the temperature required for vaporization (~250–350°C). After an additional 30–90 s, during which the sample is swept onto the column, the split valve is opened and temperature programming is initiated. If operating correctly, the baseline will drop to zero as soon as the purge is opened, allowing for the detection of relatively volatile substances. There is no discrimination of high molecular weight compounds using this method, and quantification of components is comparable to that achieved with on-column methods.

When it is necessary to analyze dilute samples, PTV injectors can be operated in solvent elimination mode. The split flow is adjusted to vent most of the solvent, and the valve is kept open during the injection with the injector temperature relatively low. The solvent vapor is vented from the system for a few seconds while the less volatile analytes remain in the injector, then the valve is closed and the injector block is heated to vaporize the analytes and sweep them onto the column. The split valve is reopened after 30–90 s to purge any sample remnants from the injector. With this method, volatile sample components may be lost during the solvent venting phase. This can be minimized by using a lower-boiling-point solvent and/or cooling the injector with a cryostat (Ravindranath 1989).

While this injection method has the advantage that injected components experience the minimum amount of heating required to move them onto the column, there are minor drawbacks. First, the partial transfer of more volatile constituents onto the column (e.g., hydrocarbons $<C_{10}$) prior to the initiation of temperature

programming may result in peak doubling or broadening. Second, septum fragments or other debris may absorb sample components, leading to sample loss or delayed elution with peak tailing (Schomburg 1990).

3.3.3.5. Cool On-Column Injection

Cool on-column injection introduces the sample directly onto the column at low temperatures. The entire sample is injected inside the column, so no loss of analyte can occur. Cool on-column injection is the method of choice for samples that begin to degrade in flash vaporization injectors because there is minimal exposure to high temperatures. The injector design is simple and often consists of only a syringe guide and a valve to stop gas flow while the injection is made. A cool air circulator at the bottom of the injector prevents the loss of highly volatile compounds during injection and allows sample introduction into the column in the liquid state so that there is no component discrimination. As with splitless injection, injections are made at a temperature approximately 20°C below the boiling point of the solvent, and after the injection the oven is rapidly heated to drive off the solvent. Since the syringe needle must fit inside the column bore, a 0.32-mm i.d. or larger column is generally used. Columns with a smaller i.d. can be accommodated by the use of smaller stainless steel needles, fused silica needles, or by connecting a retention gap (a larger bore piece of uncoated, deactivated fused silica tubing, ~1–20 m in length), between the injector and the analytical column.

Retention gaps have several other useful functions. First, they allow large injections to be made with cool on-column injection. In the absence of a retention gap, this would lead to very broad and distorted peaks because the injected sample spreads several meters or more along the column. With a retention gap of an appropriate length, as a large injection is made, the condensed liquid spreads along the inner surface of the retention gap, for a length of ~0.25–2 m/μl injected, depending on the solvent. As temperature programming begins, the solvent begins to evaporate from the inlet end of the flooded part of the retention gap, followed by the more volatile analytes, which are continually refocused by the solvent effect on the trailing edge of the evaporating solvent until they reach the front of the analytical column, where they are trapped as a narrow band on the stationary phase. Less volatile analytes remain in the retention gap until the temperature is raised to the point where they have sufficient vapor pressure to move through the uncoated retention gap to the front of the analytical column, where they also are focused as a narrow band. As the temperature continues to increase, the chromatography then proceeds as normal.

Retention gaps allow the routine injection of 5–100 μl or more of dilute samples (Grob 1991; Murphy et al. 1993). Injection of large volumes is particularly convenient in trace analysis because less sample preparation is required since concentration steps can be minimized. The retention gap technique is suitable for both volatile and nonvolatile samples. In our laboratory, we use a combination of three fused silica column pieces connected in series with press-fit connectors (see section 3.3.2). The first column is a short piece of large-bore deactivated fused silica (8.0 cm × 0.5 mm i.d.), connected between the injector and the retention gap column, permitting the use of 0.4-mm o.d. stainless steel needles with a septum injector for on-column injections. The second piece, the retention gap column, is generally 10–20 m in length, with the i.d. of the retention gap matched to that of the third piece, a standard 30-to 60-m analytical column.

A second useful function of retention gaps is to trap nonvolatile components, preserving the integrity of the analytical column. This is particularly important with cool on-column injections in which the entire sample, dirt and all, is injected into the column. As the retention gap becomes contaminated with nonvolatile residues, the first meter or so of the retention gap can be discarded, or the retention gap can be replaced entirely.

To avoid problems with making connections between retention gaps and analytical columns, Restek (Bellefonte, PA) recently began offering columns with retention gaps fused to the front of the column (Integra-Guard™ columns).

3.3.4. Columns and Stationary Phases

3.3.4.1. Column Materials

The column is the heart of the chromatographic system. Commercial capillary columns usually consist of fused silica tubing coated externally with a polyimide coating, with a thin film of stationary phase material bonded to the inner tubing surface (wall coated-open tubular [WCOT] columns). For best performance and lowest column bleed, particularly at elevated temperatures, it is crucial to obtain columns in which the stationary phase is bonded to instead of simply coated on the inside of the column. Manufacturers claim that bonded columns can be rinsed with solvent to remove accumulated nonvolatized materials collected in the column. This is done by pushing clean solvent (sequentially, methanol, methylene chloride, and hexane) through the column with 10–15 psi pressure from the clean detector end to the contaminated injector end using 4–5 ml of solvent for capillary columns, and 8–10 ml for megabore. Rinsing kits are available from several manufacturers. However, in our experience, solvent rinsing has not completely restored column performance, so that rinsed columns should not be used for demanding applications.

If a bonded phase is not available the selection of a "stabilized" phase is recommended. Stabilized phases have less temperature stability and cannot be solvent rinsed. There are also custom-made specialty phases that have been developed for unique separations. These phases are generally not commercially available, but the methods for coating and description of their unique separations have been published.

The polyimide outer coating on fused silica columns imparts strength and flexibility to the fused silica and protects it from scratches. However, these types of columns cannot be used for extended periods above 350°C without damage to the polyimide. Consequently, several manufacturers have developed fused-silica-lined, thin-walled stainless steel (J&W Restek; Chrompack, Middelburg, The Netherlands) or aluminum- or stainless steel-clad (Quadrex, New Haven, CT; Alltech) capillary columns that can be operated up to about 450°C. These columns offer the same inertness and flexibility as polyimide-clad columns, and are also more resistant to abrasion and scratches, with little risk of spontaneous breakage. These columns are suitable for the analysis of high-molecular-weight compounds such as crude oils, waxes, polymers, and triglycerides, extending the range of (thermally stable) compounds that can be analyzed well beyond C_{60}.

3.3.4.2. Column Internal Diameter

In choosing the optimum i.d. for the analytical column, four factors must be considered, column efficiency, sample capacity, the range of concentrations of analytes, and sensitivity. Narrower bore columns (0.32 mm i.d. and less) provide the best efficiency (i.e., best resolution) and sensitivity, and so are best for trace analysis and analyses requiring maximum resolution, such as the separation of isomers. However, narrow bore columns have limited sample capacity ($<< 1 \mu g$). Furthermore, with the highest efficiency, very small bore columns (0.1 mm i.d.), the volume of the sample injection that can be used is restricted due to the small carrier gas flow rate through the column (a fraction of 1 ml/min). Wider bore columns (0.45 mm and greater) provide greater sample capacity, but at a cost in decreased resolution and sensitivity. Table 3.3 illustrates the effects of column diameter on sample capacity and column efficiency.

Exceeding a column's capacity results in broadened, misshapen peaks with a jagged or trailing edge and a very sharp drop to baseline. The overloaded peak

Table 3.3. Effect of Internal Diameter on Sample Capacity and Column Efficiency, Expressed as Theoretical Plates/Meter[a,b]

Internal diameter (mm)	Film thickness (μm)	Sample capacity (ng/component)	Theoretical plates per meter
0.20	0.20	5–30	5000
0.25	0.25	50–100	4170
0.32	0.25	400–500	3330
0.53	1.0	1000–2000	1670
0.75	1.0	10,000–15,000	1170
2 mm (packed column)[b]	—	20,000	2000

[a]Values taken from Supelco, Inc. 1996 catalog (Bellefonte, PA) for non-polar phases.

[b]Sixty-meter SPB-1 capillary columns; 2-m packed column packed with SE-30 (equivalent to SPB-1). Helium carrier gas was used.

will usually elute earlier also. To avoid overloading, the sample can be diluted, the amount of sample deposited on the column can be decreased by increasing the split ratio, or the injection size can be decreased. However, if the minor components are of interest as well, then overloading the major components may be unavoidable so that the minor components can still be observed.

Other considerations may influence choice of column i.d. For example, dirty samples will contaminate and degrade the performance of a narrow-bore column more quickly than a larger bore column. The sample injection technique also merits consideration. If the concentration of the sample is approaching the detection limit of a given detector, the sample should be analyzed by splitless, direct, or on-column injection. For these injection techniques, a 0.32-mm or 0.53-mm i.d. retention gap is generally used to increase sample transfer efficiency. Finally, the type of instrument may dictate the column i.d. For example, some coupled GC-mass spectrometers have limited capacity to expel carrier gas and must be used with columns of 0.25 mm i.d. or less, where the flow rate at the optimum carrier gas velocity is less than 0.7 cm^3/min. This is within the pumping capacities of almost all mass selective detectors (MSDs), ion trap detectors (ITDs), and GC-MS systems (Kitson 1996).

3.3.4.3. Column Length

The effect of column length on a particular separation becomes less important as column length increases, because resolution is a function of the square root of column length. Thus, to double the resolution of two analytes requires a four-fold increase in column length, which may be impractical. Capillary column lengths of 20–30 m are most common, providing good resolution and reasonable analysis times. Shorter columns with slightly reduced efficiency are useful for fast screening procedures, simple mixtures, and high-molecular-weight compounds. On the other hand, columns of 100 m or longer have been used when higher resolution was needed or the sample mixture was complex, particularly with temperature programming to keep retention times within a reasonable range. Long columns (with thick films, see below) may also be useful in analysis of very volatile samples and gases. Column lengths beyond 100 m become less practical because of high backpressure, long retention times, and the diminishing increase in efficiency with increasing column length. Furthermore, increased retention times usually translates into broader peaks, which in turn results in decreased sensitivity. Thus, in most instances where improved resolution is required, it is far more profitable to investigate other stationary phases with different separation factors α, than to try to improve separations by increasing column length.

The special case of "active" compounds with polar functional groups (e.g., alcohols, amines), which are readily adsorbed onto active sites on suboptimal columns deserves mention. In this case short, thick-film columns may be advantageous because the thicker film shields active sites on the column surface and minimizes interactions with the sample compounds, and the short length reduces the possible number of active sites in the flow path.

3.3.4.4. Stationary Phases

The selection of an appropriate stationary phase depends on the chemical and physical properties of the analytes to be separated. There are two factors to be considered. First, the volatilities of the compounds will determine in general terms the parameters (e.g., temperature, stationary-phase thickness) required for analysis. Second, the interactions of the analytes with the stationary phase will determine the separation of compounds with similar volatilities. Compounds whose functional groups interact significantly with those of the stationary phase will be retained longer, while compounds with minimal interactions with the stationary phase will elute more quickly. In general terms, this suggests that more polar compounds may resolve better on columns with intermediate or polar stationary phases. In practice, however, capillary columns have such high resolving power that the majority of compounds will be resolved on either a nonpolar or polar phase. However, when analyzing new samples, particularly if they are complex or contain mixtures of very similar compounds such as isomers, analysis on at least two different columns is recommended to minimize the chance of coincidental elution. Coincidental elution can be particularly problematic with closely related compounds such as E and Z isomers and double-bond positional isomers in long chains. Monounsaturated compounds also may be poorly resolved from their saturated analogs on nonpolar stationary phases (e.g., DB-5), although they separate easily on more polar phases (e.g., DB-17).

Table 3.4 provides a cross-reference guide to the most common phases, including the different suppliers' trade names, the chemical nature of the phase, and the approximate temperature ranges over which the phase can be used. For general use, nonpolar phases have considerable advantages over polar phases. First, each stationary phase has minimum and maximum temperatures (Table 3.4), and these temperature limits tend to be broader for nonpolar phases, particularly in the higher temperature ranges, extending the range of use of nonpolar phases. Below its lower temperature limit, a stationary phase will start to behave like a solid and may exhibit poor separation properties, but the column is seldom permanently damaged. However, exceeding the upper temperature limit of a column accelerates degradation of the stationary phase.

Second, nonpolar phases are less susceptible to damage by air and water than polar phases, and consequently last longer. Generation of active sites on polar columns is particularly problematic with trace analyses, in which small amounts of polar analytes may give poor peak shapes, or worse, the peaks may completely disappear, giving false results.

Third, nonpolar phases tend to have higher efficiencies than polar phases,

Table 3.4. GC Stationary Phase Cross-Reference Guide for Commonly Used Stationary Phases

Stationary phase	J&W	Supelco	Alltech	Hewlett-Packard	Quadrex	Restek	SGE	Chrompack	Approximate temperature ranges[b] (°C)
polysiloxane phases dimethyl-polysiloxane	DB™-1	SPB™-1	AT™-1	HP™-1, Ultra 1	007-1	Rtx™-1	BP™-1	CP SIL 5CB	−60–350
5% phenyl-methyl	DB™-5, DB5-MS™	SPB™-5	AT™-5	HP™-5, Ultra 5	007-2	Rtx™-5	BP™-5	CP SIL 8CB	−60–350
50% phenyl-methyl	DB™-17	SPB™-50	AT™-50	HP™-17	007-17	Rtx™-50	—	—	40–300
14% (cyanopropyl)-phenyl	DB™-1701	SPB™-1701	AT™-1701	PAS-1701	007-1701	Rtx™-1701	BP™-10	CP SIL 19CB	−20–300
50% (cyanopropyl)-phenyl	DB™-225	SP™-2330[c]	—	HP™-225	007-225	Rtx™-225	BP™-225	CP SIL 43CB	45–240
50% trifluoro-propyl-methyl	DB™-210	SP™-2401	AT™-210	—	—	Rtx™-200	—	—	45–260
polyethylene-glycol (PEG)	DB™-WAX	Supelco wax®-10	AT™-Wax	HP™-20M	007-CW	Stabilwax®	BP™-20	CP WAX 52CB	20–260
NTPA-modified PEG	DB™-FFAP	SP™-1000	—	HP™-FFAP	—	Stabilwax®-DA	BP™-21	CP FFAP CB	40–250

[a]Many suppliers also produce several slightly different versions of the same phase for specialty applications (e.g., high-temperature or low-bleed formulations). More detailed tables for specialty and less commonly used phases can be found in Grob (1995) or in suppliers' catalogs.

[b]Upper temperatures listed are for temperature programmed runs. Upper temperature limits for sustained temperatures are ~20°C lower. Upper temperature ranges are given as approximations only, and may vary slightly between manufacturers. Furthermore, upper temperature limits for widebore/megabore columns and or thicker films may be 30°C or more lower. Manufacturers specifications for each column should be consulted.

[c]This is not an exact equivalent as it contains 80% instead of 50% of the (cyanopropyl)-phenyl material.

giving narrower peaks with better peak shapes. This small advantage can make a difference in difficult separations.

In addition to the common phases listed in Table 3.4, numerous phases are available for specialized applications. The characteristics of these phases, and their targeted applications, are detailed in suppliers catalogs. An example of the use of several specialty phases in the separation of geometric isomers is shown in Figure 3.1 (Heath & Tumlinson 1984), in which the *Z:E* isomer ratio is approximately 1:2. The 50-m OV-101, Carbowax® 20M, and SP-2340 columns were commercially available, whereas the 25-m liquid crystal column was custom made (Heath & Doolittle 1983). When the double bond was in the center of the molecule (C_7) the *Z* and *E* isomers separated on the nonpolar OV-101 column, but no separation was obtained for the isomers with the double bond in the 9 position. Increasing the polarity of the phase improved the separation of the isomers and furthermore, the elution order was reversed with the highly polar SP-2340 phase. The custom made liquid crystal phase provided the best separation of the isomers, but in general, it should be noted that liquid crystal phases have limited temperature ranges, and so are not suitable for general use.

For the neophyte, a selection of three basic bonded phase capillary columns,

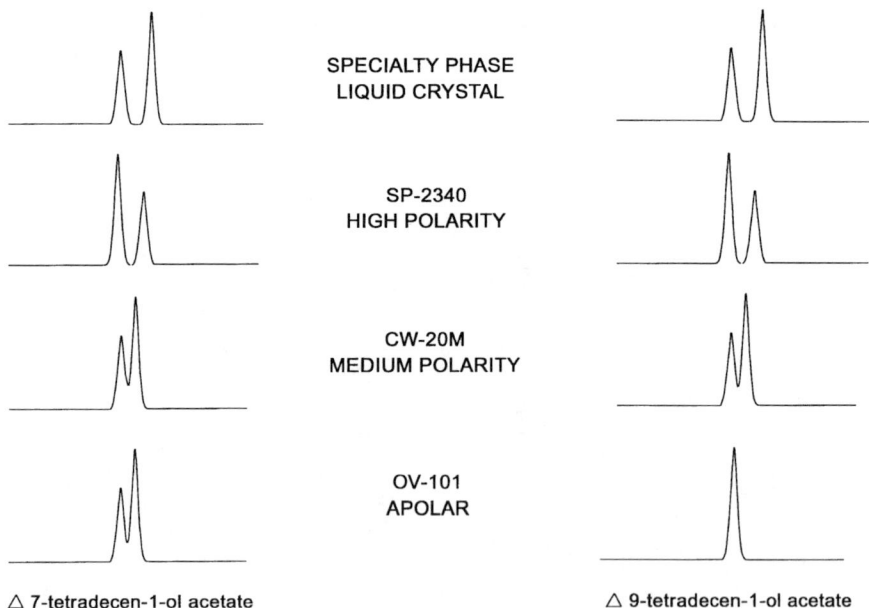

△ 7-tetradecen-1-ol acetate   △ 9-tetradecen-1-ol acetate

Figure 3.1. Separation of tetradecenyl acetates on capillary columns with four different types of stationary phases: OV-1 (dimethylsiloxane phase), Carbowax® 20-M (polyethylene glycol phase), SP®-2340 (poly(biscyanopropyl)siloxane phase), and a liquid crystal specialty phase.

spanning the range from nonpolar to polar, should be satisfactory for the vast majority of separations. These would include a (30 m × 0.25 or 0.32 mm i.d.) nonpolar dimethylsiloxane or 5% phenylmethylsiloxane column (e.g., DB-1 type or DB-5 type, respectively, Table 3.4), an intermediate polarity column such as a 50% phenylmethylsiloxane column (e.g., DB-17 type, Table 3.4), and a polar polyethylene column (e.g., Carbowax® type, Table 3.4). Other specialty columns (except for chiral separations, chapter 8) should only be considered when separation on these three phases has not been successful.

3.3.4.5. Stationary-Phase Film Thickness

The stationary-phase film thickness affects the sample capacity, column efficiency, and the retention of sample components. The standard film thickness for 0.25-mm and 0.32-mm i.d. analytical columns is 0.25 μm, while for megabore columns (0.53-mm i.d.) a thicker film (e.g., 1.5 μm) is more common. These standard film columns can be used for most routine applications. Thicker films result in increased sample capacity and retention times, at a slight cost in column efficiency for nonpolar phase films of 0.2–1 μm. Efficiency decreases more quickly with film thickness with polar phases. Thick film columns are particularly useful for two applications, the analysis of very volatile compounds such as gases and solvents, which are poorly retained on thinner-film columns, and applications requiring large sample capacity, such as micropreparative separations (section 3.5). On the other hand, very thin film columns (0.1 μm) are useful for the analysis of low-volatility, high-boiling-point compounds such as waxes, glycerides, and steroids, which may have inconveniently long retention times with films of normal thickness. Use of hydrogen as carrier gas will also be helpful in reducing retention times, as discussed in section 3.3.1.

3.3.4.6. Solid Stationary Phases

Permanent gases, solvents, and other very volatile compounds may be difficult to analyze on standard bonded phase capillary columns, even with thick films. Sometimes analyses can be done on regular thick-film columns, but only at temperatures <0°C, necessitating a GC oven with cryogenic cooling. Alternatively, these types of compounds can be analyzed on columns packed with alumina, molecular sieves, or Porapak® or Tenax®-type solid adsorbents. Capillary or megabore columns with a layer of the solid adsorbent coated on the inner walls (porous layer open tubular [PLOT] columns) are even better as they provide greater chromatographic efficiency. Conveniently, both the packed and PLOT columns are operated at normal GC temperatures (~40–250°C), avoiding the necessity for cryogenic cooling. PLOT columns are readily available from most GC suppliers, and a variety of applications examples can be found in most catalogs.

3.3.5. Column Installation and General Maintenance

3.3.5.1. Column Installation

Column installation is described in detail in instrument manuals, but a few points are worth emphasizing to obtain optimum performance from your GC system. First, the correct lengths of column extending beyond the ferrules and into the injector and detector are crucial to obtaining optimum sensitivity, and should be set accurately. The ends of the column must be cut as precisely and sharply as possible because a poorly cut column can cause a disruption of carrier gas flow at the inlet, leading to split or broadened peaks, or peak tailing due to poor sample introduction. A special capillary column cleaving tool or a diamond tipped pencil with a very fine point should always be used, and the tool should be reserved strictly for nicking columns, to preserve its point. Column ends should be cut *after* the column nuts and ferrules have been placed on the column to avoid debris from contaminating the column. When placing the column in the oven, sharp bends and twists in the column must be avoided to prevent stress that may lead to column breakage. The column should not touch the oven wall, to prevent temperature gradients. At all costs, heating of a column without carrier gas flow, which can result in permanent damage to the stationary phase, must be avoided. After installing the column in the injector, dip the detector end of the column in a vial of hexane and watch for bubbles, to ensure that there is carrier gas flow through the column. Having completed the installation, all connections should be leak checked with the system at operating pressure, particularly connections such as press-fit unions holding together retention gaps and analytical columns, which may have been loosened during installation. Leak checking should be done with a residue-free liquid, such as a few drops of methanol. All columns should be briefly preconditioned before use by one or more programmed temperature runs to clean out contaminants such as finger print residues. For analyses near the column's maximum operating temperature, more rigorous conditioning is recommended, for several hours at 10–20°C above the highest temperature that will be used, or the column's maximum isothermal limit as listed by the supplier.

3.3.5.2. Column Care and General Maintenance

Column lifetime is typically dependent on column use factors such as the nature of the sample, the injection techniques used, operating temperatures, and carrier gas purity. Water and/or oxygen contamination can cause rapid deterioration of column performance, dramatically shortening the life of a column, and oxygen and moisture traps should be used as a matter of course. Disposable retention gaps are highly recommended when running dirty samples. Periodic removal of several centimeters from each end of the column removes accumulated nonvolatiles from the front of the column and damaged areas of column that have resulted from high injector and detector temperatures. When not in use, column ends should be sealed, particularly for the less stable polar phases. This can be done with a flame, or by sticking both ends of the column into an old GC septum.

Injection port liners merit special attention, because a poorly cleaned or deactivated liner will result in poor chromatography or even the complete disappearance of peaks. Liners should be cleaned periodically to remove accumulations of nonvolatile residues and septum fragments. Light contamination can be removed by solvent rinsing and gentle scrubbing with a solvent-moistened cotton swab. More stubborn residues can be removed by scrubbing with a moistened pipe-cleaner and a nonabrasive cleanser (e.g., Softscrub®). Abrasive cleansers, and scraping with metal wires or tools should be avoided for fear of scratching the glass surface or leaving metal residues, both of which will be detrimental. If absolutely necessary, glass liners can be cleaned in Chromerge® or similar acid baths, followed by thorough rinsing in distilled water to remove all traces of acid.

After anything other than solvent rinsing, and particularly when doing trace analysis or analyses of polar compounds, injector liners will need deactivation to cover up polar silanol groups on the glass surface. This can be done in several ways. In a method recommended by Alltech (Alltech 1997), the cleaned, oven-dried (180°C for 1 h) liner can be soaked in a 5% solution of dimethyldichlorosilane (toxic and corrosive!) in toluene in a closed container in a fume hood for 1 h. The liner is then rinsed with toluene, and soaked in methanol for 1 h. After air-drying, the liner is ready for use.

For very demanding applications, an alternative method is recommended (Smith 1993). The liner is first leached overnight in a 25% vol/vol solution of hydrochloric, nitric, or sulfuric acid, followed by thorough rinsing with deionized water, and oven-drying at <150°C. The dry liner is placed in a test tube, the tube is drawn out in a flame to give a narrow neck through which a couple of drops of silanizing reagent (diphenyltetramethyldisilazane, or a 1:1 mixture of hexamethyldisilazane and trimethylchlorosilane, all of which are toxic) are added, and the tube is briefly flushed with nitrogen, then flame sealed (caution; reagents are flammable). The sealed tube is heated to at least 300°C for 3 h, preferably inside a loosely closed metal container in case of bursting. The cooled tubes can be stored indefinitely, so batchwise deactivation of liners is convenient. When the liner is needed, the tube is opened, and after rinsing thoroughly with hexane, the liner is ready for use.

3.3.6. GC Detectors

The vast majority of GC applications to chemical ecology use either the standard flame ionization detector or a mass spectrometer (mass selective detector) for detection of eluting analytes. Both of these are essentially universal detectors

(FID detectors don't detect water, most air constituents, and respond only slightly to heavily halogenated compounds and CS_2), and both have good sensitivity, well into the picogram range. However, both detectors destroy the sample during detection. FID detectors are robust and reliable, requiring little maintenance unless used with large sample sizes, in which case frequent cleaning of the soot buildup in the detector tower may be required. Mass spectrometers have the added advantage that mass spectra, particularly in conjunction with retention time matchups, provide strong evidence for a positive identification (it may still be difficult to conclusively identify closely related compounds such as isomers). For unknown compounds, a large amount of information about the structure of the compound can be obtained from the mass spectrum, and in fact, the mass spectrum is often the first piece of data used to begin the identification of a new compound. This is discussed in greater detail in chapter 4.

However, for specialized applications, a variety of other detectors warrant brief mention, and their basic characteristics are listed in Table 3.5. More detailed information on their characteristics, operation, and limitations can be found in any GC reference book (e.g. Grob 1995). Briefly, the major factors to consider when choosing a detector for a more specialized application are sensitivity, selectivity, and dynamic range. Other factors which may be of greater or lesser importance include stability, whether or not the sample is destroyed during detection, and cost.

Thermal conductivity detectors, although they have the lowest sensitivity of any of the standard GC detectors, are universal and nondestructive, making them useful for preparative applications. They have also found use in analysis of gases which cannot be detected by an FID detector (e.g., inert gases, CO_2, H_2O, O_2).

Nitrogen-phosphorous detectors (also called thermionic detectors), are, as the name implies, highly selective for compounds containing nitrogen and phosphorous, with virtually no response to hydrocarbons, and with sensitivity about an order of magnitude better than an FID detector. The high sensitivity and selectivity are enormously useful, because they enable the trace analysis of relatively crude samples in complex biological matrices. These detectors are used widely in biomedical, forensic, and environmental applications for the detection of drug residues and pesticides.

Electron capture (EC) detectors are exquisitely sensitive and selective for halogenated compounds and a few uncommon classes of compounds such as nitrates. Of themselves, these characteristics would render EC detectors useful for a few specialized applications, such as analysis of chlorinated pesticides or solvents. However, the range of use is expanded enormously by the ease with which alcohols, amines, carboxylic acids, and compounds with other polar functional groups can be derivatized with reagents such as perfluorinated acid chlorides or pentafluorobenzyl bromide (see chapter 7). The resulting derivatives have the added advantage that they are much more volatile and less polar than the parent compounds, with much better chromatographic properties. Thus, these detectors

Table 3.5. Basic Characteristics of Common GC Detectors

Detector name	Selectivity	Detection limit	Linear dynamic range
flame ionization (FID)	almost universal	low picograms	10^7
thermal conductivity (TCD)	universal	high nanogram to microgram	10^6
mass selective detector (MSD)	universal	pico-to nanogram[a]	10^5
nitrogen-phosphorous (NPD)	N, P, heteratoms	~1 picogram	10^4
electron capture (ECD)	halogens, nitrates	pico-femtogram[b]	10^4
flame photometric (FPD)	P, S	low picogram	10^3–10^4

[a]Sensitivity in full scan range is ~1 ng. With selected ion monitoring, this range is decreased to the low picograms.

[b]Sensitivity depends on the number and type of halogen atoms present. Derivatization of compounds with reagents containing five or more fluorines renders electron capture detection both selective and exquisitely sensitive.

have found widespread use in analysis of drugs and agrochemicals and their metabolites, and they represent one of the few methods available for detection of analytes at femtogram levels. However, EC detectors are finicky, and because of their high sensitivity, they are very sensitive to contamination to the extent that halogen-containing solvents cannot be used in the same room. Furthermore, the argon-methane carrier gas mixture used with these detectors results in higher analysis temperatures being required (and longer retention times) than if helium were used (see section 3.3.1).

Flame photometric detectors are selective for compounds containing phosphorous and sulfur, and have found widespread use in analysis of agrochemicals and their residues.

3.4. Retention Indices

One of the most useful characteristics of gas chromatography is that the retention times of compounds on a particular stationary phase (column) *relative to defined standards* are highly reproducible, particularly with modern capillary columns manufactured under tight quality control. Because relative retention times, quantified as retention indices (see below), can be reproduced with a high degree of fidelity on any other GC system equipped with the same type of column, GC retention indices are a valuable piece of identification information, and are commonly reported for use in subsequent identification of the same compound. The power of retention indices is further extended by their measurement on several different stationary phases, and a good match of retention indices on two or more columns provides compelling evidence for identification.

The Kovats retention index system, using linear alkanes as reference standards, is most commonly employed. Retention indices can be calculated under *isothermal* conditions from the following equation:

$$I = 100y + 100 \ (z - y) \times \frac{\log t'_{r(x)} - \log t'_{r(y)}}{\log t'_{r(z)} - \log t'_{r(y)}}$$

where t'_r is the adjusted retention time (i.e., the measured retention time minus the retention time of a nonretained compound), x is the analyte, y is a linear alkane with y carbons eluting before x, and z is a linear alkane with z carbon atoms eluting after x. Using this system, the retention index of any compound x on any type of column and stationary phase can be calculated, choosing an isothermal GC temperature to match literature data, or to provide retention times within a convenient range (e.g., 10–30 min). When determining the retention indices of several different compounds with widely varying volatilities, several runs at different isothermal temperatures may be required. If desired, a calibration plot of the logs of the adjusted retention times of the alkane reference standards versus their retention indices (e.g., decane = 1000, undecane = 1100, etc.) can be generated, and the retention indices of unknowns simply read off the graph from their retention times. Alternatively, the indices of unknowns can be calculated directly and with more precision from the equation above. Retention indices vary slightly with temperature, and markedly with the stationary phase, so all column and chromatographic variables should be reported. To calculate Kovat's indices, a mixture of linear alkanes (usually a sequential series) bracketing the required range is run, either before the sample, or spiked directly into an aliquot of the sample.

Kovat's indices can be calculated even more conveniently but with slightly less precision under temperature programmed conditions, using slow to medium, uniform programming rates (e.g., 2–8°C/min). In programmed temperature runs, the relationship between retention time and carbon number is linear rather than logarithmic. Thus, the log values in the equation above can be replaced with simple retention times, and simplifying matters even further, the actual instead of the adjusted retention times can be used (because the retention time of an unretained compound is a constant for all four terms, and so drops out). The simplified equation becomes:

$$I = 100y + 100(z - y) \times \frac{t_{r(x)} - t_{r(y)}}{t_{r(z)} - t_{r(y)}}.$$

Programmed temperature runs are obviously much more convenient when working with samples containing analytes with a range of volatilities. In practice, one must be sure that all compounds of interest elute during the temperature programming period, and the initial temperature should be well below the elution temperature of the first of the alkane standards.

Kovat's indices do vary slightly with temperature, so temperatures should always be reported, and conditions reported in the literature duplicated when trying to reproduce results. The problem is particularly acute with polar stationary phases, in which the order of elution of compounds with different functional groups sometimes changes with temperature. In general, with the precise flow and temperature control of modern GC systems, and the standardization of commercial GC columns, reproducibilities within 1–5 index units (i.e., a fraction of 1%) are obtained, with variability being highest for polar phases. A more detailed discussion of factors affecting reproducibility of Kovat's values can be found in Poole and Poole (1991).

The calculation of Kovat's indices on two or more columns of different polarity can be particularly informative, because it will provide a general idea of the polarity of the analyte under study. For example, the Kovat's index of a nonpolar compound will not change much going from a nonpolar to a polar column, whereas the Kovat's index of a polar compound will increase considerably, often by several hundred units. Furthermore, in cases where partial or complete sets of standards are available, retention time or Kovat's index changes with structure can be quantified, and used for predictive purposes. This has proven particularly useful, for example, in locating the position of methyl branches in chains (see the work of K.N. Slessor and G. Gries, for example), or the position, number, and geometry of double bonds in chains (e.g., Cork et al. 1991).

Unfortunately, much of the published Kovat's index data is in the older literature and was measured on packed or coated columns (e.g., Jennings & Shibamoto 1980; Pácákova & Feltl 1992) rather than the bonded phase columns typically used today. The problem is compounded by the fact that most compilations focus on one or a few classes of compounds. However, all is not lost, because retention indices measured on older style nonbonded capillary columns coated with squalane or methylsilicone phases (e.g., SE-30 or OV-1) are similar to those on the analogous nonpolar dimethylsiloxane bonded phases (e.g., DB-1). One useful and current compilation of Kovat's indices of essential oil compounds measured on a bonded DB-5 column can be found in Adams (1995). A commercial computerized database containing indices of several thousand compounds, measured under isothermal and temperature programmed conditions on several stationary phases, is also available (Sadtler 1985).

Other retention index systems have also been used to a lesser degree, such as the equivalent chain length (ECL) system, which is analogous to the Kovat's system except that it uses saturated fatty acid methyl esters as the reference standards.

3.5. Micropreparative Gas Chromatography

Preparative GC is an extremely useful method of collecting nanogram to multimilligram amounts of volatile and semivolatile analytes (routinely, up to ~C_{30}). In fact, it is really the only practical way of isolating submilligram samples of pure volatile compounds for spectroscopic identification methods or bioassays. Although major GC manufacturers still market GC systems capable of being used for preparative work, with the decline in the use of packed in favor of capillary columns, modern GC systems may require some adaptation for use with microscale collections. Two points must be addressed, detection of analytes, because the standard FID is a destructive detector, and getting the analytes out of the oven to a collector. Detection can be addressed in one of three ways:

1. A nondestructive thermal conductivity detector can be used, with the sample collected directly off the detector outlet. Minor modification of the detector outlet to accommodate a Swagelok® fitting, with a nut and a Teflon® or Vespel ferrule to hold a collection tube firmly (see below), may be desirable.

2. A small aliquot of the sample is run to determine retention times of compounds of interest. Then the flame gases (H_2 and air) are shut off,

a more concentrated injection is made, and the sample is collected blind (based on time alone) by placing glass collection tubes (see below) over the detector jet outlet. If the detector design makes this difficult, see the alternate collection method described below, using a spare injection port. When collecting blind like this, it must be remembered that injection of the main portion of the sample may result in much broader peaks and/or shifts in retention times. Allowances can be made for this by injection of a similar volume of a model compound, with similar retention time, while the column is still connected to the detector.

3. The column effluent can be split, with a small portion of the effluent being directed to the FID detector, and the bulk going to the collection port. The effluent is split by connecting the column to a Y- or T-splitter (fused silica press-fit, or butt connector type with a two-hole ferrule at the outlet end, section 3.3.2). The split ratio is determined by the cross-sectional areas of the uncoated fused silica capillaries (of approximately equal length) going from the Y arms to the FID and the collection port; a 0.53 megabore capillary to the collector and a 0.1-mm i.d. capillary to the detector results in a split ratio of ~28:1. If necessary, an X-shaped connector can be used, so that makeup gas can be added to sweep out the splitter quickly and preserve peak integrity.

Collections can be made using jury-rigged collection devices for onetime or infrequent collections, or with more complex devices that require some modification to the GC system. In the simplest case, as described above, the column effluent can simply be collected blind from the detector jet, with the flame turned off. Alternatively, and usually more conveniently, the column can be disconnected from the detector, and the column end directed out of the second heated injector port, with the end of the column pushed through a GC septum and protruding 1–2 mm. The septum holds the column in place, and also provides a good seal as collection tubes are placed over the end of the column to collect the effluent. This setup is also much more accessible and consequently easier than trying to collect off the detector jet at the bottom of the detector tower. Fractions are collected based on previously determined retention times, as in collections made directly from the detector.

For frequent use, a custom-built heated collection port can be built (Fig. 3.2), as described by Brownlee and Silverstein (1968), modifying their general design to fit your GC. Collection systems for multiple fractions (Wassgren & Bergstrom 1984) and automated collection have also been described (Nitz et al. 1988). In the block diagram of a typical Brownlee-Silverstein setup shown in Fig. 3.2, the aluminum heater block mounted in the wall of the GC oven can be controlled by one of the heated zone controls available on the GC system, from a detector

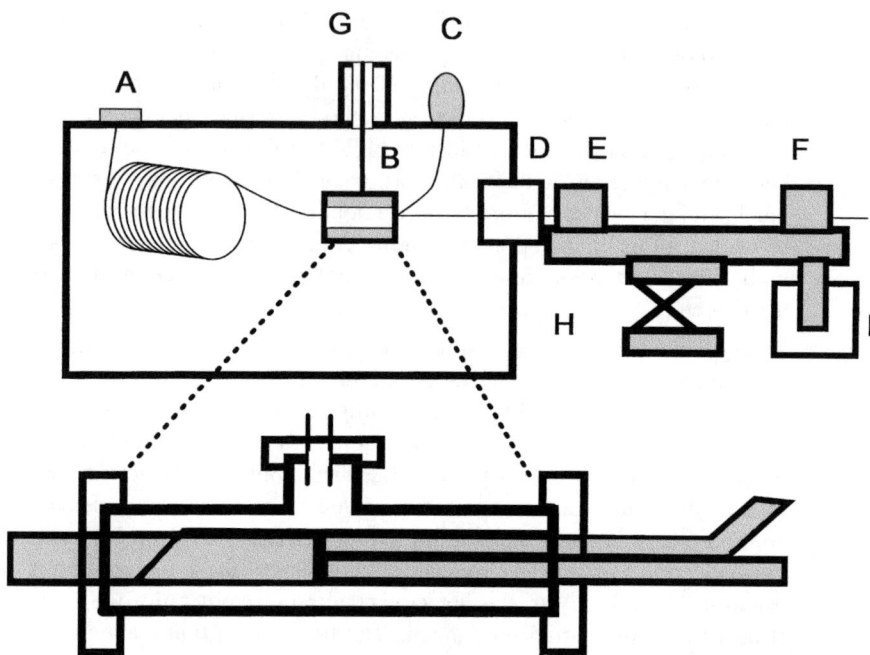

Figure 3.2. GC collector system, consisting of on-column injector (*A*) connected to retention gap and analysis column, which feeds into the glass lined union tee (*B*). The effluent is split into one of two capillaries, with one capillary connected to the FID detector (*C*) and the other capillary directed through a heated block (*D*), and connected to the collector arm. The collector consists of a heated block (*E*) and cold block (*F*), which produce a temperature gradient along the arm and condense analytes in glass collection tubes as they elute. A flow meter (*G*) monitors N_2 makeup gas into the union tee. A laboratory jack (*H*) positions the collector adjacent to the heated block of the GC. A dry ice/acetone bath (*I*) is used to chill the cold block.

or injector that is not in use. A piece of glass-lined tubing passes through the center of the block. Inside the oven, a Swagelok® union connects the large-diameter i.d. capillary tubing from the splitter to the inner end of the glass-lined tubing (Fig. 3.3). The outer end is fitted with a second 1/8-in. Swagelok® union, fitted with a nut and a Teflon ferrule. The 1-mm i.d. glass collection tubes (see below) slide into this union, and are held snugly by gently tightening the nut and ferrule. The collector block consists of a section of insulated stainless steel tubing to hold the glass collection tube, held between a heated block (at the end closest to the GC) and a cold block to produce a temperature gradient along the arm that condenses the analyte in the middle of the collection tube. Both the transfer block and the heated end of the collection arm should be maintained at a temperature 10°C above the highest column temperature for a given run to

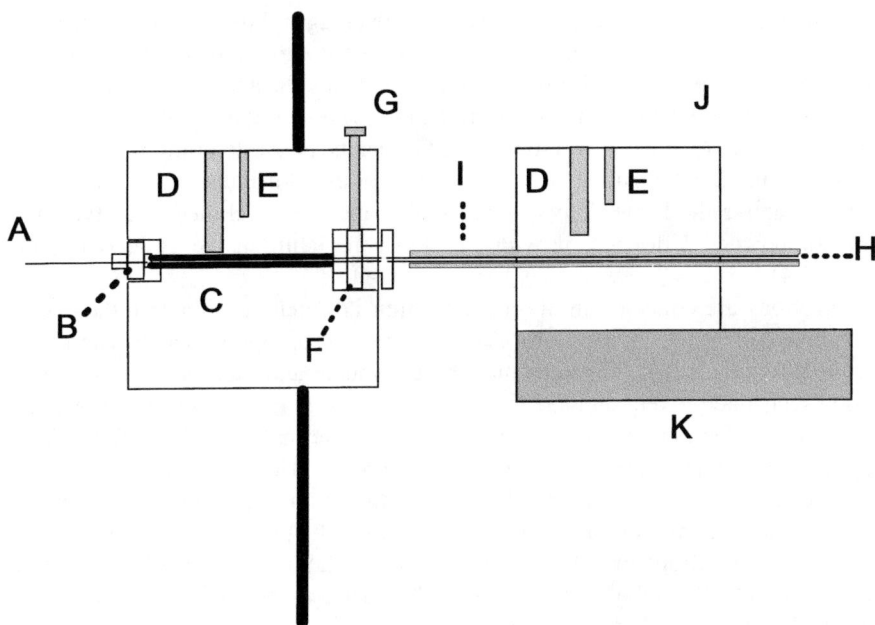

Figure 3.3. The heater block mounted in the wall of the GC oven includes a transfer capillary (*A*) from the union tee, connected with a 1/16-in. stainless steel union (*B*). To the other side of the union is attached a piece of glass lined tubing (*C*). The aluminum heater block is heated by a 70-W heater (*D*) and monitored by a PRT sensor (*E*), taken from the GC's second injector, which was not used. A 1/8-in. Swagelok® stainless steel union (*F*) attaches to the other end of the glass lined tubing and is held tightly by a retaining screw (*G*). The exposed nut on the 1/8-in. union containing a Teflon® ferrule seals the glass collection tubes (*H*) in place. The glass collection tubes fit inside a metal sheath (*I*), which passes through another heated block containing a heater and sensor. The heated block is connected to the rest of the collector system by a brace (*K*).

prevent premature condensation of analytes. The cold end of the collection unit, furthest from the GC, consists of an aluminum block through which the tube passes. A thick aluminum rod protruding from the bottom of the cooling block and immersed in a dry-ice/acetone bath or liquid nitrogen efficiently chills the whole block. In use, and with a little practice, the glass collection tubes can be changed in a matter of a couple of seconds.

When designing a collection port for modern GC systems, where electronics or pneumatics may make it difficult to build a port in the side of the instrument, a simpler solution may be to machine an aluminum block of the same size and shape as a detector block, for mounting directly in the precut hole for the second detector. A metal tube through the center of the block, terminating in a Swagelok® nut and Teflon® ferrule connection for attaching collection tubes, directs the column through the block. This has several advantages. First, it requires minimal modification of the GC since the block fits directly into the precut hole. Second, the block can be heated with the detector's cartridge heater, which is controlled directly by the GC controls, so that no extra wiring is required. Third, the block mounts and dismounts easily, so that the GC can be converted for preparative use when required, but is quickly returned to normal use. An additional dismountable bracket above the heated block, fitted with a metal tube chilled in a styrofoam cup of powdered dry ice, provides a sleeve for chilling the collection tubes (Fig. 3.4).

Fractions are collected in 30 cm × 1.0 mm i.d. melting point capillary tubes (Thomas Scientific, Swedesboro, NJ), or tubes of appropriate length can be cut from glass tube stock. The ends must be cut square and fire polished to provide good seals and prevent damage to the Teflon® ferrule in the collector. For quick and dirty collections, the tube can be chilled while making a collection simply by holding a small piece of dry ice against the center. Having collected a fraction, the tube ends should be temporarily sealed with Teflon® tape until the tube warms up to limit condensation of water inside the tube. The tube contents can be rinsed out with a few drops of solvent, directly into a vial, or into a nuclear magnetic resonance (NMR) tube, for example. Obviously, tubes should be thoroughly cleaned and oven baked before use.

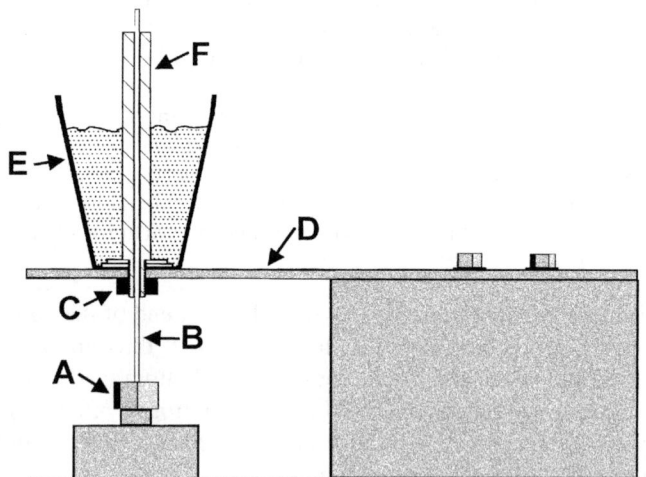

Figure 3.4. Simple cooling device for microscale collections from a heated collection port modified from an unused FID detector block: *A*, Swagelok® nut with a Teflon® ferrule to seal collection tube in place; *B*, glass collection tube; *C*, nut holding cooling rod to plywood support (*D*); *E*, styrofoam cup filled with powdered dry ice; *F*, metal cooling rod bored out to hold glass collection tube.

3.5.1. Columns for Preparative GC

Capillary, megabore, or packed columns can be used for preparative work, depending on the scale of the separation and the resolution required. Capillary columns (0.2–0.32 mm i.d.) are suitable for quantities up to a few hundred nanograms, while megabore columns can handle a few micrograms without serious overloading (see Table 3.3). Column capacity can be increased by using columns with thicker films of stationary phase, with a moderate decrease in resolution. The general characteristics of capillary and megabore columns have already been described, so the remaining discussion focuses on packed columns.

There are a wide variety of phases available for packed column chromatography, but most are for limited specialty applications. Important considerations when choosing a column include column material, stationary phase and its support matrix, and column dimensions.

Columns are made from stainless steel, nickel, glass, or more recently, stainless steel lined with fused silica (e.g., Silcosteel® columns, Restek). Glass columns are rigid, and easily chipped or broken. Stainless steel and nickel columns are durable and flexible, but steel and to a lesser extent nickel may catalyze degradation of sensitive compounds. Fused silica lined stainless steel columns represent a good compromise, with most of the advantages and few of the disadvantages of both.

Packed columns for preparative work are usually 2–5 mm in diameter, and a maximum of about 5 m in length, due to backpressure constraints. Remembering that column efficiency is proportional to the square root of column length, increasing column length to improve separations is not very effective with packed columns.

Stationary phases commonly used in preparative columns, such as dimethyl- or diphenylsilicones (nonpolar phases) or polyethers (polar, Carbowax® type phases) are often analogous to those used in capillary columns. Most separations can be accomplished with one of these three phase types. Detailed descriptions of the plethora of specialty phases can be found in suppliers catalogs (e.g., Alltech, Supelco, Restek, Chrompack), or reference manuals (e.g., Unger 1990).

For preparative work, the stationary phase is usually coated on an inert highly porous support such as Chromosorb® A or Gaschrom® Q Prep (typically 60–80 or 80–100 mesh) to provide a high surface for increased sample capacity. The support should be of the acid-washed, dichlorodimethylsilane (AW-DMCS) treated type.

With packed columns, injections are usually made directly into the column. To minimize contamination of the front of the column, disposable glass column inserts that fit into the front few centimeters of the column should be used. Furthermore, if the packing material in the front of the column is contaminated, it can be carefully dug out and replaced with fresh packing. If doing micropreparative work with capillary or megabore columns, retention gaps are often used to increase the volume of sample that can be injected (see section 3.3).

Coating stationary phases on supports and packing columns is tedious, and takes some practice to produce good columns. We do not recommend it, particularly because commercial packed columns, particularly those in standard sizes packed with off-the-shelf packings, are relatively inexpensive, and are usually of higher quality than user-prepared columns.

3.6. Chromatographic Troubleshooting

3.6.1. Isolating the Source of the Problem

Problems in gas chromatography can arise from many causes, which can at times interact to a bewildering degree. Some of the major problems include (1) electronic or mechanical failure of the instrument itself; (2) operator error; (3) contamination of critical components such as detectors, syringes, septums, or samples; and (4) leaks in the gas flow system. Certain basic procedures can make troubleshooting more efficient. Most GC instruction manuals include troubleshooting tables that can be helpful in narrowing down the search for the problem, but these tables cannot cover all potential problems. Thoughtful and careful iterative elimination of possible sources of the problem is still required. This process has the added benefit of providing you with a much more thorough knowledge of your GC system's operation, making future troubleshooting easier and faster.

The primary rule in troubleshooting is to systematically isolate the problem. Begin by carefully identifying the symptoms. Common symptoms may include (a) no or diminished peaks, (b) extraneous peaks, (c) poorly resolved or shaped peaks, (d) long retention times, or (e) other various baseline symptoms. First rule out operator error by ensuring that the operating parameters of the GC system are within the intended specifications. These include parameters such as gas flow rates, oven temperature, and program rates, and operator errors such as poor or improper sample injection technique. Improper sample injection can be the cause of a variety of chromatographic problems, which are easily cured by following a few general procedures to ensure that injections are accurate and reproducible. For example, the sample should not fill the syringe to its capacity, nor should the sample size be less than 10% of the syringe's total volume. To avoid band broadening of sample peaks, particularly with split injections, sample injection should be rapid, with a uniform motion, and the syringe should be removed quickly from the injector port immediately after the sample is injected. Additional chromatographic problems can result from variable injection volumes, which is easily solved by using an autosampler for repetitive quantitative work, or by using an internal standard in each sample injected.

When operator error can be ruled out, next prepare a new reference standard and make a test injection with a new or thoroughly cleaned syringe. Many hours can be wasted looking for instrumentation problems when in fact the problem is a contaminated standard or syringe. If the system in question consistently separates the new reference standard, the problem is related to the sample or syringe and not a mechanical or electrical problem associated with the instrument. If the problem is still not corrected, try removing and cleaning or replacing the injector liner. Next, replace the GC column with a duplicate (one that is known to provide acceptable separation under optimum conditions. If this corrects the problem, replace or service the original column (i.e., change the retention gap and cut off 1–2 m from the injector end of the column) and retest.

If the problem or symptoms persist even with a new column and injector liner, next check the system for leaks. The most common leak sites are at the septum and at the ferrules used for column connections. Improperly installed columns are a frequent source of leaks and can also be the cause of actual column breakage. Incorrectly installed or overtightened septum or column nuts can also contribute to such things as column bleed, premature failure of the septum or ferrules, and low carrier gas flow rates. The most common method for leak testing is to apply a residue-free solvent (e.g. methanol) to the connection points and watch for bubbles. If leaks are located using this method, the system should be allowed to stabilize afterward, because the liquid used can enter the GC system through an aspirator effect and cause baseline symptoms or ghost peaks. To eliminate the risk of this kind of contamination, a thermal conductivity leak detector (e.g., Gow-Mac gas leak detector, Supelco) can be used for detecting helium, hydrogen, or nitrogen leaks.

3.6.2. Troubleshooting Guides

In addition to the small sampling of GC problems mentioned above, problems can arise from a variety of other areas. Table 3.6 lists problem areas, organized by symptoms, and while not inclusive, it references some of the most common causes and possible remedies for many of GC problems. Excellent troubleshooting guides can also be found in the J&W Scientific catalog (Anonymous 1996) or in bulletins available from most GC suppliers (e.g. Alltech 1996b).

Table 3.6. Troubleshooting Guide for Common GC Problems

Problem/symptoms	Possible cause	Remedy
1. No peaks	a. Defective syringe, or sample injected in incorrect column on a multicolumn system	a. Replace defective syringe, reinject sample into proper column.
	b. FID flame not lit, detector power off	b. Check hydrogen and air flow to FID, relight flame. Check detector and electrometer settings and fuses.
	c. No carrier gas flow	c. Measure carrier gas at column exit, make sure flow is on, check for septum and column connection leaks.
2. Solvent peak only	a. Operation of GC at temperatures not appropriate for sample, column can not separate sample compounds from solvent	a. Verify that the temperature settings for the column, injector, and detector are appropriate for the sample being analyzed. Increase or decrease temperatures as necessary.
	b. Sample too dilute	b. Increase sample concentration.
3. Irregular baseline drift when operating isothermally or when temperature programming	a. Column bleed	a. Condition column or operate at a lower temperature.
	b. Poor carrier, detector gas regulation	b. Check pressure of carrier gas supply; check carrier gas regulator and flow controller for proper operation.
	c. FID only—poor hydrogen and/or air regulation	c. Check hydrogen and air regulators and flow controllers for proper operation.
	d. Detector temperature increasing or decreasing	d. Allow sufficient time for detector temperature to stabilize after temp change.
	e. Column bleed, column contaminated with high molecular weight compounds	e. Recondition column.
	f. Poor carrier gas regulation	f. Check carrier gas pressure, check gas flow controller and regulator.
4. Leading peaks (fronting or sharkfin shaped)	a. Column overloading	a. Reduce sample size.
	b. Column temperature too low	b. Increase column temperature or program rate.

5. Tailing peaks

a. Dead volume
b. Injection temperature too low
c. Active sites in column or in the injector

d. FID—cracked detector jet

a. Check all column connections, check for leaks.
b. Increase injector temperature.
c. Service or replace column, change to a bonded phase column; service injector liner.
d. Replace detector jet.

6. Split peaks

a. Gross detector overload
b. Sample flashing prior to injection/simulates two injections

a. Decrease sample size.
b. Use syringe solvent flush technique, so that sample is in barrel, between two air bubbles. Use a less volatile solvent.

7. Ghost peaks (temperature programmed/no injection)

a. Septum bleed

b. High-molecular-weight material eluting from a previous sample injection
c. Carrier gas impurities

a. Replace septum with high-temperature septum, reduce injector temperature, check septum purge.
b. Recondition column, allow sufficient time for previous sample injection to elute.
c. Install, regenerate, or replace carrier gas filter(s).

8. Ghost peaks with injection

a. High-molecular-weight material eluting from a previous sample injection
b. Carrier gas impurities
c. Sample decomposition
e. Contamination from syringes, glassware, etc.

a. Recondition column, allow sufficient time for previous sample injection to elute.
b. Install, regenerate, or replace carrier gas filter(s).
c. Reduce injector temperature. Try different column.
e. Clean glassware and syringe, prepare fresh sample.

9. Square or flat top peaks

a. Detector saturated (normal for solvent peak)

a. Reduce sample size.

10. Baseline stepping

a. Grounding problem

a. Check all electrical connections and cables.

11. Unresolved peaks

a. Column temperature too high
b. Column too short
c. Incorrect column for application
d. Carrier gas flow too high
e. Poor injection technique

a. Reduce column oven temperature.
b. Use longer column.
c. Change columns to match application.
d. Reduce carrier gas flow rate.
e. Review standard injection techniques.

Continued next page

123

Table 3.6. Continued

Problem/symptoms	Possible cause	Remedy
12. Broad solvent peak	a. Dead volume in injection port due to poor column installation	a. Check column installation, change to on-column injection if possible.
	b. Injection port temperature too low	b. Increase injection port temperature.
	c. Sample solvent reacts with detector or is retained by column	c. Change to more volatile, less retained solvent.
13. Spikes on baseline	a. Dirty detector	a. Clean detector.
	b. Sample condensing in detector	b. Raise detector temperature.
	d. Contaminated sample/solvent	d. Prepare new sample, use fresh solvent.
	c. Contaminated carrier or flame gases	c. Replace purifiers or gas cylinders if used.
	d. Electronic problem (especially if some of the spikes are negative)	d. Check electronic cables, connections, and switches.
14. High background noise	a. Contaminated carrier gas, detector gases	a. Replace carrier gas tank, replace or regenerate carrier gas filter.
	b. Incorrect carrier, detector gas flow rates, leaking	b. Adjust flow rate to proper levels, check for leaks.
	c. Bad ground connection or electronics failure	g. Check electrical ground, electronics.
15. Decreasing retention times, loss of resolution	a. Carrier gas flow increase	a. Check and reduce carrier gas flow if necessary.
	b. Loss of stationary phase	b. Check that column is not being operated above its maximum operating temperature.
	c. Incorrect split ratio	c. Check split ratio.
	d. Increase in column temperature	d. Check temperature controllers.
16. Retention times increasing	a. Leak in system	a. Check all connections for leaks, including septum.
	b. Carrier gas flow decreasing, carrier gas tank running dry	b. Check carrier gas flow and pressure at tank, increase if necessary.

3.7. References

Adams, R.P. 1995. Identification of Essential Oil Components by Gas Chromatography/ Mass Spectrometry. Allured Publishing, Carol Stream, IL.

Alltech. 1996a. Capillary Instruction Manual, Bulletin no. 242. pp. 1–18, Alltech Associates, Deerfield, IL.

Alltech. 1996b. GC Troubleshooting Guide, Bulletin no. 792. pp. 1–53, Supelco, Inc., Supelco Park, PA.

Alltech. 1997. Gas Chrom® Newsletter, Bulletin #363, pp. 4–5, Alltech Associates, Deerfield, IL.

Anonymous. 1996. High Resolution Chromatography Products, pp. 5–26, J&W Scientific, Folsom, CA.

Bemgard, A. & C. Ostman. 1992. High temperature and high pressure stable gluing of press-fit connectors for fused silica and metal capillary tubing. J. High Res. Chromatog. **15**:131–133.

Brownlee, R.G. & R.M. Silverstein. 1968. A micro preparative gas chromatograph and a modified carbon skeleton determinator. Anal. Chem. **40**:2077–2079.

Cork, A., M. Agyen-Sampong, S.J. Fannah, P.S. Beevor & D.R. Hall. 1991. Sex pheromone of female African white rice stem borer, *Maliarpha separatella* (Lepidoptera: Pyralidae) from Sierra Leone: identification and field testing. J. Chem. Ecol. **17**:1205–1219.

Grob, K. 1986. Classical Split and Splitless Injection in Capillary GC. Alfred Huthig, Heidelberg.

Grob, K. 1991. On-Column Injection in Capillary Gas Chromatography, 2nd ed. Huthig, Heidelberg.

Grob, R.L. 1995. Modern Practice of Gas Chromatography, 3rd ed. Wiley Interscience, New York.

Heath, R.R. & R.E. Doolittle. 1983. Derivatives of cholesterol cinnamate: a comparison of the separations of geometrical isomers when used as gas chromatographic stationary phases. H.R.C.-C.C. **6**(1):16–19.

Health, R.R. & J.H. Tumlinson. 1984. Techniques for purifying, analyzing, and identifying pheromones. *In:* Techniques in Pheromone Research, eds. H.E. Hummel & T.A. Miller, pp. 287–322, Springer-Verlag, New York.

Jennings, W. 1987. Analytical Gas Chromatography. Academic Press, New York.

Jennings, W. & T. Shibamoto. 1980. Qualitative Analysis of Flavor and Fragrance Volatiles by Glass Capillary Gas Chromatography. Academic Press, San Francisco.

Kitson, F.G. 1996. Gas Chromatography and Mass Spectrometry: A Practical Guide. Academic Press, San Diego.

McMinn, D.G. & H.H. Hill. 1992. Detectors for Capillary Chromatography. John Wiley & Sons, New York.

Murphy, R.E., R.R. Heath & J.G. Dorsey. 1993. The optimization of capacity and efficiency when coupling open tubular columns in gas chromatographs. Chromatographia **37**(1/2):65–72.

Nitz, S., F. Draweert, M. Albrecht & U. Gellert. 1988. Micropreparative system for enrichment of capillary GC effluents. J. High Res. Chrom. **11**:322–327.

Liddle, P.A.P. & A. Bossard. 1985. A case of artefact formation when using hydrogen as a carrier gas in capillary gas chromatography. J. High Res. Chromatog. Chromatog. Commun. **7**:646–647.

Pácákova, V. & L. Feltl. 1992. Chromatographic Retention Indices. An Aid to Identification of Organic Compounds. Ellis Horwood, New York.

Poole, C.F. & S.K. Poole. 1991. Chromatography Today. Elsevier Science Publishing, New York.

Ravindranath, B. 1989. Principles and Practice of Chromatography. John Wiley & Sons, New York.

Sadtler. 1985. The Sadtler Standard Gas Chromatography Retention Index Library. Sadtler Research Laboratories, Philadelphia, PA.

Sandra, P. ed. 1986a. Sample Introduction in Capillary Gas Chromatography, Vol. 1 GC. Alfred Huthig, Heidelberg.

Sandra, P. ed. 1986b. Sample Introduction in Capillary Gas Chromatography, Vol. 2 GC. Alfred Huthig, Heidelberg.

Schomburg, G. 1990. Gas Chromatography: A Practical Course. VCH. New York.

Scott, R.P.W. 1960. Gas Chromatography. Proceedings 3rd Int. Symp. Butterworths, London.

Smith, K. 1993. Chemical deactivation of injector liners. Separation Times **7**:2. J & W Scientific.

Smith, R.M. 1988. Gas and Liquid Chromatography in Analytical Chemistry. John Wiley & Sons, New York.

Unger, K. 1990. Packings and Stationary Phases in Chromatographic Techniques. Marcel Decker, NY.

Wassgren, A.-B. & G. Bergstrom. 1984. Revolving fraction collector for preparative capillary chromatography in the 100 µg to 1 ng range. J. Chem. Ecol. **10**:1543–1550.

Wesen, C. & H. Mu. 1992. Re-use of press-fit connectors and splitters for GC capillary columns. J. High Res. Chromatog. **15**:136.

4

Mass Spectrometry

Francis X. Webster, Jocelyn G. Millar,
and David J. Kiemle

4.1. Introduction

Mass spectrometry is based on a simple idea: a compound or mixture of compounds is ionized and broken into fragments, the ions are separated (or analyzed) on the basis of mass/charge ratio, and the relative abundance of each ion is recorded as a spectrum. However in practice, mass spectrometry has become a bewildering array of instruments and experiments, most of which are highly specialized for different types and sizes of compounds and mixtures. A complete review of mass spectrometry would cover hundreds of monographs and many thousand journal articles. The objectives of this chapter are rather more modest. We attempt to describe those instruments and experiments that are likely to be most useful to natural products chemists and chemical ecologists. We further try to provide a realistic assessment of the advantages and limitations of each method or experiment, and the information obtainable from each experiment. Thus, the intent of this chapter is to provide a guide to the optimal use of mass spectrometry, particularly with limited amounts of sample. Readers are referred to standard reference books such as McLafferty and Turecek (1993) for background on mass spectral interpretation.

The field of mass spectrometry is covered by instrument type rather than by specific experiment or compound class. The type of instrument is broken down by (1) ion-separation methods and (2) ionization methods. In general, the method of ionization is independent of the method of ion separation and vice versa, although there are obvious exceptions. Thus, the classification scheme is artificial, but it should nonetheless provide a working understanding of mass spectrometry and its limitations.

4.2. Background

Mass spectrometry is usually used for two main purposes in chemical ecology. First, if the compounds under study are known and characterized and their mass spectra are known, then the mass spectrometer becomes an analytical tool whose sensitivity and specificity are unsurpassed. In this regard, the coupling of mass spectrometers to chromatographic instruments for use as detectors is routine and commonplace. The two most common types of coupled instruments are the gas chromatograph-mass spectrometer (GC-MS) (Karasek & Clement 1988) and the liquid chromatograph-mass spectrometer (LC-MS) (Brown 1990).

Also useful in studies with known compounds is tandem mass spectrometry or MS-MS ("MS squared"), and, with certain types of instruments (ion traps), MS to the nth ($MS^{(n)}$) is possible where $n = 2$ to 9. In practice, n rarely exceeds 2 or 3. With MS-MS, a "parent" ion from the initial fragmentation (the initial fragmentation gives rise to the conventional mass spectrum) is selected and allowed or induced to fragment further, thus giving rise to "daughter" ions. In complex mixtures, these daughter ions provide unequivocal evidence for the presence of a known compound. Thus, by simply ionizing a crude sample, selectively fishing out an ion characteristic of the compound under study, and obtaining the diagnostic spectrum of the daughter ions produced from the first ion, a compound can be unequivocally detected in a crude sample, with no prior chromatographic (or other separation steps) being required. Thus, MS-MS can be a very powerful screening tool. This technique alleviates the need for complex separations of mixtures for many routine analyses. For instance, crude urine samples from humans (or animals) can be analyzed for drugs or drug metabolites by MS-MS without prior purification. For unknown compounds, these daughter ions can provide structural information as well.

The other use for mass spectrometry in chemical ecology studies is in structure elucidation of either completely unknown compounds or compounds whose structures are unknown but that are related to known compounds. In either case, mass spectrometry can provide crucial structural information. In some cases, the structure can be determined outright from mass spectrometry, but more often, the MS data complement information obtained from other spectroscopic methods such as nuclear magnetic resonance (NMR).

Regardless of the instrument or experiment, interpretation of a mass spectrum requires the recognition of the molecular ion (M^+) or ions related to the molecular ion (e.g., MH^+ or $M+Na^+$, etc.) and an understanding of fragmentation modes. Interpretation of mass spectra is not the goal of this chapter and those interested are referred to either an introductory text (e.g. Silverstein & Webster 1998) or more advanced treatises such as McLafferty and Turecek (1993) or Splitter and Turecek (1994). These sources provide access to a wealth of literature and databases to aid the interpretation process.

The question of sensitivity as it relates to the various forms of mass spectrometry is both important and elusive. A simple albeit vague statement is that mass spectrometry is a very sensitive analytical method, during which the sample is destroyed. The problem with giving absolute numbers is that the sensitivity is dependent on the compound and the instrumentation employed. One can generalize that, for example, a modern ion trap coupled with gas chromatography can detect picogram quantities of most classes of volatile and semivolatile organic compounds. LC-MS has equally impressive sensitivity, although with proteins, numbers are usually quoted on a molar rather than a weight basis. To provide a rough comparison, rough estimates of the sensitivities obtainable with various mass analyzers and/or ionization methods are given in Tables 4.1 and 4.2. These numbers should be used with caution because sensitivity varies widely between instrument types and manufacturers.

Sample preparation for mass spectrometry is usually simple. Many if not most analyses are carried out with the mass spectrometer coupled to some type of chromatograph. In these cases, sample preparation is the same as that required for the requisite type of chromatography. One obvious corollary is that the compounds of interest do not require rigorous purification as long as an MS method coupled with chromatography is being used.

However, there are MS experiments that do not use chromatographic effluent as their source of analyte, such as fast atom bombardment (FAB) and matrix-

Table 4.1. Comparison of Characteristics of Mass Analyzers

Mass analyzer	Mass range	Resolution	Sensitivity	Advantages	Disadvantages
Magnetic sector	1–15,000 m/z	High, > 100,000	Low	High resolution	Low sensitivity Very expensive High technical expertise
Quadrupole	1–5000 m/z	Unit	High	Easy to use Inexpensive High sensitivity	Low resolution Low mass range
Ion trap	1–5000 m/z	Unit	High	Easy to use Inexpensive High sensitivity Tandem MS (MS^n)	Low resolution Low mass range
Time of flight	Unlimited	Unit	High	High mass range Simple design	Very low resolution
Fourier transform	Up to 70 kDa	High, 100,000	High	Very high resolution and mass range	Expensive High technical expertise

Table 4.2. Comparison of Characteristics of Ionization Methods

Ionization method	Ions formed	Sensitivity	Advantages	Disadvantages
Electron impact	M^+	ng–pg	Database searchable Lots of structural information Couple with chromatographic methods	M^+ often absent
Chemical ionization	$M+1$, $M+29$, etc.	ng–pg	$M+1$ quasimolecular ion usually present Couple with chromatographic methods	Little structural information
Field desorption	M^+	μg–ng	Nonvolatile compounds	Specialized equipment Pure sample required
Fast atom bombardment	$M+1$, M^+ cation, $M+$ matrix	μg–ng	Nonvolatile compounds Some sequencing information	Matrix interference Difficult to interpret
Plasma desorption	M^+	μg–ng	Nonvolatile compounds	Pure sample required Matrix interference
Laser desorption	$M+1$, $M+$ matrix	μg–ng	Nonvolatile compounds Burst of ions	Pure sample required Matrix interference
Thermospray	$M+1$	μg–ng	Nonvolatile compounds	Pure sample required Outdated
Electrospray	$M+1$, $M+2^{2+}$, etc.	ng–pg	Nonvolatile compounds Couple with liquid chromatography Forms multiply charged ions	Limited classes of compounds Little structural information

assisted laser desorbtion ionization (MALDI), making them exceptions to the general rule of simple sample preparation. For these experiments, the compound of interest is dissolved in a matrix and ionized from its surface. These methods require pure samples (although some solvent or buffer might be tolerated or even necessary) because the spectrum obtained will contain ions from everything in the sample. Mixtures can be analyzed deliberately, but the resulting spectrum will be a composite, which may be of limited or no value when trying to identify an unknown compound; in fact the composite spectrum may serve to confuse rather than clarify the identification. As the various experiments are discussed, sample preparation is highlighted only if it is critical to that particular experiment.

Instruments are classified first by the method of ion separation and then by the method of ion generation (ionization). The methods for ion detection are not considered, nor are they important to understanding the type of information that can be obtained from the mass spectrum. The mass spectrometer is viewed as an instrument in which the important components can be mixed and matched. In concept, this is a useful idea and indeed, in practice, many commercial manufacturers sell instruments with "interchangeable" ionization modules. The hardware for separating and detecting the ions is invariably restricted to a single type for a given instrument.

4.3. Mass Analyzers

The mass analyzer, which separates the mixture of ions generated into ions of individual masses to obtain a spectrum, is the heart of each mass spectrometer, and there are several different types with different characteristics. Each of the major types of mass analyzer is described below. This section concludes with a brief discussion of tandem MS and related processes.

4.3.1. Magnetic Sector Mass Spectrometers

The magnetic sector mass spectrometer uses a magnetic field to deflect moving ions into a curved path (Fig. 4.1). Magnetic sector mass spectrometers were the first commercially available instruments, and they remain important. Ions are separated based on their mass/charge ratios, with lower mass ions being deflected more than higher mass ions with the same charge. Resolution depends on each ion entering the magnetic field (from the source) with the same kinetic energy, accomplished by accelerating the ions (which have charge z, where $z = $ the number of excess electronic charges per molecule) with a voltage V. Each ion acquires kinetic energy $E = zV = mv^2/2$. When an accelerated ion enters the magnetic field (B), it experiences a deflecting centrifugal force (Bzv), which bends the path of the ion orthogonal to its original direction. The ion is now traveling in a circular path of radius r, given by $Bzv = mv^2/r$, or rearranged, $r = mv/Bz$. The two equations can be combined to give the familiar magnetic sector

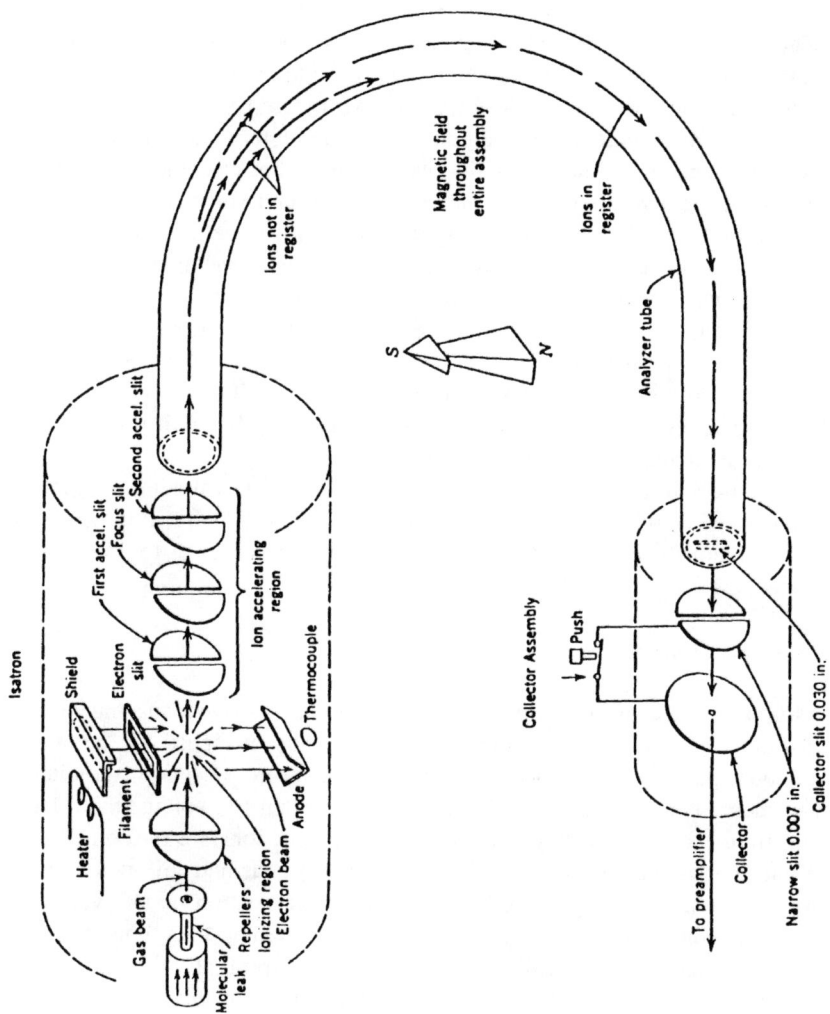

Figure 4.1. Schematic diagram of a single-focusing, 180° magnetic sector mass analyzer. The magnetic field is perpendicular to the page.

133

equation: $m/z = B^2r^2/2V$. Because the radius of the instrument is fixed, the magnetic field (or the accelerating voltage V) is scanned to bring the ions sequentially into focus. As these equations show, a magnetic sector instrument separates ions on the basis of momentum (mv) rather than mass alone. Therefore, ions of the same energy but different mass will come into focus at different points.

An electrostatic analyzer (ESA) can greatly reduce the energy distribution of an ion beam by forcing ions of the same charge (z) and kinetic energy, regardless of mass, to follow the same path. A slit at the exit of the ESA further focuses the ion beam before it enters the magnetic field. The combination of an ESA and a magnetic sector is known as double focusing, because the two fields counteract the dispersive effects each has on direction and velocity.

The resolution of a double-focusing magnetic sector instrument can be as high as 100,000 through the use of narrow slit widths. This very high resolution enables the measurement of "exact masses," which unequivocally provide a molecular formula, an enormously useful piece of data. However, because of the narrow slits, high-resolution instruments sacrifice a great deal of sensitivity; by comparison, slits allowing an energy distribution for about 5000 resolution give at least 0.5 m/z accuracy across the entire mass range, i.e., the "unit resolution" that is used in a standard mass spectrometer.

The upper mass limit for commercial magnetic sector instruments is about m/z 15,000. Raising this upper limit is theoretically possible but impractical. For the vast majority of chemical ecology samples, a mass range to 15,000 is entirely adequate.

4.3.2. Quadrupole Mass Spectrometers

The quadrupole mass analyzer is much smaller and less expensive than a magnetic sector instrument. A quadrupole, as shown schematically in Figure 4.2, is comprised of four rods, 10–20 cm long and hyperbolic in cross section, mounted parallel to each other at the corners of a square. A constant DC voltage modified by a radio frequency (RF) voltage is applied to the rods. Ions are introduced at one end into the "tunnel" down the middle of the four rods and follow a spiral trajectory down the axis (Fig. 4.2). For any given combination of DC voltage and modifying voltage applied at the appropriate frequency, only ions with a given m/z value have a stable trajectory and are able to spiral all the way down the quadrupole to the detector. All other ions with different m/z values have unstable or erratic paths and collide with one of the rods or pass outside the quadrupole. Thus, quadrupole mass analyzers can be regarded as tunable mass filters. In practice, the mass filtering occurs at a very fast rate so that the entire mass range (e.g., m/z 30–800) can be scanned in less than a second.

With respect to resolution and mass range, quadrupoles are generally inferior to magnetic sector instruments. For instance, the current upper mass limit is generally less than m/z 5000. On the other hand, sensitivity is generally high

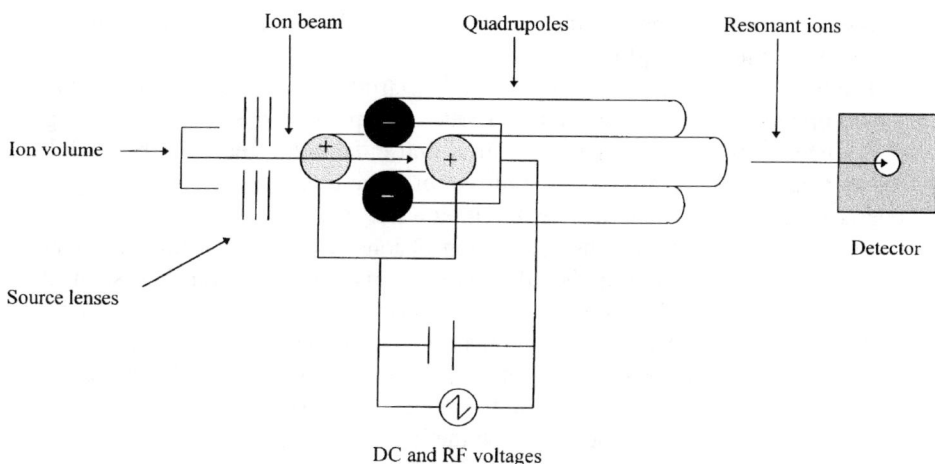

Figure 4.2. Schematic diagram of a quadrupole mass spectrometer.

because there is no need for resolving slits, which remove a portion of the ions. Another advantage of quadrupoles is that they operate most efficiently with ions of low velocity, which means that their ion sources can operate close to ground potential (i.e., low voltage). Since the ions generally have energies of less than 100 eV, the quadrupole mass spectrometer is ideal for interfacing to LC systems for atmospheric pressure ionization (API) techniques such as electrospray (section 4.4.3.2), which work best with ions of low energy.

4.3.3. Ion-Trap Mass Spectrometers

The ion trap is sometimes considered as a variant of the quadrupole because the appearance and operation of the two are related. However, the ion trap can be much more versatile and clearly has greater potential for development. Historically, ion traps have lagged in popularity because earlier versions had serious problems and gave inferior results compared to quadrupoles. For example, spectra were often "concentration dependent"; larger sample sizes often gave many (ion + 1) peaks, rendering the resulting spectra difficult to interpret and unsearchable with standard mass spectral databases. The manufacturers state that these problems have been resolved and the electron impact (EI) spectra are now fully searchable with commercial databases. Furthermore, ion traps are more sensitive than quadrupole instruments, and the ion trap is routinely configured to carry out tandem experiments, with no extra hardware needed.

In one sense, an ion trap is aptly named because, unlike the quadrupole, which merely acts as a mass filter, it can "trap" ions for relatively long periods of time, with important consequences. The simplest use of the trapped ions is to sequentially eject them to a detector, producing a conventional mass spectrum.

Before other uses of trapped ions are briefly described, a closer look at the ion trap itself would be helpful.

The ion trap generally consists of three electrodes, one ring electrode with a hyperbolic inner surface and two hyperbolic endcap electrodes at either end (Fig. 4.3). The ring electrode is operated with a sinusoidal radio frequency field while the endcap electrodes are operated in one of three modes. The endcap may be operated at ground potential, or with either a DC or an AC voltage.

The mathematics describing the motion of ions within an ion trap are given by the Mathieu equation. Details and discussions of three-dimensional ion stability diagrams can be found in either March and Hughes (1989) or Nourse and Cooks (1990). The beauty of the ion trap is that by controlling the three parameters of RF voltage, AC voltage, and DC voltage, a wide variety of experiments can be run quite easily (for details see March & Hughes 1989).

There are three basic modes in which the ion trap can be operated. First, when the ion trap is operated with a fixed RF voltage and no DC bias between the endcap and ring electrodes, all ions above a certain cutoff *m/z* ratio will be trapped. As the RF voltage is raised, the cutoff *m/z* is increased in a controlled manner and the ions are sequentially ejected and detected. The result is the standard mass spectrum, and this procedure is called the "mass-selective instability" mode of operation. The upper mass range in this mode is limited by the maximum RF potential that can be applied between the electrodes. Ions of mass beyond the upper limit are removed after the RF potential is brought back to zero.

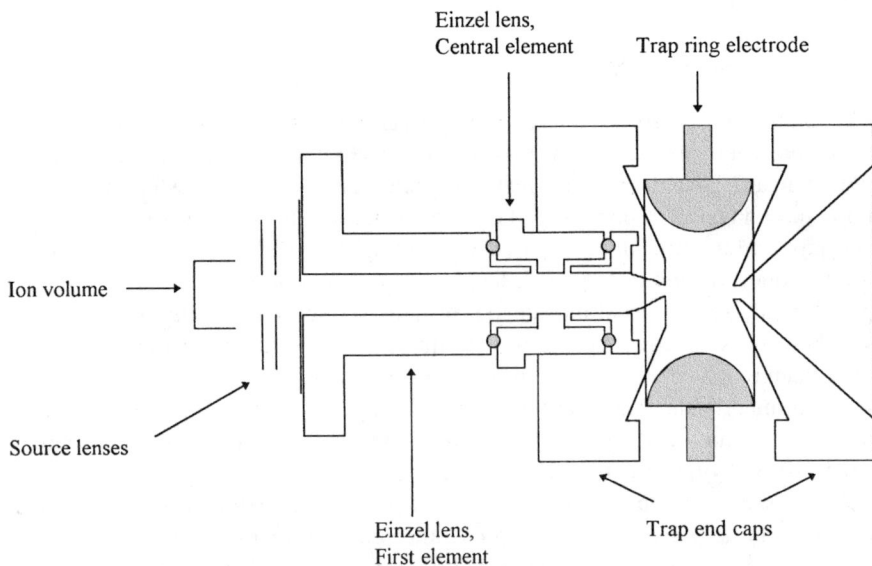

Figure 4.3. Schematic diagram of an ion-trap mass spectrometer.

The second mode of operation uses a DC potential across the endcaps; the result is that there is now both a low and high end cutoff (*m/z*) of ions. The possibilities of experiments in this mode of operation are tremendous, and most operations with the ion trap use this mode. As few as one ion can be selected. One use of this mode of operation would be for selective ion monitoring, but there is no practical limit on the number of ions that can be selected.

The third mode of operation is similar to the second, with the addition of an additional oscillatory field between the endcap electrodes, which results in adding kinetic energy selectively to a particular ion. With a small amplitude auxiliary field, selected ions gain kinetic energy slowly, during which time they usually undergo a fragmenting collision; the result can be a nearly 100% MS-MS efficiency. If the inherent sensitivity of the ion trap is considered along with the nearly 100% MS-MS efficiency, ion traps used for tandem MS experiments greatly outperform triple quadrupole instruments (see section 4.3.6).

Another way to use this kinetic energy addition mode is to selectively reject unwanted ions from the ion trap. These could be ions derived from solvent or from the matrix in FAB or liquid secondary ionization mass spectrometry (LSIMS) experiments (section 4.4.2.2). A constant frequency field at high voltage during the ionization period will selectively reject a single ion. Multiple ions can also be selected in this mode.

4.3.4. Time-of-Flight Mass Spectrometer

The concept of time-of-flight (TOF) mass spectrometers is simple (Cotter 1994). Ions are accelerated through a potential (*V*) and are then allowed to "drift" down a tube to a detector. Assuming that all of the ions arriving at the beginning of the drift tube have the same energy, given by $zeV = mv^2/2$ (z = number of charges; e = electronic charge), then ions of different mass will have different velocities: $v = (2zeV/m)^{1/2}$. For a drift tube of length L, the time of flight for an ion is given by $t = L/v = (L^2m/2zeV)^{1/2}$, from which the mass for a given ion can be easily calculated.

The critical aspect of this otherwise simple instrument is the production of ions at an accurately known start time and position. These constraints necessitate that TOF spectrometers use pulsed ionization techniques, which include plasma and laser desorption (e.g., MALDI, section 4.4.2.4).

The resolution of TOF instruments is usually less than 1000 because some variation in ion energy in unavoidable. Also, because the difference in arrival times at the detector can be less than 10^{-7} s, fast electronics are necessary for adequate resolution. On the positive side, the mass range of these instruments is unlimited, and like quadrupoles, they have excellent sensitivity due to lack of resolving slits. Thus, the technique is most commonly used with large biomolecules.

4.3.5. Fourier Transform Mass Spectrometer

Fourier transform mass spectrometers are not yet common because of their expense; in time, they may become more widespread as advances are made in the manufacture of superconducting magnets. In a Fourier transform (FT) mass spectrometer, ions are held in a cell with an electric trapping potential within a strong magnetic field. Within the cell, each ion orbits in a direction perpendicular to the magnetic field, with a frequency proportional to the ion's m/z ratio. A radio frequency pulse applied to the cell brings all of the cycloidal frequencies into resonance simultaneously to yield an interferogram, conceptually similar to the free induction decay signal in NMR or the interferogram generated in Fourier transform infrared (FTIR) experiments. The interferogram, which is a time domain spectrum, is Fourier transformed into a frequency domain spectrum, which then yields the conventional m/z spectrum.

Because the instrument is operated at fixed magnetic field strength, high-field superconducting magnets can be used, which in turn enables detection of very high masses because mass range is directly proportional to magnetic field strength. Furthermore, since all of the ions from a single ionization event can be trapped and analyzed, the method is both very sensitive and works well with pulsed ionization methods. The most compelling aspect of the method is its high resolution, making FT mass spectrometers an attractive alternative to other mass analyzers. The FT mass spectrometer can be coupled to chromatographic instrumentation and used with various ionization techniques, which means that it can be easily used with small molecules. Further information on FT mass spectrometers can be found in the book by Gross (1990).

4.3.6. Tandem Mass Spectrometry

As briefly described in section 4.2, two or more mass analyzers can be linked in series to produce an instrument capable of selecting a single ion which is then further fragmented and analyzed. For instance, three quadrupoles can be linked (a so-called "triple quad") to produce what was at one time a relatively inexpensive tandem mass spectrometer. In this arrangement, the first quadrupole selects a specific ion for further analysis, the second quadrupole functions as a collision cell to fragment the selected ion (collision-induced decomposition [CID]), and the third quadrupole separates the product ions, to produce the spectrum of daughter ions. The field of tandem mass spectrometry is well developed with good reference books available (Benninghoven et al. 1987; Wilson et al. 1989).

In order for an instrument to carry out MS-MS, it must be able to do the three operations outlined above. However, ion-trap systems capable of MS-MS and $MS^{(n)}$ do not use a tandem arrangement of mass analyzers at all. Rather, a single ion trap is used for all three operations simultaneously. As was stated above, these ion-trap MS-MS experiments are very sensitive, are free of the earlier problems, and are user-friendly. A modern ion trap brings the capability of doing tandem MS experiments to most chemical ecologists, but it must still be emphasized that these experiments may be of limited usefulness in the identification of unknown compounds.

4.4. Sample Introduction and Ion Production Methods

The large number of ionization methods, some of which are highly specialized, precludes complete coverage. The most common ones in the three general areas of gas-phase, desorption, and evaporative ionization are described below.

Ionization methods are in many cases inextricably linked with sample introduction methods. In the simplest (and oldest) case, utilizing a direct insertion probe, the sample is coated onto the inside of a glass capillary inserted into the end of the probe. The probe is then introduced slowly into the ion source through several vacuum locks. The probe can be flash heated to speed vaporization of less volatile samples. However, the sample must be pure because there is no separation step prior to introduction into the ion source.

With GC-MS, the GC effluent can empty directly into the ion source when using narrow-bore capillary columns with low flow rates. For larger flows, several types of separation device are available which selectively strip off most of the carrier gas so that the high vacuum inside the mass spectrometer can be maintained. In LC-MS, comparatively large volumes of solvent must be stripped off, and several different techniques have been developed. These include mechanical interfaces, in which the sample in solution is deposited on a moving belt or wire and the solvent is evaporated off before the belt enters the ion source. Alternatively, with ionization methods such as thermospray (section 4.4.3.1) and electrospray (section 4.4.3.2), the sample solution is nebulized into a vacuum, resulting in very rapid evaporation of the solvent.

Desorption ionization methods used for large, nonvolatile, or ionic compounds (section 4.4.2), in which the sample is dissolved in a matrix and then introduced into the mass spectrometer on a probe, are not readily coupled to chromatographic methods. Thus, these methods, like the direct insertion probe, also require relatively large amounts (~1 μg or more) of pure compound. Details of sample preparation and introduction for these various methods are described in the following sections.

4.4.1. Gas-Phase Ionization Methods

Gas-phase methods for generating ions for mass spectrometry are the oldest and most popular methods. They are applicable to compounds that have a vapor pressure of ~10^{-6} Torr at a temperature at which the compound is stable, that is, to the majority of nonionic organic molecules with molecular weights <1000. This range can be further extended by derivatization of nonvolatile, highly functionalized compounds such as sugars.

4.4.1.1. Electron Impact Ionization

Electron impact is the most widely used method for generating ions for mass spectrometry. Vapor phase sample molecules are bombarded with high energy electrons (generally 70 eV), ejecting an electron from a sample molecule to produce a radical cation, known as the molecular ion. Because the ionization potential of typical organic compounds is less than 15 eV, the bombarding electrons impart 50 eV (or more) of excess energy to the newly created molecular ion. This energy is dissipated in part by the breaking of covalent bonds, which have bond strengths between 3 and 10 eV. Bond breaking is usually extensive and critically, both highly reproducible and characteristic of a particular compound. Furthermore, fragmentation is to a considerable extent predictable, providing the source of the powerful structure elucidation potential of mass spectrometry. Fragmentation is frequently extensive, which often results in a mass spectrum with no discernible molecular ion. Reduction of the ionization voltage reduces fragmentation, and can be used to obtain a molecular ion, but with the disadvantages that the spectrum changes and so cannot be compared to standard literature spectra, and that sensitivity is greatly reduced. As a corollary, when comparing EI spectra with literature spectra, it is crucial to check the ionization voltage that was used, particularly in the older literature where energies of less than 70 eV were frequently used to enhance the molecular ion.

To many, mass spectrometry is synonmous with EI MS, for several reasons. First, historically EI was available before other ionization methods, so most early work used EI MS. Second, the major libraries and databases of mass spectral data, some of which contain over 100,000 spectra and which are relied on heavily for identifications and cited frequently, are of EI mass spectra. Third, the wealth of fragments produced in a highly reproducible pattern from a given organic compound by EI MS provides a mass spectrum, which is usually diagnostic for that compound. However, some caution is necessary, because closely related compounds such as stereoisomers or other isomers can give very similar spectra. A case in point is the plethora of isomeric sesquiterpene hydrocarbons, some of which are difficult if not impossible to distinguish solely on the basis of mass spectra. In such cases, additional data such as GC retention indices or other types of spectra are crucial for an unambiguous identification. In general, however, the relative complexity and uniqueness of the EI spectra of many compounds coupled with the great sensitivity of the method make EI MS a powerful and popular analytical tool, particularly when coupled with chromatographic methods.

4.4.1.2. Chemical Ionization

Electron impact ionization often results in such extensive fragmentation that no molecular ion is observed. One way to circumvent this problem is to use "soft" ionization techniques, of which chemical ionization (CI) is the most important (Harrison 1983; Morris 1981). In CI, a reagent gas (usually methane, isobutane, or ammonia, but others are used) is introduced into the ion source and ionized. Sample molecules (M) collide with the reagent gas ions (e.g., CH_5^+, $C_4H_9^+$, etc.), and undergo secondary ionization, usually by proton transfer from the reagent gas ions, producing MH^+ ($M+1^+$) ions. However, hydride abstraction (producing $M-H^+$ ions) and charge transfer (producing M^+ ions) are not uncommon, so that a cluster of ions around the molecular ion rather than a single ion is often seen. Furthermore, electrophilic addition of reagent gas ions to the parent compound usually produces diagnostic ions at $M+15^+$ and/or $M+29^+$ for methane, $M+43^+$ for isobutane, and $M+18^+$ for ammonia, which further confirm the assignment of the molecular ion. The choice of reagent gas depends on the types of compounds being analyzed and their propensity to fragment. Of the three common reagent gases (ammonia, isobutane, and methane), ammonia transfers the least excess energy to the sample molecules. This results in comparatively large MH^+ ions and little fragmentation. In fact, the NH_4^+ reagent gas ion binds a proton so strongly that it will not ionize compounds with a low affinity for protons, such as saturated hydrocarbons. Conversely, most organic molecules have a stronger proton affinity than the CH_5^+ reagent gas ion generated from methane, so significant amounts of energy are transferred to the sample molecule during protonation, resulting in less abundant quasimolecular ions and more fragmentation. Isobutane is intermediate between these two.

In general, the excess energy transferred to the sample molecules during CI is small, generally less than 5 eV. There are several consequences. First, the protonated molecular ion ($M+1^+$) is usually abundant and is often the base peak (i.e., the largest peak in the spectrum). Second, much less fragmentation takes place than in EI, and so CI spectra are much simpler than and usually lack most of the interpretable structural information contained in EI spectra. Thus, CI MS is usually not useful for structure elucidation. Third, because the total ion current is concentrated into a few ions, CI can result in greater sensitivity than EI. Fourth, because of the simplicity of the spectra, CI MS is not useful for peak matching (either manually or by computer). In short, in chemical ecology, the main use of CI MS is in the detection of molecular ions, and hence determination of molecular weights.

Because the interaction with reagent gas takes place in the vapor phase, CI is not very useful for non-volatile compounds. A variant of the CI method called desorption chemical ionization (DCI) utilizes a wire onto which the solid sample has been deposited. The wire is introduced into the plasma of a CI source and heated. Ionized sample molecules are released and detected. However, as the molecular weight of the sample increases, sensitivity decreases and pyrolitic processes predominate.

4.4.2. Desorption Ionization Methods

Desorption ionization describes those techniques in which ionized sample molecules are emitted directly from a condensed phase into the vapor phase. The primary use is for large, nonvolatile, or ionic compounds. There can be significant disadvantages. First, desorption methods generally do not use available sample efficiently because only a fraction of the applied sample is actually utilized. Second, the information content of the resulting spectra is often limited and can be complicated by interactions of the sample with the matrix in which it is dissolved (see below), and by matrix ions and oligomers. For unknown compounds, these methods are used primarily to provide molecular weight information, and in some cases to obtain an exact mass (i.e., molecular formula). However, even for this purpose, desorbtion methods should be used with caution because the molecular ion or quasimolecular ion may not be apparent.

4.4.2.1. Field Desorption Ionization

In the field desorption method, the sample is applied to a metal emitter, the surface of which is covered with carbon microneedles. The microneedles activate the surface, which is maintained at the accelerating voltage and functions as the anode. Very high voltage gradients at the tips of the needles strip electrons from the sample, and the resulting cation is repelled away from the emitter. The ions generated have little excess energy so there is minimal fragmentation, and the molecular ion is usually the only significant ion seen.

Field desorption was eclipsed by the advent of FAB (section 4.4.2.2). Despite the fact that the method is often more useful than FAB for nonpolar compounds and does not suffer from the high level of background ions that are found in matrix-assisted desorption methods, it has not become as popular as FAB, probably because manufacturers have strongly marketed and supported FAB.

4.4.2.2. Fast Atom Bombardment Ionization

Fast atom bombardment uses high-energy xenon or argon atoms (6–10 keV) to bombard samples dissolved in a liquid of low vapor pressure (e.g., glycerol or nitrobenzyl alcohol). The matrix protects the sample from excessive radiation damage. A related method, liquid secondary ionization mass spectrometry, is similar except that it uses somewhat more energetic cesium ions (10–30 keV).

In both methods, positive (by cation attachment, $M+H^+$ or $M+Na^+$) and negative (by deprotonation, $M-1^+$) ions are formed; both types of ions are usually singly charged. FAB is used primarily with large nonvolatile molecules, particularly to determine molecular weight. For most classes of compounds, the rest of the spectrum is less useful, particularly because the lower mass ranges may be dominated by matrix ions. With some instruments, FAB can be used in high-resolution mode to obtain molecular formulas. Furthermore, for certain classes of compounds that are composed of "building blocks," such as polysaccharides and peptides, some structural information may be obtained because fragmentation usually occurs at the glycosidic and peptide bonds respectively, thereby affording a method of sequencing these selected classes of compounds.

The upper mass limit for FAB (and LSIMS) is between 10 and 20 kDa, and FAB is really most useful up to about 6 kDa. FAB is seen most often with double-focusing magnetic sector instruments where it has a resolution of about 0.3 *m/z* over the entire mass range; FAB can however be used with most types of mass analyzers. The biggest drawback to using FAB is that the spectrum always shows a high level of matrix-generated ions, which limit sensitivity and which may obscure important fragments.

4.4.2.3. Plasma Desorption Ionization

Plasma desorption ionization is a highly specialized technique used almost exclusively with time-of-flight mass analyzers. The fission products from californium 252 (^{252}Cf) with energies in the range of 80–100 MeV are used to bombard and ionize the sample. Each time a ^{252}Cf nucleus fissions, two particles are produced moving in opposite directions. One of the particles hits a triggering detector and signals a start time. The other particle strikes the sample matrix, ejecting sample ions into the TOF MS. The sample ions are most often released as singly, doubly, or triply protonated moieties. These ions are of fairly low energy so that structurally useful fragmentation is rarely observed and, for poly-saccharides and polypeptides, sequencing information is not available. The mass accuracy of the method is limited by the TOF MS, but it can be used with compounds up to at least 45 kDa in mass.

4.4.2.4. Laser Desorption Ionization

A pulsed laser beam can be used to ionize MS samples. Because this method of ionization is pulsed, it must be used with either a TOF or a Fourier transform mass spectrometer. Two types of lasers have found widespread use: a CO_2 laser which emits radiation in the far infrared region, and a frequency-quadrupled neodymium/yttrium-aluminum-garnet (Nd/YAG) laser which emits radiation at 266 nm, in the ultraviolet (UV) region. Without matrix assistance, the method is limited to low molecular weight molecules (<2 kDa).

The power of the method is greatly enhanced by using matrix assistance (matrix-assisted laser desorption ionization). Two matrix materials, nicotinic acid and sinapinic acid, which have absorption bands coinciding with the laser employed, have found widespread use and samples of several hundred kilodaltons mass have been successfully analyzed. A few picomoles of sample are mixed with the matrix compound, followed by pulsed irradiation, which causes sample ions (usually singly charged monomers but occasionally multiply charged ions and dimers have been observed) to be ejected from the matrix into the mass analyzer. The ions have little excess energy and show little propensity to fragment. For this reason, the method can be used for mixtures. The mass accuracy is low when used with a TOF MS, but very high resolution can be obtained with an FT MS. As with other matrix assisted methods, MALDI suffers from background interference from the matrix material, which is further exacerbated by formation of adducts between the matrix and sample molecules. Thus, the assignment of the molecular ion of an unknown compound can be uncertain.

4.4.3. Evaporative Ionization Methods

There are two important methods in which ions or less often neutral compounds in solution (containing formic acid) have their solvent molecules stripped by evaporation, with simultaneous ionization leaving behind the ions for mass analysis. These methods have become popular, particularly when coupled with liquid chromatography.

4.4.3.1. Thermospray Mass Spectrometry

In the thermospray method, the sample in solution is introduced into the mass spectrometer by means of a heated capillary tube. The tube nebulizes and partially vaporizes the solvent, forming a stream of fine droplets that enter the ion source. The solvent is removed by evaporation and the remaining sample ions are analyzed. This method can handle high flow rates and buffers and consequently it was a popular solution to the problem of interfacing mass spectrometers with aqueous liquid chromatography. The method has largely been supplanted by electrospray.

4.4.3.2. Electrospray Mass Spectrometry

The electrospray ion source is operated at or near atmospheric pressure, so-called atmospheric pressure ionization. A solution of the sample enters the ion source through a fused silica capillary maintained at a high potential with respect to a counterelectrode. The potential difference produces a field gradient of up to 3 kV/cm. As the solution exits the capillary, an aerosol of charged droplets forms. The droplets shrink as the solvent evaporates, thereby concentrating the charged sample ions. When the electrostatic repulsion between the charged sample ions reaches a critical point, the droplet undergoes a so-called "Coulombic explosion," which releases individual sample ions into the vapor phase. These ions are then focused through a number of sampling orifices into the mass analyzer.

Electrospray MS has undergone an explosion of activity since about 1990, particularly for samples that have multiple charge bearing sites such as proteins, in which ions with multiple charges are formed. Since the mass spectrometer measures mass to charge ratio (m/z), rather than mass directly, these multiply charged ions are recorded at apparent mass values of ½, ⅓, ¼, etc. of their actual masses. Large proteins can have 40 or more charges so that molecules of up to 100 kDa can be detected in the range of conventional quadrupole, ion-trap, or magnetic sector mass spectrometers (Desiderio 1991). The appearance of the spectrum is a series of peaks increasing in mass, which correspond to pseudomolecular ions possessing sequentially one less proton and therefore one less charge.

Determination of the actual mass of the ion requires that the charge of the ion be known. If two peaks differing by a single charge can be identified, the calculation is reduced to simple algebra. Recall that each ion of the sample molecule (M_s) has the general form $(M_s + zH)^{z+}$, where H is the mass of a proton (1.0079 Da). For two ions differing by one charge, $m_1 = [M_s + (z + 1)H]/(z + 1)$ and $m_2 = [(M_s + zH)/z]$. Solving the two simultaneous equations for the charge z, yields $z = (m_1 - H)/(m_2 - m_1)$. A simple computer program automates this calculation for every peak in the spectrum and calculates the actual mass directly.

Many manufacturers have introduced inexpensive mass spectrometers dedicated to electrospray, for two reasons. First, the method has been very successful while remaining fairly user friendly. Second, the analysis of proteins, peptides, and nucleic acid fragments has grown in importance and they are probably analyzed best by the electrospray method, particularly when coupled with liquid chromatography (Brown 1990).

4.5. Mass Spectral Database Searching

The ease and speed with which a computer can search the relatively inexpensive commercial MS databases to produce likely matches with an unknown spectrum is seductive. Specifically because computerized database searching is so easy and nearly universally employed, several strong cautionary statements are in order, particularly for neophytes.

Routine computer searches can be carried out only on EI mass spectra. Thus, all other ionization methods are precluded unless a specialized database is available. The method of ion separation or mass analysis is theoretically irrelevant, but in practice needs to be considered, because different types of mass analyzers do produce slightly to considerably different spectra. In particular, older ion traps used with an EI source may not produce spectra that match magnetic sector or quadruple generated spectra. This is discussed in greater detail in section 4.6.

There are several points to consider when reviewing the results of an MS database search. First, the most widely used database is the NBS/Wiley collection of mass spectra, which contains well over 100,000 compounds and is routinely sold with EI mass spectrometers. Other specialized databases are available (users can create their own readily), but they will not be considered here. Second, the algorithm or searching routine used to search the database should be known because the reliability of the match can depend critically on the searching method. Information on the searching algorithm is available from the manufacturer. Third, the mass spectrometer must be tuned and operated with standard parameters, because specialized manual tuning for specific experiments, or operation at energies other than the standard 70 eV used in EI will change the resulting spectra. Fourth, some spectra are stored in databases in abbreviated form, with only a subset of the ions being listed. Thus, when comparing the spectrum of an unknown compound to a retrieved database spectrum, it is not uncommon for the spectrum of the unknown compound to contain ions that do not appear in the database spectrum (some of these ions may also be due to contaminants or incompletely separated compounds). This is usually less of a cause for concern than the converse. That is, if the database spectrum contains significant ions that are not present in the spectrum of the unknown compound, then the unknown compound is almost certainly not the same compound. Even this statement is not absolute, however. For example, a partial mass spectrum obtained from very small amounts of sample may be missing significant ions. Furthermore, it is rare, but not unheard of, for database spectra to contain ions from background or impurities, for example, from GC column bleed.

A common instance in which database searching and matching to determine the identifications of unknown compounds must be used with extreme caution occurs with the analysis of compounds for which there are numerous isomers, such as the analysis of mixtures containing mono- and sesquiterpene hydrocarbons. Spectra of several or many of these isomers can be virtually indistinguishable, so in most cases, it is absolutely crucial to use other types of data to confirm tentative identifications made on the basis of mass spectral matches with database spectra. Unfortunately, this has not stopped some researchers from publishing identifications made primarily on the basis of matches with database spectra, with the result that significant portions of the literature, on essential oil analysis, for example, may be of questionable reliability.

Overall, there is no substitute for careful comparison by eye of the spectrum of an unknown compound and a retrieved database spectrum. The human eye is better at discerning differences between the complex patterns of peaks present in mass spectra than the probability-based algorithm used in the computer search. Furthermore, even with extremely close matches, the identification must be considered tentative until it can be checked by comparison of the spectra (and if possible, other properties such as GC or LC retention times, NMR spectra, etc.) of the unknown compound with those of a standard of known structure. As emphasized above, this is particularly true for classes of compounds for which there are many known and common isomers.

4.6. Information Content of Mass Spectra

A detailed review of mass spectral interpretation is not possible in the limited space available here. Our discussion is limited to a consideration of the content of practical information in a mass spectrum, particularly for the identification of an unknown compound.

The information content of mass spectra varies widely based on the method used to obtain the mass spectrum and the particular type of sample. In general, the mass spectrum is often one of the first pieces of data obtained when beginning the identification of an unknown compound, for two reasons. First, the sample requirements are relatively small, varying from <1 ng to a microgram or more, depending on the MS method(s) used. Second, in many cases, the mass spectrometer is coupled with a chromatographic method (e.g., GC-MS or LC-MS), so that the mass spectrum is often obtained before the compound is fully isolated.

Mass spectra of completely unknown compounds are usually difficult or impossible to interpret, so that it is unusual to identify a structure solely by mass spectrometry. The problem is exacerbated by the fact that even if a tentative structural assignment can be made, isomers may give very similar spectra. However, several useful pieces of information can usually be gleaned even from the most uninformative mass spectrum as follows.

The molecular weight of a compound is obtained from the molecular ion from the standard EI mass spectrum, or if necessary, from the MH^+ ions produced by "softer" ionization methods such as CI or FAB. In EI spectra, the suspected molecular ion can usually be confirmed or rejected based on some of the high mass fragments; typical fragments would include losses of 15, 29, 43, . . . (loss of methyl, ethyl, propyl, . . .), 18 (loss of water), and losses corresponding to other simple cleavages or losses of small neutral molecules (e.g., methanol, acetic acid, HCN, etc.). If the first fragments seen would be due to unlikely or impossible losses from the putative molecular ion (e.g., loss of 4–14, 21–25 atomic mass units), then the putative molecular ion is actually a fragment ion, and the real molecular ion is too unstable to be seen. In such cases, CI, FAB, or another soft ionization method will be required. If the molecular ion is of reasonable intensity, the characteristic pattern of M+1 and M+2 ions due to ^1H and ^{13}C isotope contributions to the molecular weight should also be visible (see below).

Chemical ionization results in far less fragmentation so that the MH^+ ion is seen in the large majority of cases, and it is often the base peak. CI is frequently used to confirm molecular ions tentatively assigned by EI. Using the common reagent gases (methane, isobutane, or ammonia), CI usually produces MH^+ ions from transfer of a proton to the molecule, and often M^+ and $M-1^+$ ions as well, resulting in a cluster of ions centered around the molecular ion. Diagnostic adducts are also produced with each reagent gas (methane, M+29; isobutane, M+43; ammonia, M+18), which confirm the molecular ion assignment. Other ionization methods such as FAB, field desorption, or electrospray usually produce a molecular ion or a predictable quasimolecular ion, depending on the particular method and sample matrix used.

Determination of the unit resolution molecular weight by any of the above methods allows a list of possible molecular formulas to be generated. This can usually be narrowed down to a short list of about six or less formulas based on other information gleaned during extraction and purification (e.g., nonpolar compounds are unlikely to have large numbers of heteroatoms present, compounds with minimal UV absorption are unlikely to be highly unsaturated, etc.). It is even more useful to run the sample on a high-resolution instrument to obtain the exact mass, which unambiguously determines the exact molecular formula, an extraordinarily useful piece of information in the identification of an unknown.

Several other pieces of information can be gleaned from the unit resolution molecular ion. First, for compounds containing only C, H, O, or N (i.e., many natural products), if the molecular weight is odd, then the compound must contain an odd number of nitrogen atoms; if even, then the compound must contain either zero or an even number of nitrogens. This is known as the nitrogen rule.

Second, careful examination of the cluster of peaks around the molecular ion is often informative. For those elements with more than one isotope of significant natural abundance (^{35}Cl:^{37}Cl, 76:24; ^{79}Br:^{81}Br, 50.5:49.5; ^{32}S:^{34}S, 95:4.2), mass spectra with unmistakable M:M+2 ratios are obtained (with more than one of these atoms per molecule, the M+, M+2, M+4, . . . patterns become complicated, but are completely predictable, and consequently diagnostic). Even for compounds that do not contain any of these atoms, careful examination of the M:M+1:M+2 ratios can provide a good indication of the ratios of carbon and hydrogen in a molecule, due to the small contributions from ^{13}C and deuterium (see Silverstein & Webster 1998 for details).

In more general terms, a large molecular ion in EI spectra is usually indicative of a more compact or highly conjugated system, such as a polycyclic or aromatic structure, while weak or absent molecular ions are indicative of more easily fragmented structures, such as those containing branch points; those that can easily lose small neutral molecules (e.g., molecules containing one or more heteroatoms); or those that contain long, extended chain structures rather than compact, multicyclic structures.

One further piece of data can be deduced from the molecular formula, that is, the number of rings plus unsaturations. Given the generic formula $C_xH_yN_zO_n$, the number of rings + unsaturations can be calculated from the simple formula:

$$\text{Rings + unsaturations} = x - y/2 + z/2 + 1.$$

More generally, x equals the number of tetravalent atoms (C or Si), y is the number of monovalent atoms (H, F, Cl, Br, I), z is the number of trivalent atoms (usually N, P), and n is the number of divalent atoms (usually O or S). There are several points to note. First, it must be remembered that several atoms have more than one common valence state (e.g., S can be di-, tetra-, or hexavalent; P can be tri- or pentavalent) so several different numbers of rings + unsaturations may have to be considered for compounds containing these atoms. Furthermore, modifications may have to be made for ionized compounds (e.g., tetravalent ammonium nitrogen) or unusual compounds such as *N*-oxides. Second, both the rings and double bonds in aromatic compounds must be counted (e.g., benzene formally contains a ring plus three double bonds). Third, $C\equiv C$, $C\equiv N$, and $N\equiv N$ triple bonds each count for two sites of unsaturation. Overall, the number of rings + unsaturations, in combination with other snippets of information such as the polarity of the compound and the UV spectrum, can be helpful in determining whether the compound is likely to be (poly)cyclic or aromatic, and provide limited information about possible functional groups (e.g., a carbonyl represents one site of unsaturation, whereas an ether or alcohol does not).

The pieces of data described above (i.e., molecular weight and formula, odd or even numbers of nitrogen, presence of sulfur or halogens, number of rings + unsaturations) are obtained from any MS method that gives a molecular ion (or quasimolecular ion). However, from this point, the information content of spectra of unknown compounds obtained by various ionization methods varies widely. At one extreme, spectra produced by methods such as FAB and electrospray generally contain little further useable information, particularly as the spectra may be largely composed of matrix ions or adducts of the matrix with the sample. Attempted interpretation of such spectra is more likely to cause confusion than enlightenment. Chemical ionization spectra lie somewhere in the middle. They may contain limited fragmentation information from facile cleavages, particularly when using reagent gases such as methane which impart considerable energy to the sample molecules. Thus, for example, favored cleavages of alkyl groups from carbons alpha to heteroatoms in rings may produce high-abundance fragments in CI spectra. At the other extreme are EI spectra, which generally contain a wealth of fragmentation information. However, with an unknown compound, interpretation of this information may be difficult or impossible, even for experts, because of the multitude of different possible identities for each fragment. Furthermore, EI spectra often do not show molecular ions, and in cases where there is a highly favored fragmentation, the spectrum may consist primarily to almost exclusively of the fragment(s) from this single cleavage. Whereas this single large fragment may be diagnostic for a particular subunit (e.g., the *m/z* 85 ion characteristic of 2-alkyltetrahydropyrans), the virtual absence of other significant fragments results in little or no information about the rest of the molecule.

For further information on the interpretation of mass spectral fragmentation patterns, the reader is referred to standard monographs (e.g., McLafferty & Turacek 1993) or spectral interpretation handbooks (e.g., Silverstein & Webster 1998). However, a few general comments about spectral interpretation protocols can be summarized as follows:

1. Collate all information obtained during the extraction and purification of the compound, such as polarity, possible presence of functional groups, volatility, stability to air and heat, and so forth.

2. If at all possible, conclusively determine the molecular ion, and hence possible molecular formulas. An exact measurement giving the molecular formula unequivocally is even more helpful.

3. Check the molecular ion cluster for diagnostic signatures of chlorine, bromine, or sulfur. The M:M+1:M+2 ratios may also be informative.

4. Check the general appearance of the spectrum, such as the relative size of the molecular ion, the amount of fragmentation, and the presence or absence of series of ions separated by 14 mass units (indicative of alkyl chains).

5. Look for ions due to loss of characteristic fragments, such as alkyl groups ($- CH_3$, $- C_2H_5$, etc.) or small neutral molecules (water, methanol, acetic acid, ammonia, etc.; see tabulation in, e.g., Silverstein & Webster 1998).

6. Look for diagnostic ions, characteristic of a specific structural unit, such as a heteroatom-containing ring (see tabulation in Silverstein & Webster 1998).

7. If you have any suspicion that an unknown compound may be similar to a known compound, check the spectrum of the known compound against that of the unknown compound. This is particularly useful, for example, in the identification of compounds of related structure, such as the pheromones of insects within the same genus.

The proper selection of instrument and experiment may seem a daunting challenge, and indeed, it often boils down to trial and error and/or available equipment. However, a few simple guidelines may be useful.

First, and most obvious, if the compounds are volatile, then the selection of GC-MS as the method of choice, with its tremendous resolving power and high sensitivity, is neither difficult nor contentious, with the exception of compounds with limited thermal stability. Furthermore, GC-MS is inexpensive, user-friendly, and ubiquitous and little else needs to be said apart from a few comments on the different types of GC-MS systems. To date, quadrupole GC-MS units have been the workhorses of GC-MS analysis, and they are excellent systems, with good sensitivity, ease of use, and minimal maintenance. However, the choice of ion-trap GC-MS, which was at one time controversial, should also be considered now that the earlier problems with this method have been overcome. The modern ion-trap GC-MS deserves a second look for several reasons, the most compelling being that the ion trap is exquisitely sensitive. A convenient measure of sensitivity is to compare the sensitivity of a modern ion trap to that of a GC flame ionization detector (FID). Under almost all conditions, the ion trap is at least one and up to several orders of magnitude more sensitive. Thus, any analysis that can be conducted by FID can certainly be carried out by ion-trap GC-MS, resulting in full-scan mass spectra on subnanogram quantities. Other modern GC-MS systems, while being able to provide full-scan spectra on nanogram to subnanogram quantities, are probably not as sensitive. Other significant reasons to consider ion-trap GC-MS are the facts that EI and CI spectra can be obtained from a single run, and that MS-MS (or MS^n) and other modes of operation are standard capabilities.

Thermally stable compounds of insufficient volatility to be analyzed by GC-MS, but sufficient volatility to be vaporized by flash heating under the high-vacuum conditions inside a mass spectrometer may be amenable to analysis by direct insertion probe. However, the sample must be pure, because the spectrum obtained will be a composite from all molecules present in the sample, and comparatively large amounts of sample (about tens of nanograms to micrograms) may be required. More generally, for compounds of limited or no volatility, or thermally unstable compounds, there are a number of choices, which need to be carefully considered to maximize the information obtained from a minimum amount of sample. The first decision is whether to use an LC-MS method, for example, when the pure compound has not yet been isolated. The LC-MS method also has the advantages that it requires less sample than the matrix-assisted methods, and uses the sample more efficiently, but only some ion production methods are available with LC-MS. An LC-MS with electrospray is versatile and experimentally simple, and for those compounds that can accommodate multiple charges, electrospray has the added advantage of allowing very high mass analysis.

If a pure sample is in hand in microgram amounts, the matrix-assisted methods such as FAB and MALDI may be satisfactory in providing molecular weight information, particularly with instruments capable of exact mass determinations so that exact molecular formulas are obtained. However, except in some special cases, the useful information content of the remainder of the FAB or MALDI spectra may be limited. The special cases would include compounds with repeating units, such as oligosaccharides or peptides, from which some sequence information may be obtainable.

In summary, with astute choices of methods and careful compilation and interpretation of the resulting data, mass spectrometry is an indispensible component in the identification of biologically active compounds of unknown structure. However, it is not without its problems and pitfalls, and neophytes and occasional users are urged to consult or collaborate with more experienced users to minimize the frustration and confusion that accompanies obtaining ambiguous or uninterpretable results. Furthermore, it must always be remembered that mass spectra are not necessarily unique to a particular compound, and final confirmation of the structure of an unknown compound can only be obtained by comparison with an authentic standard.

4.7. References

Benninghoven, A., F.G. Rüdenauer & H.W. Werner. 1987. Secondary Ion Mass Spectrometry: Basic Concepts, Instrumentation Aspects, Applications and Trends, Vol. 86. Wiley, New York.

Brown, M.A. ed. 1990. Liquid Chromatography/Mass Spectrometry: Applications in Agricultural, Pharmaceutical, and Environmental Chemistry. American Chemical Society, Washington, DC.

Cotter, R.J. ed. 1994. Time-of-Flight Mass Spectrometry. American Chemical Society, Washington, DC.

Desiderio, D. M. 1991. Mass Spectrometry of Peptides. CRC Press, Boca Raton, FL.

Gross, M.L. ed. 1990. Mass Spectrometry in the Biological Sciences: A Tutorial. Kluwer Academic Publishers, Norwell, CN.

Harrison, A.G. 1983. Chemical Ionization Mass Spectrometry. CRC Press, Inc., Boca Raton, FL.

Karasek, F.W. & R.E. Clement. 1988. Basic Gas Chromatography-Mass Spectrometry: Principles and Techniques. Elsevier, New York.

March, R.E. & R.J. Hughes. 1989. Quadrupole Storage Mass Spectrometry. Wiley, New York.

McLafferty, F.W. & F. Turecek. 1993. Interpretation of Mass Spectra, 4th ed. University Science Books, Mill Valley, CA.

Morris, H.R. ed. 1981. Soft Ionization Biological Mass Spectrometry. Heyden & Son Inc., Philadelphia, PA.

Nourse, B.D. & R.G. Cooks. 1990. Aspects of recent developments in ion-trap mass spectrometry. Anal. Chim. Acta **228**:1–21.

Silverstein, R.M. & F.X. Webster. 1998. Spectrometric Identification of Organic Compounds, 6th ed. Wiley, New York.

Splitter, J.S. & F. Turecek. eds. 1994. Applications of Mass Spectrometry to Organic Stereochemistry. VCH, New York.

Wilson, R.G., F.A. Stevie, & C.W. Magee. 1989. Secondary Ion Mass Spectrometry. Wiley, New York.

5

Structure Elucidation by NMR

Francis X. Webster and David J. Kiemle

5.1. Introduction

The phenomenal growth and progress of nuclear magnetic resonance spectrometry, universally abbreviated as NMR, has revolutionized structure determinations in the past two decades. Progress in three areas accounts for most of the improvements: magnetic field strength increases, other hardware additions and improvements, and new experiments. This chapter focuses on these three areas in the context of natural products chemistry and chemical ecology.

Spectrometric methods, especially NMR, historically have close ties to chemical ecology. There was a clearly defined, discrete transition from the classical chemical methods utilized by Butenandt et al. (1959) in their pioneering work on the pheromone of the silkworm moth, *Bombyx mori,* and the instrumental methods outlined and developed by Silverstein and his colleagues (and others) on his work with bark beetles, and in particular, with the publication of his classic textbook, *Spectrometric Identification of Organic Compounds.* For the chemical ecologist, NMR spectrometers and mass spectrometers remain the workhorse instruments for chemical analysis. Mass spectrometry enables the chemist to detect compounds whose structures have already been determined in routine analyses, at nanogram and subnanogram levels. For completely new compounds, however, mass spectrometry usually provides limited albeit important structural information. NMR spectrometry, on the other hand, enables the determination of the structures of completely new compounds; the amount needed to carry out these NMR analyses is now approaching the nanogram level.

The first part of this chapter gives a snapshot view of hardware (and related software) as of late 1996. Progress in NMR spectrometry has always been partially fueled by hardware advances. The second area covered in this chapter is the current status of useful experiments for structure determination. The number of experiments that could be included in this section (along with correspondingly confusing acronyms) is prohibitively long; in actual practice, many of these experiments are either redundant or superfluous in the structure determination process. Only those experiments that give unique, useful information are presented here. For a more complete list of experiments and their pulse sequences and for more experimental detail, consult one or more of the excellent books listed in the references (Chandrakumar & Subramanian 1987; Derome 1987; Ernst et al. 1987; Farrar 1987; Sanders & Hunter 1987; Brey 1988; Kessler et al. 1988; Martin & Zektzer 1988; Schraml & Bellama 1988; Friebolin 1993; Croasmum & Carlson 1994; Gunther 1995; Silverstein & Webster 1998). It should be pointed out that the experiments covered in this chapter can be run and processed routinely on modern instruments without the advice or intervention of an NMR specialist.

The chapter ends with a brief treatment of gradient NMR, an area that has been developing over the last decade. The coverage of gradient NMR includes no "new" experiments, merely improved ways of running them. Magnetic resonance imaging (MRI) and diffusion experiments are not covered in this chapter.

5.2. Sample Preparation Considerations

Sample handling and preparation are always important considerations but especially so when working with small quantities of nearly irreplaceable compounds. The worst or most difficult situation arises when working with small quantities of relatively volatile material. An example of such a situation might be a 1- to 10-μg sample of a C-10 to C-20 compound. In cases where there is much less than 1 μg of sample, specialized microprobes are usually necessary to carry out any NMR experiment; these probes can be obtained either from the instrument manufacturer (e.g., Bruker or Varian) or from third-party probe manufacturers (e.g., Nalorac). These probes are not routinely available and are not considered further.

There are three factors that are important in preparing a small sample of a volatile compound for NMR study. The first factor is sample purity. As stated elsewhere in this chapter, purer is better. While this statement is succinct, it may not completely convey the level of purity necessary. An often used criteria for purity of a volatile compound is a single peak on one or more capillary gas chromatography (GC) columns. This criteria is also useful for NMR, but in addition, the sample must be completely free of solvents and water. About the only way to achieve this level of purity is to collect the sample from the effluent of a gas chromatograph. Many methods have been devised to carry out such collections, the best of which when working with very small quantities is the gradient-cooling method of Brownlee and Silverstein (1968). For compounds of low volatility (caution), high-vacuum evacuation for prolonged periods can completely remove water and solvents. This method is fraught with numerous difficulties and potential sources of contamination.

With a suitably pure compound in hand, the second experimental factor of importance is the choice of an appropriate deuterated NMR solvent. Again, the question of purity is paramount. The solvent must be of the highest isotopic purity and the highest chemical purity, which includes the exclusion of water. The solvent of choice is "100%" deuterobenzene (C_6D_6) in sealed ampoules instead of the usual NMR solvent, deuterochloroform ($CDCl_3$). Chloroform apparently cannot be purified to the same extent that benzene can, and chloroform is subject to light-induced degradation to hydrogen chloride, dichlorocarbene, and phosgene. The exclusion of water from both the sample and the solvent is never absolute but must be minimized. Any water that is present will be found at about δ 0.5 in the proton spectrum in benzene, a region generally free of resonances (there are exceptions, such as the protons on cyclopropane rings), while water is generally found at about δ 1.5 in deuterochloroform, which may interfere substantially with the compound of interest.

The final important experimental factor in sample preparation is the choice of NMR tube. For routine NMR experiments, a thin-walled 5-mm tube designed to fit a standard 5-mm probe is filled with ~0.5 ml of solvent and sample. The volume observed in an NMR experiment is actually much smaller, so the sample is generally dissolved in much more solvent than is absolutely necessary. When sample size is severely limiting, the concentration of the sample falls to unacceptably low levels. To compensate, tube manufacturers have developed special tubes that minimize the volume of solvent, thus raising the concentration of sample.

There have been a number of successful approaches to tube design for lowering the volume of solvent for experiments in a standard 5-mm probe, and a variety of microcells are available (e.g., Wilmad Glass, Buena, NJ). The tubes are of two kinds: the cavity cell type and the insert or coaxial tube type. Cavity cells consist of a thick-walled tube with a spherical or oblong cavity to contain the sample. Coaxial tubes consist of a capillary sample tube mounted in a holder inside a standard 5-mm NMR tube. A major drawback of the cavity cells is the difference in dielectric constant between the thick glass (or polytetrafluoroethylene [PTFE]) wall and the solvent. The coaxial tubes are probably more useful and used more often. However, the sample tube must be mounted parallel with the outer tube, and several attempts at proper alignment may be necessary before optimal conditions are found. Hybrids of the two tube types are also made. Regardless of which tube is selected, prior experimentation with the system using a suitable standard is strongly recommended before working with a precious sample.

5.3. Background

The presentation of this chapter presumes a basic understanding of both ^{1}H and ^{13}C NMR spectrometries. Any good introductory textbook on spectroscopic methods (see, e.g., Silverstein & Webster 1998) will thoroughly cover the interpretation of these standard experiments. Those completely unfamiliar with NMR are encouraged to consult one of these texts to familiarize themselves with basic concepts before proceeding further in this chapter.

NMR spectrometers came into widespread use as analytical tools when the field strength limit[1] was 60 MHz and experiments were conducted in continuous wave (CW) mode. Since those early days, the drive has always been for higher field strength, which results in, among other things, improved sensitivity and greater resolution. The improved sensitivity translates into successful analyses conducted on ever smaller quantities of material. The greater resolution obtained at higher field is probably the more important factor and certainly the more complicated one. Greater resolution simplifies the spectrum by reducing overlap of signals and by simplifying the patterns that arise from spin-spin coupling. Simple patterns are called "first order"; in theory, all patterns can be reduced or simplified to first-order patterns by going to high enough field strengths.

As technology improved, the 60-MHz permanent magnets gave way to electromagnets that could record ^{1}H NMR spectra at 100 MHz. Along with new magnets came the introduction of a computer interfaced with the spectrometer. The computer interface both controlled the spectrometer and permitted a fundamentally new method of acquiring and processing NMR data. The method is called Fourier transformation or simply FT (Williams & King 1990). The Fourier transformation method has also found widespread use in other types of spectrometry.

---

[1]The convention of describing field strength in terms of frequency (e.g., 60 MHz) is based on the fundamental NMR equation, $v=\gamma \mathbf{B}_0/2\pi$, which relates frequency, v and field strength $\mathbf{B}_0$. Although technically incorrect, the convention has been universally adopted.

The FT method allowed chemists for the first time to record routine ^{13}C NMR spectra on unenriched samples; the method also opened the door to the field of two-dimensional (2D) NMR or correlation NMR, which is covered in some detail below.

Electromagnets became impractical as the push for higher field strengths continued, but the advent of superconducting magnets signaled the beginning of a slow but inexorable march to higher fields. The current "world record" for a commercial NMR spectrometer is 800 MHz, a record that surely will not last long.

As spectrometer field strengths increase, there has been a concomitant improvement in probe design. For instance, Bruker reports a signal-to-noise ratio (rms) = 1234:1 on 0.1% ethylbenzene for their 800-MHz AVANCE spectrometer with a TXI triple-gradient probehead. The sensitivity increase over a lower field instrument is greater than the linear increase in field strength. With similar types of probes, Bruker claims signal/noise ratios of 605:1 at 500 MHz and 730:1 at 600 MHz, respectively.

The other major area (besides gradients) of hardware improvements for NMR spectrometers over the past two decades has been the attached computer. Larger, faster computers make it easier to "Fourier transform" in two or more directions (dimensions) on bigger data sets. For example, average 2D data sets now range from 4 to 8 megabytes while large data sets can be as large as 64 megabytes for a single experiment! Larger data sets also require improved "graphical resolution" in order to visualize the results.

These hardware advances, by themselves, are not the most exciting changes that have been occurring over the last decade. The FT method of obtaining and processing data has allowed the development of correlation NMR experiments, very often referred to as, but not limited to, 2D experiments.[2] In a correlation experiment, the chemical shift of one type of nucleus (usually ^1H or ^{13}C) is allowed to evolve or correlate with another magnetic parameter such as the same or another nucleus' chemical shift (again, ^1H or ^{13}C) by way of dipole-dipole interactions. The bulk of this chapter outlines the useful experiments in this category and how to interpret them. In addition, limitations are discussed as these experiments relate to chemical ecology. Many of the experiments have been used most often on compounds of less than 1000 Da, but some are applicable to proteins, polysaccharides, nucleic acids, and other high-molecular-weight compounds.

The approach taken in this chapter is to apply each experiment to a single, rather complex compound so that the reader can see directly the outcome of each experiment. This approach is taken so that those readers who are unfamiliar with the experiments outlined can develop a strategy for elucidating structures without resorting to a haphazard trial and error approach.

---

[2]Experiments requiring Fourier transformation in three or more directions (e.g., a 3D experiment) are conceptually similar but not included here. It is our opinion that these experiments are not yet routine and certainly not needed for the purposes of this chapter.

5.4. Example: Quinine

The compound selected to illustrate these methods in quinine (shown below), a well-known, naturally occurring compound possessing many interesting structural features. No attempt is made to "prove" the structure; rather one or more structural features are illustrated with each experiment. Our intent, however, is to cover *every* experiment that might be needed to completely identify an unknown structure, except its absolute stereochemistry. That is not to say that every structure can be determined unequivocally by these methods, only that these methods cover the type of information that is available by NMR experiments for structure elucidation.

Before the results can be discussed, some experimental details and concepts are delineated. First and foremost, the purity of the sample should be assessed. While it is possible to use these experiments with some impurities present, it is always safe to assume that purer is better. Often the purity of the sample and the total amount of the sample are inversely related, leading to the second important consideration: total amount of sample. For each experiment outlined, some consideration is given to the amount of sample needed, but no absolute declarations are possible, partly because actual amounts depend on the instrument available.

All of the experiments shown in this chapter were run on an 8-year-old Bruker 300 AMX spectrometer, which operates at 300 MHz for ^1H. This instrument was selected to show that it is not necessary to have the newest equipment or the highest field to run these experiments. In fact, all but one of these experiments are "routine experiments" on this instrument, with the INADEQUATE experiment (see section 5.11), which requires a large sample size and a long (48 h or more) data acquisition time, being the exception.

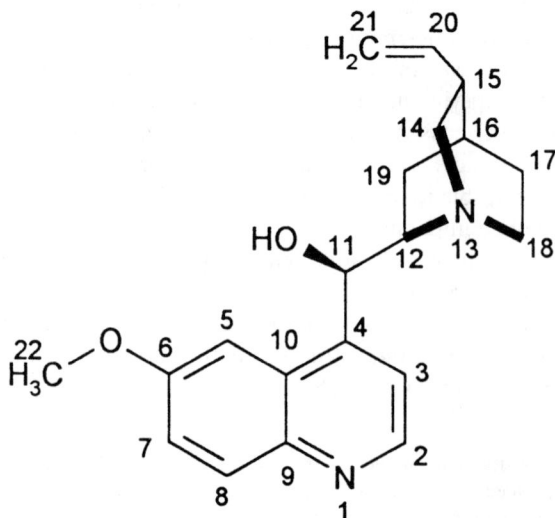

5.5. ¹H and ¹³C Spectra

The place to begin in any structure elucidation problem is with ¹H, ¹³C, and distortionless enhancement by polarization transfer (DEPT) spectra. Figure 5.1 displays the 300-MHz ¹H NMR spectrum of quinine and Figure 5.2 contains the ¹³C and DEPT spectra of quinine (discussed in next section). A limitation already arises concerning the ability to obtain a ¹³C NMR spectrum with limited material. On the 300-MHz instrument used for this discussion, less than ~1 mg precludes the direct observation of a ¹³C NMR spectrum. This problem can be overcome by indirect methods available through correlation experiments. The point is considered again after a discussion of ¹H-¹³C correlation.

Simple proton and carbon spectra form the basis for most structure determinations; indeed, many structures can be determined from these alone. Clearly, however, not all structures can be determined from this limited information. In such cases and depending on availability of sample, more sophisticated experiments can produce exquisitely detailed structural information.

Results from these three experiments are given in Table 5.1. The chemical shift values and multiplicities will greatly simply the discussion of other experiments in

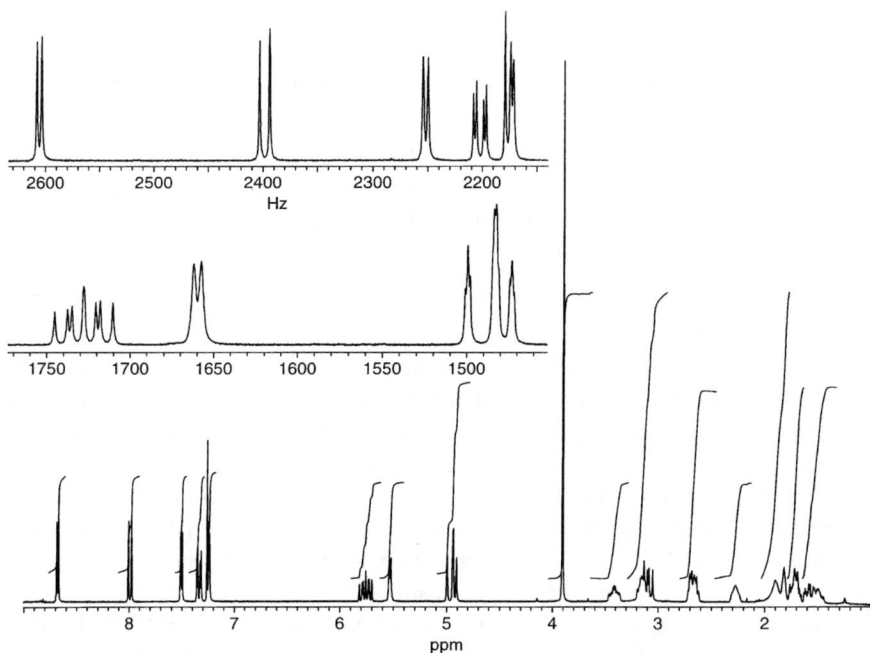

Figure 5.1.   The 300-MHz ¹H NMR spectrum of quinine in CDCl₃. Greater detail can be seen for the downfield resonances in the inset.

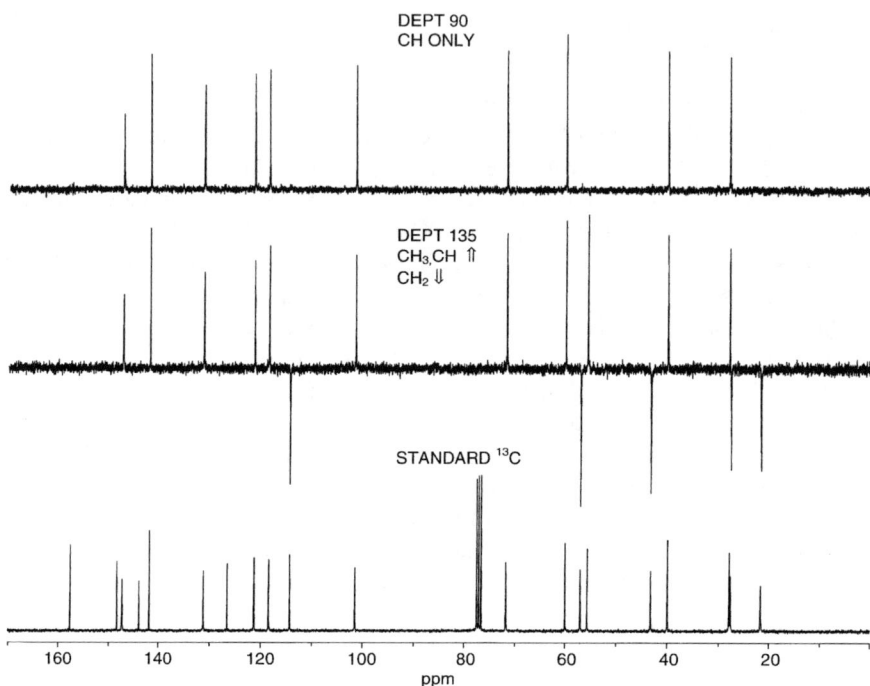

Figure 5.2. The 75.2-MHz ^{13}C and DEPT spectra of quinine in CDCl$_3$.

this chapter. It should be noted that the values in Table 5.1 were not determined solely on the basis of these three experiments, but instead after careful consideration of all of the experiments outlined below.

5.6. DEPT

The DEPT spectra require some explanation. Along with a fully decoupled ^{13}C NMR spectrum, it is customary to determine whether each carbon resonance is due to a methyl group, a methylene, a methine group, or to carbon that is directly bonded to no hydrogen atoms, usually referred to as a "quaternary" carbon. Obsolete experiments formerly used to obtain the same information are the off-resonance ^{1}H decoupled experiment, the attached proton test (APT), and more recently, the INEPT experiment (*i*nsensitive *n*uclei *e*nhancement by *p*olarization *t*ransfer). The DEPT experiment has distinct advantages over the other three experiments. Off-resonance decoupled spectra often suffer from overlap of multiplets, while results of the APT experiment are sometimes difficult to phase and interpretation can be ambiguous. A serious drawback of the INEPT experiment is the fact that triplets and other multiplets are not immediately recognizable

Table 5.1. Basic Chemical Shift and Coupling Information for Quinine[a]

Position in structure	Chemical shift for ^1H	Multiplicity	Chemical shift for ^{13}C	Type of carbon
2	8.68	d	147.2	CH
3	7.50	d	118.4	CH
4	—	—	148.3	C
5	7.23	d	101.4	CH
6	—	—	157.6	C
7	7.33	dd	121.3	CH
8	7.98	d	131.1	CH
9	—	—	143.9	C
10	—	—	126.5	C
11	5.53	d	71.7	CH
12	3.15	m	59.9	CH
14	3.09	dd	56.9	CH_2
14′	2.64	m	—	CH_2
15	2.27	m	39.9	CH
16	1.82	m	27.8	CH
17	1.54	m	27.6	CH_2
17′	1.50	m	—	CH_2
18	3.41	m	43.1	CH_2
18′	2.69	m	—	CH_2
19	1.72	m	21.6	CH_2
19′	1.59	m	—	CH_2
20	5.76	dq	141.8	CH
21	4.94	tt	114.2	CH_2
22	3.90	s	55.6	CH_3

[a]See text for numbering of the structure.

because intensities are distorted. A bonus of the DEPT experiment is that it is inherently more sensitive than a standard ^{13}C experiment (up to four times more) because the DEPT benefits from polarization transfer.

The results of the DEPT experiment are very straightforward. In the first spectrum (upper spectrum Fig. 5.2), only the ^{13}C signals from methine (CH) carbons are seen. In the second spectrum (middle spectrum, Fig. 5.2), methyl and methine carbon signals are above the axis, while methylene carbon signals are inverted. Finally, quaternary carbons appear only in the standard ^{13}C spectrum (bottom spectrum, Fig. 5.2), so any carbon signal which appears only in the standard spectrum and not in either of the two DEPT spectra must be due to a quaternary carbon. Thus, for our model compound quinine, this data is summarized in the right-hand column of Table 5.1.

5.7. COSY

One of the first successful 2D experiments is called COSY (*c*orrelation *s*pectroscopy); COSY spectra correlate ^1H-^1H spins that are coupled (also called *J*-

coupling). A useful way to consider a COSY spectrum is as a series of homonuclear decoupling experiments in which all coupled protons are found. Of course, there are no decouplings in the experiment and COSY is not a replacement for homonuclear decoupling.

5.7.1. DQF-COSY

Quinine provides a good example to illustrate the power of the COSY experiment. The proton spectrum of quinine (Fig. 5.1) is rather complex; there are 19 chemical-shift-distinct resonances. Splitting of many of the resonances is evident and represents coupling of proton spins. Figure 5.3 displays the double-quantum-filtered COSY (DQF-COSY) spectrum of quinine. Double-quantum filtering is a refinement for COSY that reduces "clutter" on the diagonal of the 2D spectrum.

The DQF-COSY spectrum of quinine is a typical 2D spectrum and illustrates most of the important features of 2D correlation spectra.[3] There are two frequency axes (labeled F1 and F2) that represent two orthogonal Fourier transformations of the data. These two axes are the "dimensions" in a 2D spectrum. In the case of COSY, these axes are both for ^{1}H; two identical axes produce a symmetric appearance and a diagonal. For convenience, a proton spectrum is provided along each axis. These reference spectra are not actually part of the COSY spectrum; these spectra do, however, provide indispensable markers of chemical shift. The information in this 2D spectrum (as in most 2D spectra) is represented by a series of contour plots just as a hill is represented by series of contours on a contour map.

The contours on the diagonal are of little interest because these contours are nothing more than the peaks that are found in the 1D proton spectrum. The off-diagonal or cross peaks, which are symmetrically disposed horizontally and vertically about the diagonal, represent the new and interesting information. The easiest way to interpret a COSY spectrum is to select any cross peak and to draw perpendicular lines to each of the proton spectra on the orthogonal axes. These absorptions or multiplets are coupled. An expanded view of the downfield portion (from 7.0–8.8 ppm) of the COSY spectrum containing the aromatic proton peaks can be found in Figure 5.4.

Included in this figure are the perpendicular lines to show the correlations. For instance, the proton at about δ 8.7 correlates with the proton at about δ 7.5; these protons (on C-2 and C-3, respectively) are therefore coupled. The other correlations are easily found, although their assignments are not yet trivial.

Figure 5.5 displays the upfield portion of the COSY spectrum. This region of

---

[3]There are 1D analogues for most 2D correlation experiments. These 1D analogues are generally considered "inferior" to the 2D experiments. The 1D HOHAHA, an analogue of the 2D TOCSY, is considered later in the chapter.

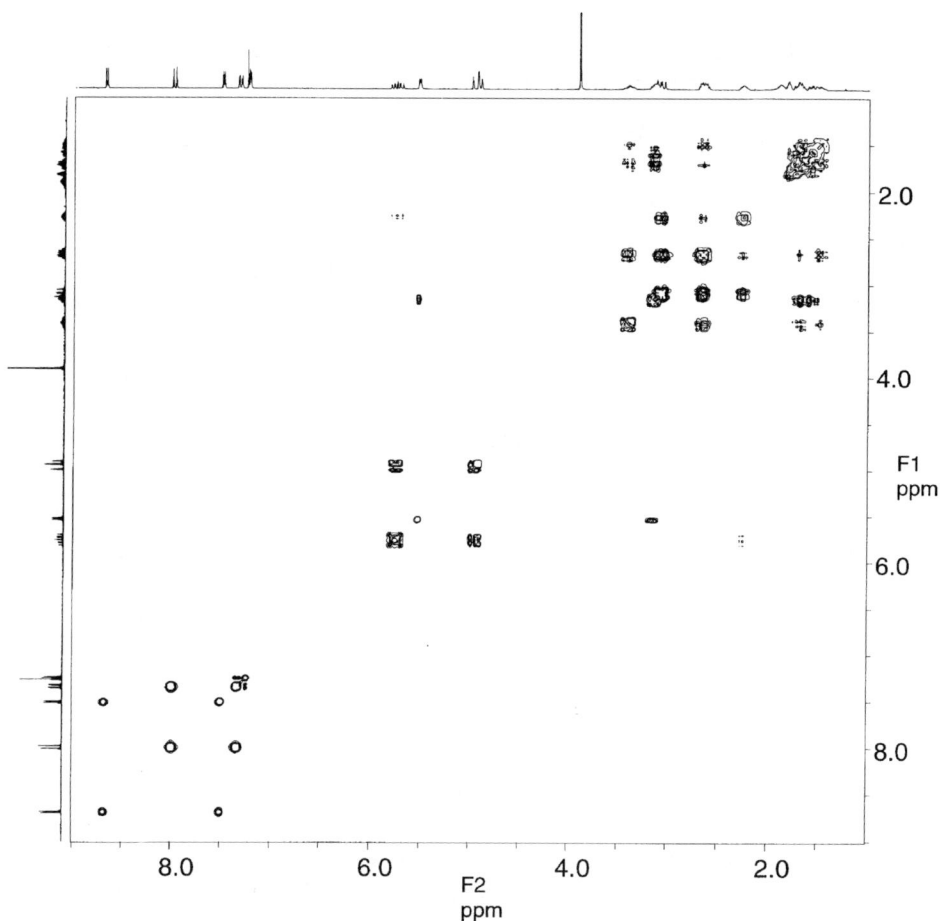

Figure 5.3. The 300-MHz DQF-COSY spectrum for quinine in CDCl₃. None of the correlations are shown in the figure.

the spectrum contains overlapping peaks and it is obviously more complicated. Only two correlations are shown in this figure. The other ones can easily be drawn in and tabulated. Although a complete "interpretation" can not yet be made, the COSY spectrum contains a large amount of valuable information from proton-proton couplings.

A more powerful version of COSY is the phase-sensitive DQF-COSY experiment. The results of this experiment give both positive and negative contours, which are usually shown with different colors. The results of the experiment enable one to measure (among other things) both the sign and the magnitude of the coupling constants.

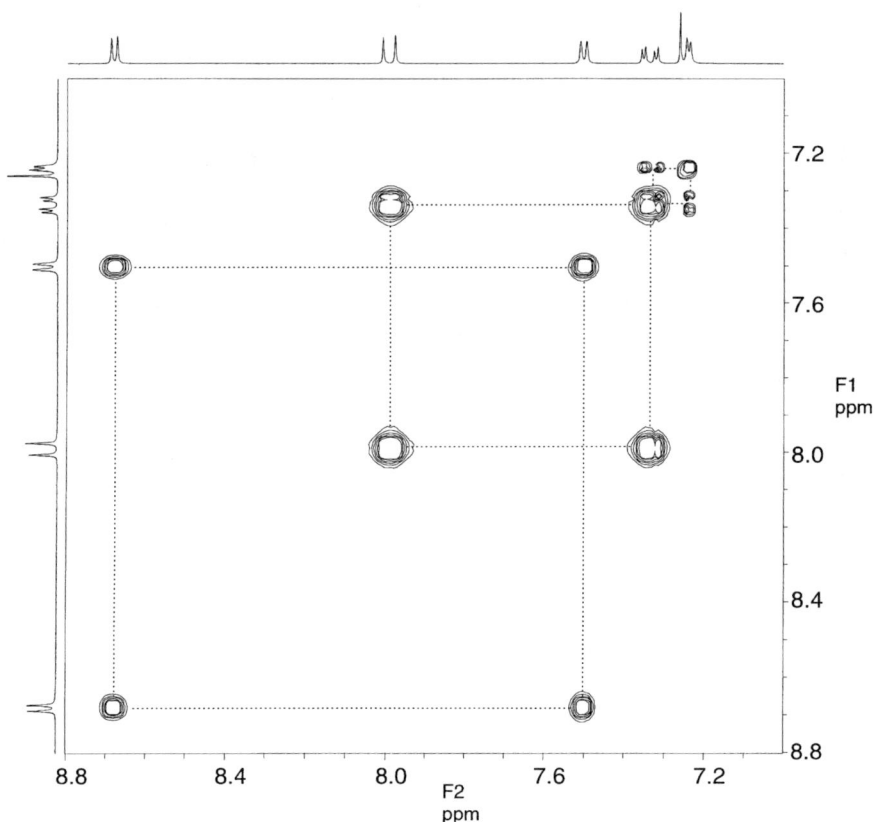

Figure 5.4. Expanded view of the downfield portion of the DQF-COSY spectrum of quinine (see Fig. 5.3). The dotted lines show correlations; all correlations are shown.

5.8. TOCSY

There are other types of ^{1}H-^{1}H correlations that might be considered in certain instances, depending on the structural problem at hand. One important experiment is TOSCY (*to*tally *c*orrelated *s*pectroscopy), which is particularly useful in structures with distinct, isolated spin systems such as carbohydrates, peptides, and nucleic acids. The experiment is similar to COSY with an additional feature known as "spin locking" during which there is a "mixing period." During this mixing period, magnetization is relayed from one spin to its neighbor and then to its next neighbor and so on. The longer the mixing period, the further the relay, which, in theory, can extend throughout the entire spin system.

The 2D TOCSY spectrum of quinine is shown in Figure 5.6. The appearance of this spectrum is similar to COSY; there is a proton axis in both dimensions

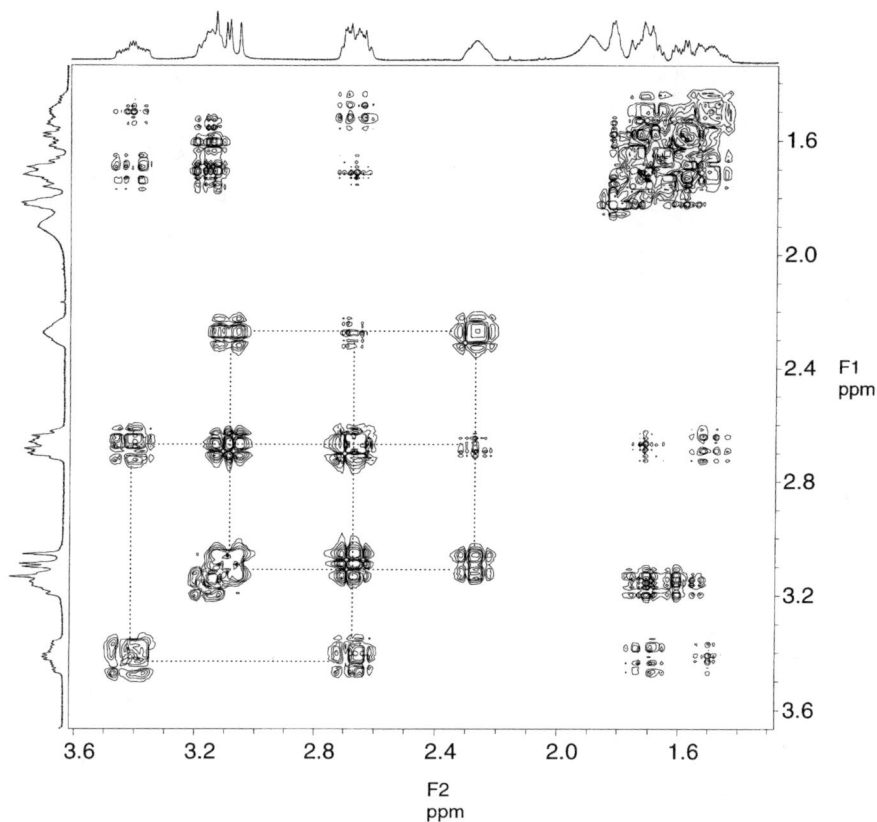

Figure 5.5. Expanded view of the upfield portion of the DQF-COSY spectrum of quinine (see Fig. 5.3). Some, but not all, correlations are drawn in.

and a proton spectrum provided for clarity. There is a diagonal and the cross peaks again are symmetrically disposed about the diagonal. In fact, the COSY interactions (correlations) are a subset of the TOCSY cross peaks. The additional cross peaks arise from magnetization transfer, allowing connectivity beyond two adjacent carbons to be determined. These additional peaks are found the same way as in COSY.

5.8.1. HOHAHA

A very interesting and useful variation of the 2D TOCSY is the 1D version, which is usually called HOHAHA (*h*omonuclear *h*artmann-*h*ahn). For this experiment, a proton resonance is selected and irradiated with a shaped pulse which provides a high degree of selectivity. An appropriate mixing time is applied so that magnetization can be relayed (only from the selected resonance) to its neighboring

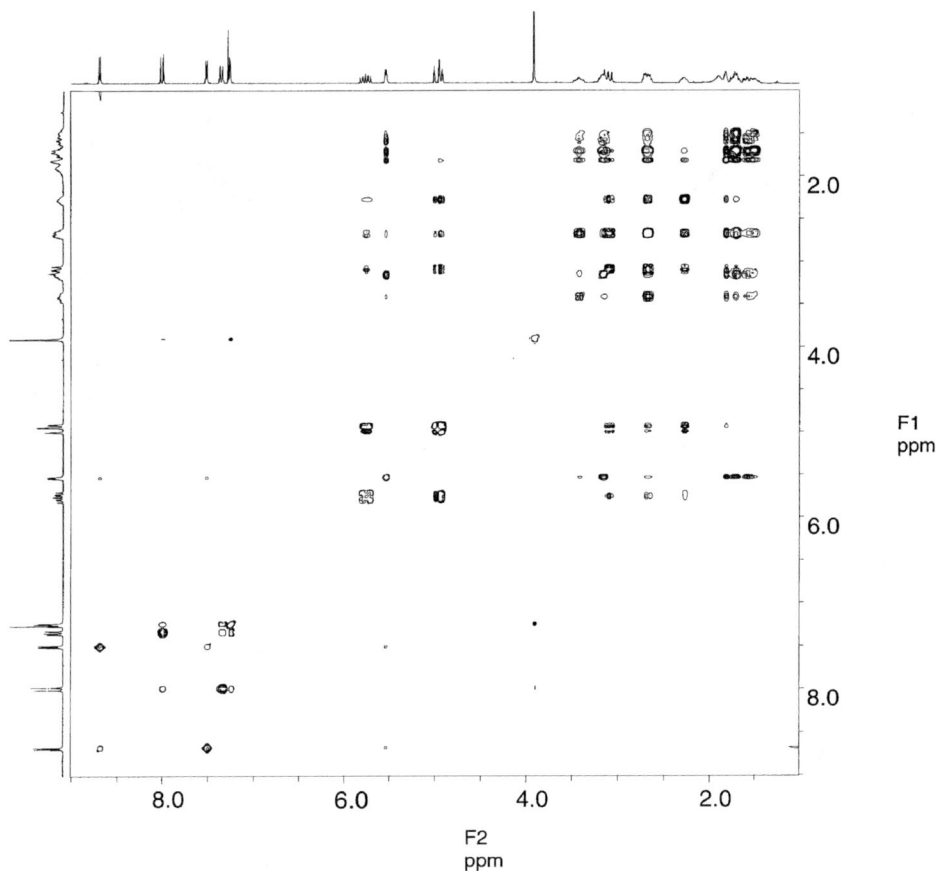

Figure 5.6. The 300-MHz 2D TOCSY spectrum of quinine in CDCl$_3$. None of the correlations are shown.

spins. At the end of the mixing time, a 1D spectrum is acquired. The only signals that are recorded in this spectrum are those to which magnetization has been transferred. All other proton resonances are absent.

Three 1D HOHAHA spectra are displayed in Figure 5.7; the three spectra have different mixing times while the same proton resonance, the methine proton on C-11 (δ 5.5), is irradiated. At 20 ms mixing time, the magnetization is relayed only to the C-12 methine proton (δ 3.15). At a longer mixing time of 120 ms, magnetization has been further transferred to the proton on C-3 (δ 7.50) and the methylene protons on C-19 (δ 1.59, 1.72). After the longest mixing time for this experiment of 230 ms, magnetization also has been transferred to the proton at C-2 (δ 8.68), and there is evidence of transfer to protons on C-15, C-16, and C-18. As can be seen, the signal strength dissipates with longer mixing times.

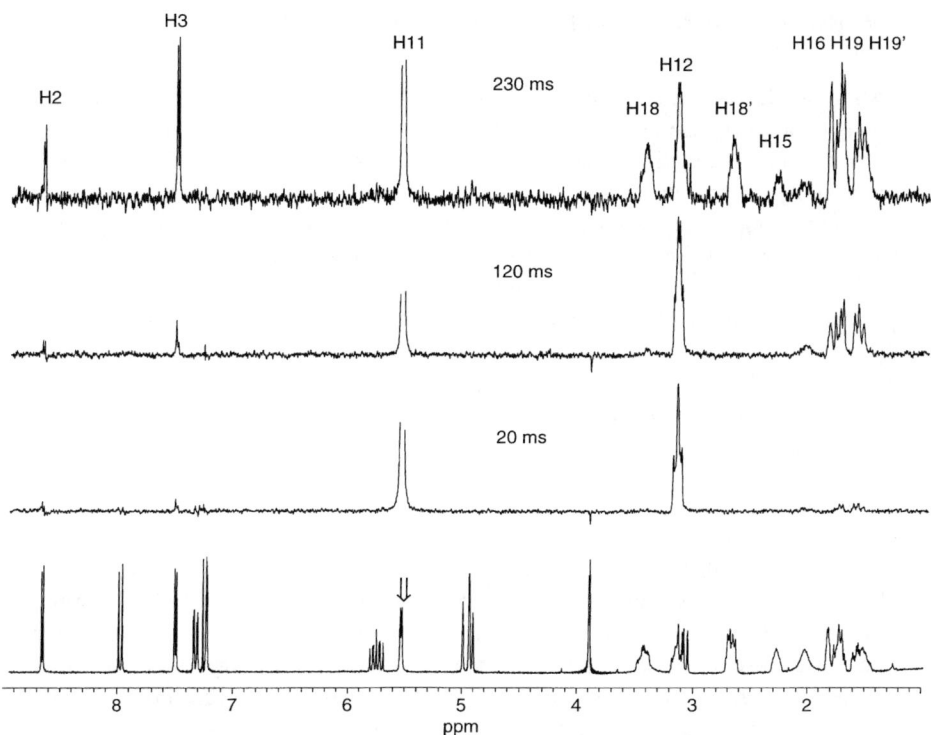

Figure 5.7. The 300-MHz 1D HOHAHA spectra of quinine in CDCl₃. The first line is the standard ¹H spectrum for comparison. The arrow indicates which resonance was irradiated. Mixing time is listed in milliseconds. The decreasing signal/noise ratio is a function of increasing mixing time.

This 1D experiment can be used to great advantage on systems in which proton signals overlap yet the protons are not coupled to each other. For instance, imagine a trisaccharide in which the ring protons for the three different sugar residues overlap. Selection of the anomeric proton of one of the residues for irradiation and a systematic increase of the mixing time gives a series of ¹H edited spectra, each containing another of the ring protons further removed from the anomeric position. For most hexoses, magnetization can be transferred as far as the methylene group at C-6. It is easy to imagine similar experiments for peptides and nucleic acids. In many cases, the 2D experiment gives too much information at one time, which precludes the *sequential* matching of signals. Thus, even though the series of experiments can take more instrument time and more operator assistance, this is one case where the 1D experiment is often more helpful than the 2D experiment.

5.9. Experiments Based on the Nuclear Overhauser Effect

So far, we have discussed interactions between nuclei that take place through bonds connecting the nuclei. However, there also can be through-space interactions between magnetic nuclei, which can increase (or decrease) the intensities of the signals of the nuclei being observed. These interactions, which are independent of bonding, occur when one nucleus is irradiated at its resonance frequency, and nuclei which are nearby in space experience enhanced relaxation, which slightly increases the size of the observed signals. This effect, known as the nuclear Overhauser effect (NOE), is very short range (~2–4 Å), falling off as the inverse sixth power of the distance between the irradiated and observed nuclei. NOE forms the basis of several useful experiments which provide information complementary to the information obtained from other NMR experiments based on connectivity.

5.9.1. NOESY and ROESY

For large molecules such as peptides and nucleic acids, a 2D experiment called NOESY (Nuclear Overhauser Enhanced Spectroscopy) provides detailed information about the tertiary structures of these macromolecules by revealing which nuclei are close together in space. These measurements are extraordinarily useful because they are obtained in (aqueous) solution so that the macromolecules are in their native conformations, while similar information from x-ray crystallography can only be obtained in the solid crystalline state. Interactions of large molecules with small molecules can also be studied by NOESY, such as the interaction of a receptor with a ligand.

The NOESY experiment may give weak signals with small molecules. However, a variation of the NOESY experiment, called ROESY (Rotating Frame Nuclear Overhauser Effect Spectroscopy), utilizes a technique known as "spin locking" to obtain through space information on small molecules in a 2D format, and the ROESY experiment is recommended for small molecules (MW < 1000). These through space 2D experiments are compared to the COSY spectrum because many of the COSY cross peaks are found in the NOESY and ROESY experiments.

5.9.2. NOE Difference Spectra

A 1D NOE version for through space correlations by difference spectrometry can be quite easily carried out. An NOE difference spectrum is obtained by irradiating one resonance (at less than full power) followed immediately by acquisition of a ^{1}H spectrum. The signal of any proton close in space (~2–4 Å) to the one that was irradiated will be enhanced by the nuclear Overhauser effect. The spectrum obtained is then subtracted from the original, thereby producing a difference spectrum which shows only those signals which were enhanced. An

NOE difference spectrum is very effective in showing small differences in signal intensity (usually on the order of a few percent) that result from NOE enhancement.

NOE difference spectrometry is especially useful for distinguishing relative stereochemistry or positional isomers. It is important to realize that NOE cannot be used to determine absolute stereochemistry, at least directly.[4] Elucidation of a structure such as quinine does not benefit greatly from this type of correlation but useful information can nonetheless be obtained. Application to quinine can yield the position of the methoxy group, which cannot be determined by other ^{1}H-^{1}H correlations, although the question can be answered readily by ^{1}H-^{13}C correlation methods (see below).

The NOE difference spectrum of quinine, for which the methyl group of the methoxy group has been irradiated, is shown in Figure 5.8. The difference spectrum shows that the signals from the protons on C-5 (δ 7.23) and C-7 (δ 7.33) are, on average, equally close to the methyl group because they both show large and similar enhancement in the difference spectrum. The through-space correlation therefore places the methoxy group between these two protons.

NOE correlations can be applied to various other structural questions. For instance, in trisubstituted olefins, the stereochemistry can be determined indirectly by chemical shift of the single olefinic proton. If one of the groups on the olefin is a methyl, the chemical shift of the methyl group may be diagnostic, and occasionally long range coupling can provide the needed information. These assignments are risky, however, especially when only one isomer is available. Alternatively, NOE difference spectrometry can solve the problem directly without the need for both isomers, because there is usually some enhancement to the methyl (or alkyl) group on the same side of the double bond as the lone olefinic proton.

Because NOE difference spectra show only a small subset of the multiplets in the total proton spectrum, NOE difference spectra also can be used to fish out a particular multiplet from an overlapped group of multiplets. That is, the shape of multiplets are unchanged in the NOE difference experiment; only the intensities of the few signals from protons close in space to the irradiated proton are enhanced. These few multiplets are then cleanly displayed in the difference spectrum, free from overlapping interferences.

5.10. Heteronuclear Correlated Experiments

5.10.1 HETCOR and COLOC

Historically, for the natural products chemist working with very small quantities of sample, correlations involving ^{1}H-^{13}C interactions were never considered be-

---

[4]NOE difference spectrometry and similar types of experiments can be used to determine absolute stereochemistry after derivatization with a compound of known stereochemistry. Relative stereochemistry along with the absolute stereochemistry of the derivative gives the desired information.

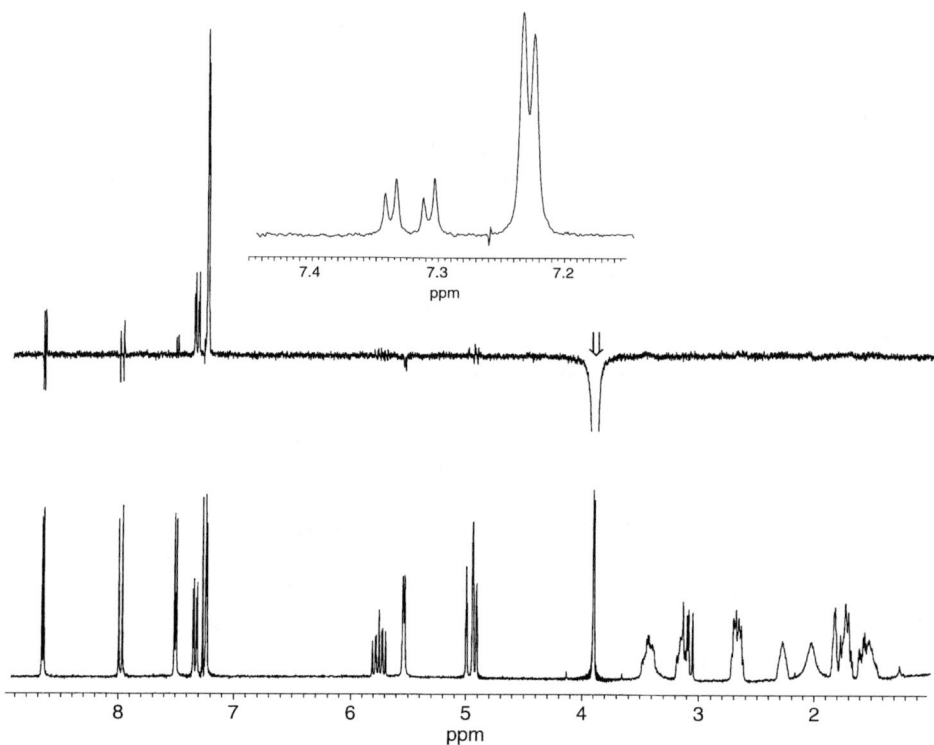

Figure 5.8. The 300-MHz NOE difference spectrum of quinine in $CDCl_3$. The arrow indicates which resonance was irradiated. The first line is the standard ^1H spectrum for reference. The inset for the NOE difference spectrum shows the pattern more clearly. Apparent peaks at δ 5.0, 5.6, 5.8, 8.0, and 8.7 are dispersive peaks and are not true enhancements.

cause they traditionally required direct observation of the ^{13}C nucleus. The two most important experiments in this category are HETCOR (*het*eronuclear *cor*relation) and COLOC (*co*rrelation spectroscopy via *lo*ng-*r*ange *c*oupling). The HETCOR experiment is a 2D experiment showing directly bonded ^1H-^{13}C couplings ($^1J_{CH}$). COLOC is a similar type of 2D experiment that shows long range ^1H-^{13}C couplings, i.e., two-bond ($^2J_{CH}$) and three bond ($^3J_{CH}$) couplings. For structure elucidation, these experiments are powerful tools.[5] The obvious limitation is that usually more than 10 mg of sample is required. To overcome the problem, "inverse" or "indirect" methods have been developed.

In principle, in a ^1H-^{13}C correlation experiment, one should be able to observe

---

[5]The HETCOR experiment remains a popular and useful technique when the amount of sample is not limiting. The COLOC experiment has been largely replaced by HMBC.

either the ^{1}H nucleus or the ^{13}C nucleus. In practice, there are various technical problems in using the ^{1}H nucleus for observation. Therefore, most of the original experiments devised for ^{1}H-^{13}C correlation actually observed ^{13}C, even though it is only about 0.000175 times as sensitive as the ^{1}H nucleus. Today, the problems have been completely overcome with the development of "inverse probes" and better pulse sequences. It is therefore possible to run heterocorrelation (^{1}H-^{13}C) experiments on the same amount of sample as needed for homocorrelation (^{1}H) experiments. To put it another way, the HMQC and HMBC experiments, which are described below, require about as much sample as a COSY, which is ~ 0.5 mg for a 300-MHz instrument, and less for higher field instruments.

5.10.2. HMQC

The proton detected version of HETCOR, in which one-bond or directly bonded carbon-hydrogen couplings are correlated, is the HMQC (*h*eteronuclear *m*ultiple *q*uantum *c*oherence) experiment. The HMQC spectrum of quinine is shown in Figure 5.9.

The most obvious difference in this 2D spectrum when compared to the other 2D spectra so far in this chapter is that there is no diagonal and no symmetry; this will always be true whenever the F1 axis and the F2 axis represent different nuclei. In this presentation, the F1 axis (carbon) is along the right side and the F2 axis (proton) is along the bottom. Opposite these axes are found the corresponding 1D spectra, which again are given as a convenience and are not part of the actual 2D spectrum. Interpretation of this spectrum is straightforward. A horizontal line is drawn from any carbon until a cross peak is encountered. Another line is drawn perpendicular to the first line to find the proton or protons with which it correlates. (Interpretation could start on the proton axis to obtain the same results. In cases of overlap in the proton spectrum, it will not always be possible to find all of the proper starting points. Overlap is less likely on the carbon axis.)

There are only three possible cases for a carbon atom in an HMQC spectrum. First, if the horizontal line encounters no cross peaks, then the carbon has no attached protons. Second, if the drawn line encounters only one cross peak, then the carbon may have either one, two, or three protons attached; if two protons are attached, then they are either chemically shift equivalent or they fortuitously overlap. Third, if the line encounters two cross peaks, then the special case of diastereotopic protons of a methylene group has been found. Much of this information will already be available from the DEPT spectra (see above); indeed, the HMQC spectrum should, whenever possible, be considered along with the DEPT.

In the HMQC spectrum of quinine (Fig. 5.9), two correlations are shown. The carbon resonance at about δ 72 gives a single cross peak with the proton at about δ 5.53. This interaction is the C-11 carbon and its attached methine proton. The carbon resonance at about δ 43, which corresponds to a methylene group (CH_2) from the DEPT 135 spectrum (Fig. 5.2), correlates with two proton resonances

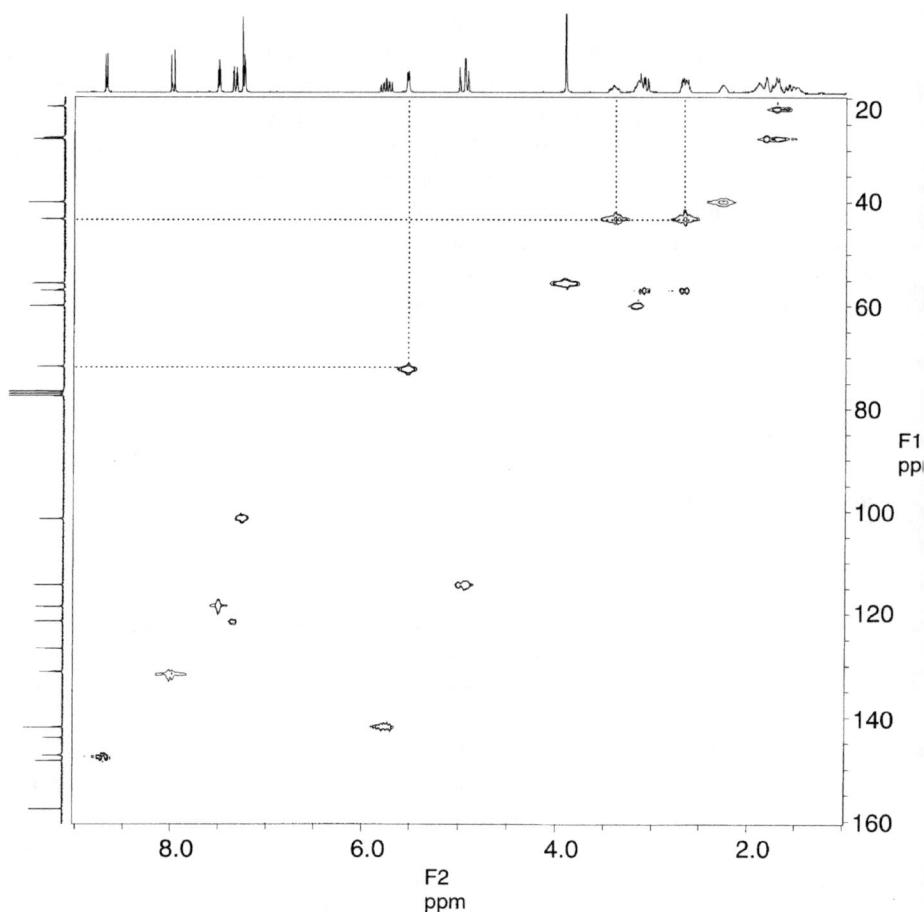

Figure 5.9. The HMQC spectrum of quinine in CDCl$_3$. Two correlations have been shown as examples.

at about δ 2.64 and 3.41 on the proton axis. The carbon resonance is for C-18 while the two proton resonances represent the two diastereotopic hydrogens.

The information derived from HMQC raises the COSY spectrum to a new level. In other words, the various correlation spectra build on each other in a synergistic manner. For example, the fact that the two protons on C-18 are coupled can be easily gleaned from the COSY spectrum. It is not certain that the two protons are on the same carbon and diastereotopic until the information from the HMQC is available (however, the magnitude of the coupling constant from the 1D spectrum may suggest that they are geminal). The interplay among the various spectra can become quite elaborate. It is helpful to tabulate certain types of data; for certain experiments, as can be seen below, it becomes essential to look at the data in tabular form to minimize confusion due to data overload!

5.10.3 HMBC

In the HMQC experiment described above, long range (i.e., two- and three-bond) proton-carbon couplings were not observed by appropriate choices for parameters in the pulse sequence while the directly attached (i.e., one-bond) couplings were preserved and correlated in a 2D experiment. The HMBC (*h*eteronuclear *m*ultiple *b*ond *c*oherence) experiment, on the other hand, capitalizes on these two- and three-bond couplings, yielding an extremely useful (although sometimes cluttered) spectrum. HMBC is a proton-detected, long-range ^{1}H-^{13}C heteronuclear correlation 2D experiment. This experiment indirectly gives carbon-carbon (although not ^{13}C-^{13}C) correlations. In addition, quaternary carbon atoms show correlations to nearby protons, something not found in HMQC spectra. Since both $^2J_{CH}$ and $^3J_{CH}$ couplings are preserved, interpretation is often tedious. It is helpful to keep in mind the HMQC correlations.

The HMBC spectrum for quinine is shown in Figure 5.10. In some ways, the HMBC and HMQC spectra appear similar; however, the HMBC spectrum clearly shows more correlations. Closer inspection reveals that the one-bond correlations are gone. Interpretation of HMBC spectra is more difficult than interpretation of those from HMQC. Also, one or more expected correlations may be absent. In particular, while the two-bond correlations ($^2J_{CH}$) are almost always found, the three-bond ($^3J_{CH}$) correlations are occasionally absent. The variations in the correlations that are found are due to the variations in the magnitude of $^2J_{CH}$ and $^3J_{CH}$ coupling constants.

In principle, interpretation is as above for HMQC. One set of correlations are drawn in. Before considering the "real" correlations, a point about an interesting and common artifact should be discussed. Intense proton peaks, usually those due to methyl groups, sometimes give ^{13}C satellites in HMBC spectra. These artifact peaks are easy to spot because they show up as two "cross peaks" that do not line up (correlate) with any protons in F2. In the HMBC for quinine, ^{13}C satellites of the methoxy group are evident and should be ignored. They are found straddling the vertical line drawn in at about δ 3.9. (In F1, they are in line with the methoxy carbon at about δ 56 and are indicated with arrows in Fig. 5.10.)

The methoxy protons have no two-bond correlations, and only one three-bond correlation, to C-6, thus establishing a direct connectivity between the methoxy group and the aromatic ring. HMBC correlations from C-6 (shown with perpendicular lines in Fig. 5.10) can be quickly and easily found. The new correlations found are to protons on C-5, C-7, and C-8; these correlations are important because C-6 is quaternary and would otherwise be difficult to assign. The process can be continued, but the mental bookkeeping quickly becomes a taxing problem

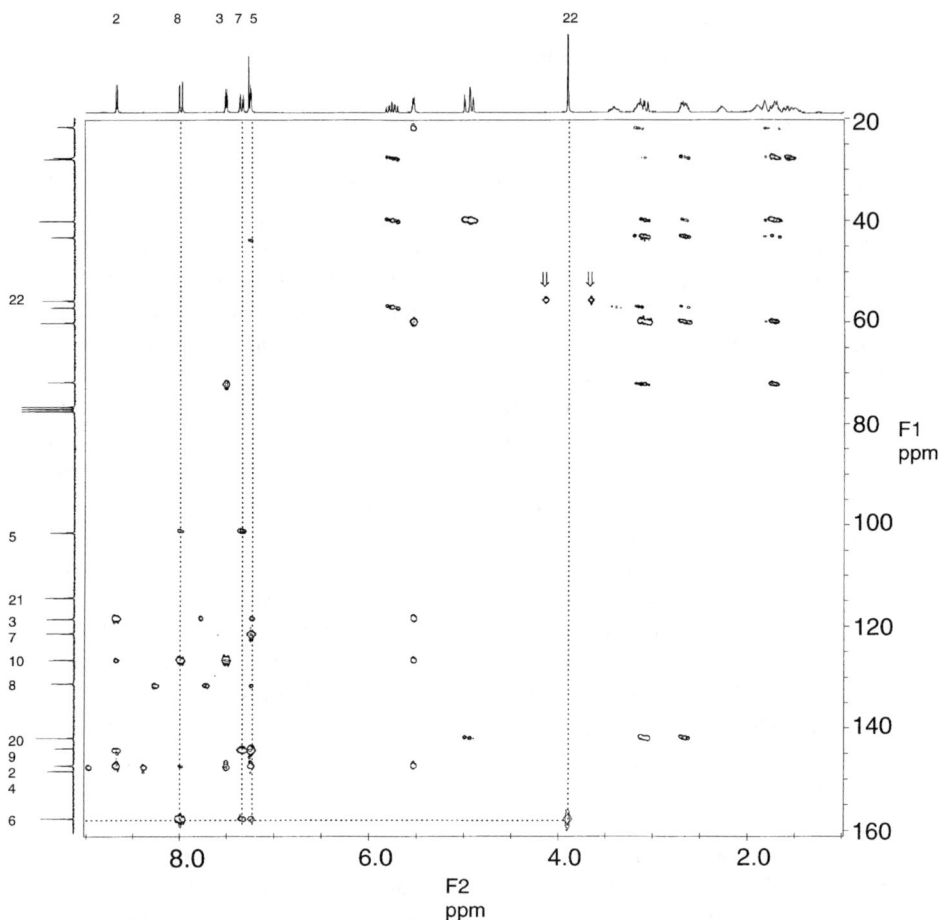

Figure 5.10. The HMBC spectrum of quinine in CDCl₃. One correlation is shown as an example. The numbers on the left side next to the carbon spectrum correspond to carbon numbers in the structure and in Table 5.1.

and confusion often results. A good way to keep track of these data is to construct a table listing the carbon resonances in one direction and the proton resonances in the other. In Table 5.2, the carbons are given across the top and protons along the side. The numbering for quinine is shown in the structure; the primed numbers refer to diastereotopic pairs of protons. It is usually easier to read this table from the top, in other words, from the carbon resonances, because overlap is less likely in carbon resonances. From either direction, the results are identical.

The entries in Table 5.2 can be of four types. If a given proton and carbon are bonded, "Bonded" is entered. These entries are the exact correlations found in the HMQC spectrum. There are no correlations for these entries in the HMBC

Table 5.2. Expected HMBC Interactions[a]

										Carbon 13 Signals										
	2	3	4	5	6	7	8	9	10	11	12	14	15	16	17	18	19	20	21	22
2	Bonded	α	β					β	β											
3	α	Bonded	α						α											
5			β	Bonded	α	β		β												
7				β	α	Bonded	α	β												
8					β	α	Bonded	α	β											
11		β	α						β	Bonded	α						β			
12			β							α	Bonded						α			
14												Bonded	α	β		β		β		
14'												Bonded	α	β		β		β		
15												α	Bonded	β	β	β	β	α	β	
16												β	α	Bonded	α	β	α	β		
17													β	α	Bonded	α	β			
17'													β	α	Bonded	α				
18											β	β	β	β	α	Bonded				
18'											β	β	β	β	α	Bonded				
19										β	α		α	α	β		Bonded			
19'										β	α		β	α	β		Bonded			
20												β	β	β				Bonded	α	
21																		α	Bonded	
21'																		α	Bonded	
22					β															Bonded

[a]See text for detailed explanation.

except the ^{13}C satellites noted above. If there is no entry in the table for a given proton-carbon pairing, then the proton and carbon are separated by four or more bonds and no correlation is expected.[6] If an "α" is entered for a given proton-carbon pair, then there are two bonds separating the pair. If a "β" is entered for a given proton-carbon pair, then there are three bonds separating the pair. For both an "α" and a "β," a correlation is expected. It must be emphasized that there is no difference between the "α" and "β" entries as far as the spectrum is concerned; they are used in the table merely to sort the interaction types.

A complete analysis of the HMBC spectrum for quinine is a large, but rewarding, undertaking, which considered in the context of the HMQC and the COSY unambiguously supports the structure for quinine. Closer inspection of the HMBC spectrum and the associated table yields exquisitely detailed structural information that can be deciphered with a methodical approach. The carbon skeleton, the backbone of the structure, is deduced by indirect evidence. The HMBC experiment provides direct correlations to quaternary carbons, and equally as important, provides a way to "see through" heteroatoms such as oxygen and nitrogen.

5.11. INADEQUATE

There is a direct method of determining the carbon skeleton by ^{13}C-^{13}C correlations. From ^{1}H-^{1}H correlations to ^{1}H-^{13}C correlations, the next logical step utilizes ^{13}C-^{13}C correlations. The best of the experiments exploiting ^{13}C-^{13}C correlations is the 2D experiment INADEQUATE (*i*ncredible *n*atural *a*bundance *d*ouble *qua*ntum *t*ransfer *e*xperiment). For the elucidation of the structures of organic compounds, this experiment is, without question and without exception, the most powerful and the least ambiguous method available. The results of the experiment are easy to interpret as well. The problem with the experiment is simple: sensitivity. The natural abundance of ^{13}C in unenriched samples is about 1%. The probability that any two adjacent carbon atoms will both be ^{13}C atoms (independent events) is 0.01×0.01 or 0.0001; in rounded whole numbers, that is about 1 in 10,000 pairings.

For the chemical ecologist isolating a small amount of a semiochemical, the prospects of having enough sample for INADEQUATE are remote. A discussion of the experiment is nonetheless included because it is really the only way of obtaining direct carbon-carbon correlations. Furthermore, there is a strong likelihood that hardware and pulse sequence improvements will soon reduce the need for large samples.

The ability to record only the occurrences of ^{13}C-^{13}C pairings, a seemingly impossible task, is accomplished through the technique of double quantum filter-

---

[6]Four-bond correlations are quite rare but occasionally do occur. In the case of quinine, a four-bond correlation is found between C-5 and the proton attached to C-8.

ing. Thus, only transitions from systems with two spins (i.e. two ^{13}C bonded to each other, either AB or AX) and higher are detected during acquisition. The signal from such transitions is invariably weak so that the amount of sample required for INADEQUATE on a modern high field spectrometer is about 70–300 mg dissolved in 0.5 ml of an appropriate deuterated solvent. (If possible, the sample should contain a catalytic amount of paramagnetic chromium [III] acetyla-cetonate to shorten the T1 relaxation times. See section on ^{15}NMR spectrometry.)

The 2D INADEQUATE spectrum of quinine is presented in Figure 5.11. The

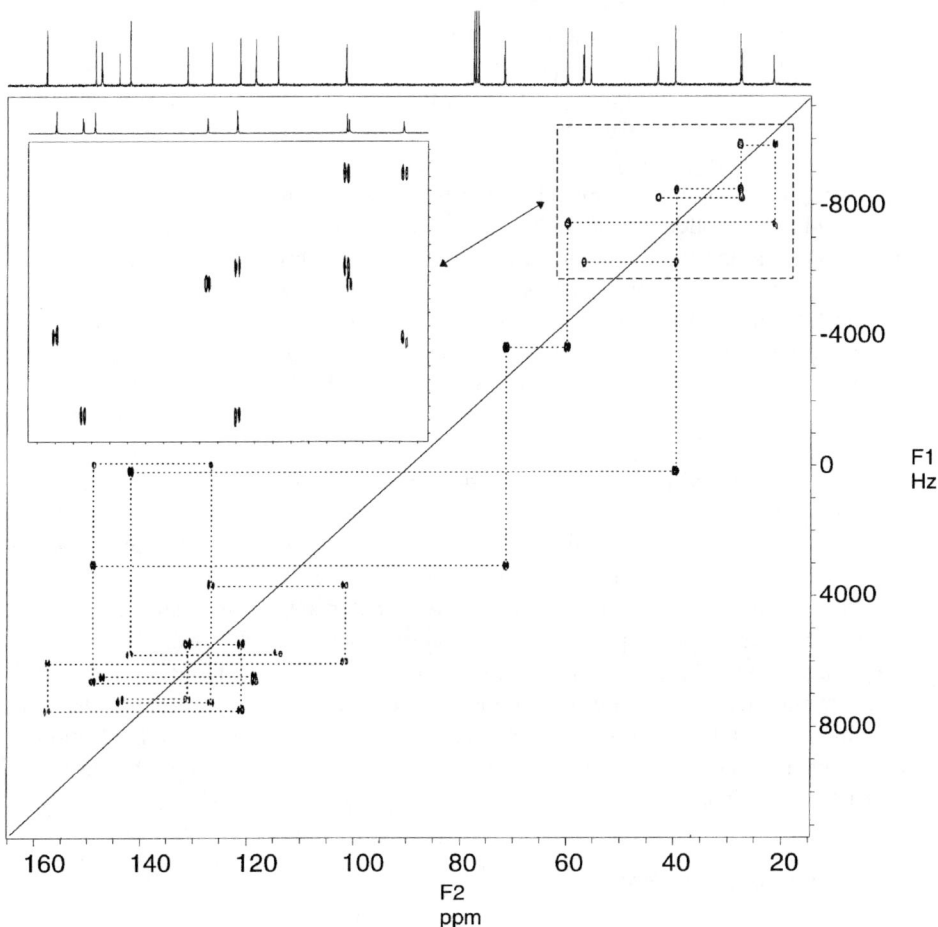

Figure 5.11. The 2D INADEQUATE spectrum of quinine in $CDCl_3$. One expansion is shown to illustrate the doublets for each AX pair. All correlations are shown. See text for explanation of the F1 axis.

F2 axis is the familiar carbon axis, which, of course, is established during acquisition. The F1 axis is not the usual carbon axis and requires further explanation.

During the period in which the carbon-carbon spins are interacting, the frequencies that evolve are not the chemical shifts of the coupled nuclei as they are in a typical COSY. Instead, the sum of the chemical shifts of the coupled nuclei evolves during this period, and because of the double-quantum filtering, only the two-spin AB and AX systems contribute significantly to signal intensity in the INADEQUATE spectrum. Proper selection of certain delays in the pulse sequence selects the larger one-bond couplings ($^1J_{CC}$), thus ensuring that the signals are due to directly bonded carbon-carbon correlations. The F1 axis is usually given in hertz and it is two times the range in F2.

In the 2D INADEQUATE spectrum of quinine, cross peaks or correlations are found at ($v_A + v_X$, v_A) and at ($v_A + v_X$, v_X) in the (F1, F2) coordinate system for a given AX system. The actual cross peaks themselves are doublets (see the expanded inset, Fig. 5.11) with a spacing equal to the ($^1J_{CC}$) coupling constant. The midpoint of the line connecting the two sets of doublets is (v_A, v_X), ($v_A + v_X$)/2; thus, the collection of midpoints for all of the pairs of doublets lie on a line running along the diagonal. This is an important observation because it can be used to distinguish genuine cross peaks from spurious peaks and other artifacts.

The INADEQUATE spectrum in Figure 5.11 shows both the line connecting the midpoints and all of the "connectivities." For instance, connectivities from C-11 (δ 71.7) join it with C-4 (δ 148.3) and C-12 (δ 59.9). C-12 shows further connectivity to only C-19 (δ 21.6), while C-4 connects to C-3 (δ 118.4) and C-10 (δ 126.5). C-3 is, in addition to C-4, connected to C-2 (δ 147.2) whereas C-10 is also connected to C-5 (δ 101.4) and C-9 (δ 143.9), and so on. Tracing the remaining connectivities gives the entire carbon skeleton but not the entire structure. There are no connectivities for the two nitrogen atoms or for the oxygen atom. Even with these limitations, the power of the method is evident.

The experiments outlined thus far in this chapter are outlined in Table 5.3, which summarizes the types of information obtained, limitations of the experiment, sample size, and instrument time. The information in the table is not meant to be comprehensive nor are the sample sizes and instrument times meant to be absolutes. Instead, the information is intended to provide guidelines for selection of the most fruitful experiments to run with the available sample and with possible restrictions on instrument time.

5.12. Heteronuclear NMR

Another class of experiments that are generally out of the realm of natural products isolation and identification (therefore only briefly treated here) is heteronuclear NMR, or the NMR analysis of nuclei other than protons and carbon.

Table 5.3. Summary of the Experiments Outlined in this Chapter

Experiment	Information obtained	Limitations of experiment	Minimum/typical sample size	Instrument time required for minimum/typical sample size

Standard 1D ^{1}H	1. Integration—ratio of protons of each type
	2. Chemical shift—type of functional groups and chemical environment
	3. Coupling
	  a. Patterns (e.g., doublets)—number of adjacent nuclei
	  b. Magnitude of coupling in hertz—stereochemical information
		Chemical shifts overlap
		
		Non-first-order coupling
			1 µg/5–20 mg	16 h/5 min

Standard 1D ^{13}C	1. Chemical shift—functional groups and their chemical environment 2. Number of signals—number of nonequivalent carbons	Chemical shifts overlap	1mg/10–15 mg	16 h/1 h
DEPT	1. Type of ^{13}C for each signal (i.e., C, CH, CH$_2$, CH$_3$)		1 mg/10–50 mg	8 h/30 min
2D DQF COSY	1. Complete coupling information at a glance 2. Coupling constants that were unobtainable by 1D methods can be determined		50 µg/5–20 mg	24 h/2 h
2D TOCSY	1. Complete spin networks can be determined. 2. Spectral edited slices of 2D clean overlapped regions	Not selective enough to give order of propagation	20 µg/5–20 mg	16 h/1 h
1D HOHAHA	1. Complete spin networks determined 2. Coupling constants, patterns of overlapped regions 3. Spectral editing of overlapped regions		20 µg/5–20 mg	16 h/2 h
2D NOESY or ROESY	1. Complete through space coupling information at a glance 2. Through-space distance calculated		20 µg/5–20 mg	16 h/2 h

Continued

Table 5.3. Continued

Experiment	Information obtained	Limitations of experiment	Minimum/typical sample size	Instrument time required for minimum/typical sample size
1D NOE difference	1. Through-space coupling, or distance measurements up to 4.5 Å		10 μg/5–20 mg	16 h/1 h
HMQC	1. Connectivity of ^{1}H bonded directly with ^{13}C 2. Diastereotopic protons 3. Separation of ^{1}H signals according to ^{13}C help with overlapped ^{1}H signal identification	Quaternary signals are not seen	100 μg/5–20 mg	24 h/2 h
HMBC	1. Quaternary signals are seen if there is a ^{1}H within three bonds 2. ^{1}H to two- and three-bond ^{13}C observed 3. Correlations go through oxygen and nitrogen	Not all signals observed ^{13}C satellites observed	100 μg/5–20 mg	64 h/4 h
INADEQUATE	1. Complete carbon skeleton constructed, except isolated carbons	Solubility	70 mg/300 mg	72 h/16 h

There are some notable exceptions, such as the NMR of nitrogen in protein analysis, and the NMR of phosphorus in nucleic acid work. Because quinine contains two nitrogen atoms, a very brief description of ^{15}N NMR is presented. Each nucleus is unique in terms of NMR and each must be considered separately.

5.12.1 ^{15}N NMR

Nitrogen has two magnetically active isotopes accessible by NMR, ^{14}N and ^{15}N. The most abundant isotope of nitrogen, ^{14}N, which represents greater than 99% of nitrogen's natural abundance, possesses a spin of 1 and hence an electric quadrupole moment. This nucleus has an inherent low sensitivity and a very broad line due to quadrupolar relaxation. The NMR of the ^{14}N nucleus finds use in many instances but will not be considered further here.

The other isotope of nitrogen, ^{15}N, also has an inherent low sensitivity, which when coupled with a very low natural abundance, leads to an extremely low absolute sensitivity. Modern instrumentation has largely overcome the problem of sensitivity in cases where the amount of sample is not the limiting factor, unlike most chemical ecology studies.

There are two important experimental factors that must be considered to run successful ^{15}N experiments. The ^{15}N nucleus tends to relax very slowly; T_1 relaxation times of greater than 80 s have been measured. Thus, either long pulse delays must be incorporated into the pulse sequence, or alternatively, another route for spin relaxation can be provided. A common procedure is to add a "catalytic" amount of paramagnetic chromium (III) acetylacetonate, whose unpaired electrons efficiently stimulate relaxation.

The other experimental factor that must be considered is the nuclear Overhauser effect. The magnetogyric ratio for ^{15}N is small and *negative* ($\gamma = -2712$). A quick calculation shows that the maximum NOE enhancement for ^{15}N is $\gamma_H/(2)\gamma_N$ [26,753/ $(2) \times (-2,712)$], which is equal to -4.93. For the general case, a spin-one-half nucleus with a positive magnetogyric ratio gives positive NOE enhancement with proton decoupling, while a spin-one-half nucleus with a negative magnetogyric ratio gives negative NOE enhancement.

For the ^{15}N nucleus, the maximum enhancement is $-4.93 + 1$ or -3.93. In the case where ^{15}N-^{1}H dipolar spin-lattice relaxation overwhelms, the signal is inverted (negative) and its intensity is nearly four times what it would be in the absence of ^{1}H irradiation. However, since ^{15}N dipolar relaxation is only one of many relaxation mechanisms for ^{15}N, proton decoupling can lead to NOEs ranging from 0 to -4.93 or a signal ranging from $+1$ to -3.93. The experimental downside to this is that any NOE between 0 and -2.0 lowers the absolute intensity of the observed signal. In fact, an NOE of exactly -1.0 produces no signal at all!

The ^{15}N NMR spectrum of quinine is shown in Figure 5.12. Because of proton decoupling, the spectrum has two singlets representing the two very different nitrogen atoms. Tables of chemical shifts can be found in Levy and Lichter

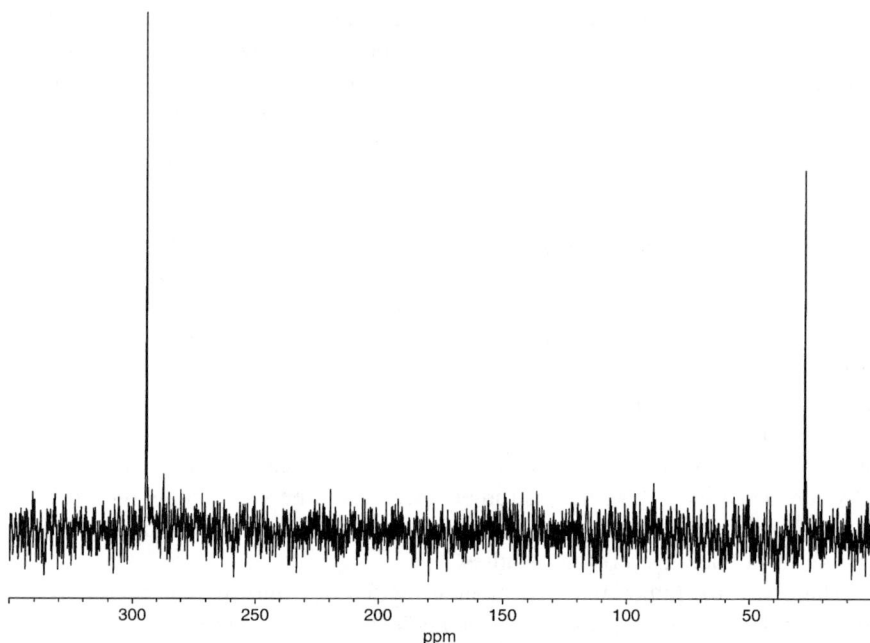

Figure 5.12. The 30.4-MHz proton-decoupled ^{15}N NMR spectrum of quinine in CDCl$_3$. The chemical shifts are given in terms of liquid ammonia.

(1979). The take-home message is that one of the nitrogen atoms is a tertiary amine, while the other is quinoline. In a real structure determination problem, such information could be tremendously useful. However, as mentioned, the problem of large sample requirements will preclude the widespread application of ^{15}N NMR to most chemical ecology studies.

5.13. Gradient Field NMR

One of the fastest growing areas in NMR in the past decade has been the use of "pulsed field gradients," or PFG NMR. It is ironic to consider that so much effort has been expended over so many years to avoid "gradients" in the field, to maintain the magnetic field as homogeneous as possible. Today, most modern high-field NMR spectrometers are routinely equipped with hardware (coils) that rapidly ramps the field along one of three mutually orthogonal axes. These magnetic field gradients are incorporated into the pulse sequence in a large range of applications. A brief discussion of gradients is included here because there are many applications to correlation experiments. Gradients are treated in a general manner; for a more technical treatment and for a myriad of other applications consult the review by Price (1996).

Pulse sequences and phase cycling are important considerations in any correlation experiment. The details of phase cycling are beyond this treatment (see Derome 1987) but it easy to appreciate one of their negative aspects: time. In correlation experiments, anywhere from 4 to 64 phase cycles must be summed to produce one Free induction decay curve (FID). If the signal-to-noise ratio is poor, then the identical cycle is repeated until sufficient signal is acquired. These phase cycles are wasteful of spectrometer time and account for at least one reason why 2D experiments take so long.

In a PFG experiment, the pulse sequence can be rewritten so that phase cycling can be eliminated altogether. Thus, if the signal-to-noise ratio is sufficient, each pulse can be saved as one FID in a 2D experiment. The saving in instrument time is enormous; experiments that previously took several hours to an entire day can now be run in a matter of minutes. One reassuring point to realize is that even though the experiment is run differently, the results and hence the interpretation remain the same.

Even though the initial interest in gradient NMR stemmed from the fact that familiar experiments can be run more quickly and more efficiently, the most exciting features of gradients are the experiments that cannot be run without them. Gradients allow for improved magnetic resonance imaging, improved magnetic resonance microscopy, better solvent suppression (especially water), and entirely new areas of inquiry such as diffusion measurements. A look into the future of nuclear magnetic resonance finds gradient fields playing an increasing role, with increasing importance.

5.14. References

Brey, W. ed. 1988. Pulse Methods in 1D and 2D Liquid-Phase NMR. Academic Press, New York.

Brownlee, R.G. & R.M. Silverstein. 1968. A micro-preparative gas chromatograph and a modified carbon skeleton determinator. Anal. Chem. **40**:2077–2080.

Butenandt, V.A., R. Beckmann, D. Stamm & E. Hecker. 1959. Uber den Sexuallockstoff des Seidenspinners *Bombyx mori*. Reindarstellung und Konstitution. Z. fur Naturforsch. **14b**:283–284.

Chandrakumar, N. & S. Subramanian. 1987. Modern Techniques in High Resolution FT NMR. Springer-Verlag, New York.

Croasmum, W.R. & R.M.K. Carlson. 1994. Two-Dimensional NMR Spectroscopy, 2nd ed. VCH, New York.

Derome, A. 1987. Modern NMR Techniques for Chemistry Research. Pergamon, Oxford.

Ernst, R.R., G. Bodenhausen & A. Wokaum. 1987. Principles of Nuclear Magnetic Resonance in One and Two Dimensions. Clarendon, Oxford.

Farrar, T.C. 1987. An Introduction to Pulse NMR Spectroscopy. Farragut, Chicago.

Friebolin, H. 1993. Basic One- and Two-Dimensional Spectroscopy, 2nd ed. VCH, New York.

Gunther, H. 1995. NMR Spectroscopy, 2nd ed. Wiley, New York.

Kessler, H., M. Gehrke & C. Griesinger. 1988. Two-Dimensional NMR Spectroscopy, Angew. Chemie Int. Ed. **27**:490–536.

Levy, G.C. & R.L. Lichter. 1979. Nitrogen-15 Nuclear Magnetic Resonance Spectroscopy. Wiley, New York.

Martin, G.E. & A.S. Zektzer. 1988. Two-Dimensional NMR Methods for Establishing Molecular Connectivity. VCH, New York.

Price, W.S. 1996. Gradient NMR. *In:* Annual Reports on NMR Spectroscopy, Vol. 32, ed. G.A. Webb, Academic Press, London. pp. 51–142.

Sanders, J.K.M. & B.K. Hunter. 1987. Modern NMR Spectroscopy. Oxford University Press, Oxford.

Schraml, J. & J.M. Bellama. 1988. Two-Dimensional NMR Spectrometry. Wiley, New York.

Silverstein, R.M. & F.X. Webster. 1998. Spectrometric Identification of Organic Compounds, 6th ed. Wiley, New York.

Williams, K.R. & R.W. King. 1990. The Fourier transform in chemistry—NMR. J. Chem. Ed. **67**:125–138.

6

Infrared and Ultraviolet Spectroscopy Techniques

Walter Soares Leal

6.1. Introduction to Infrared Spectroscopy

The infrared (IR) region of the electromagnetic spectrum extends from the red end of the visible spectrum out to the microwave region, encompassing wavelengths between 0.78 and 1000 μm or, in wavenumbers (the reciprocal of wavelength, proportional to frequencies), between 10 and 12,800 cm^{-1}. The majority of analytical applications for compound identification are confined to the middle of the infrared region extending from 670 to 4000 cm^{-1} (15 to 2.5 μm).

Infrared absorption spectroscopy measures the twisting, bending, rocking, and vibrational motions of atoms in molecules. On interaction with broad-spectrum infrared radiation, portions of the incident radiation are absorbed when the energy of the incident radiation coincides with the energy required to excite a particular vibrational mode. The multiplicity of vibrations occurring simultaneously produces a highly complex absorption spectrum, which is uniquely characteristic of both the functional groups comprising the molecule and the overall configuration of the atoms. The infrared spectrum of an organic compound usually provides a unique "fingerprint" distinguishable from the spectra of all other compounds (except enantiomers, which have the same IR spectra). Furthermore, by interpretation of the spectrum, it is possible to determine the presence or absence of specific functional groups. It is beyond the scope of this chapter to discuss the basics of infrared spectroscopy; the reader is referred to such standard reference sources as Roeges (1994), Skoog (1985), Silverstein et al. (1991), or Willard et al. (1981). This chapter focuses mainly on vapor-phase infrared, and the characteristic functional group frequencies that are useful for microscale analysis and structure determination in the field of chemical ecology, particularly in pheromone research.

6.2. Coupled Gas Chromatography-Fourier Transform Infrared Spectroscopy Instrumentation

In coupled gas chromatography-Fourier transform infrared (GC-FTIR), IR measurements are conducted on the GC effluent after chromatographic separation. Several methods are used. In the first, the effluent is passed through a heated flow cell, or "light pipe." A light pipe basically consists of a glass tube coated internally with gold. Since vapor-phase spectra are obtained as the compounds in the GC effluent travel through the light pipe, the time of residence in the cell restricts the number of scans that can be obtained for each point on the FTIR chromatogram. Theoretically, the minimum identifiable quantity of a strongly absorbing sample that could be characterized with a light pipe GC-FTIR interface would be about 400 pg in a completely optimized system (Griffiths et al. 1986), but further decreases are unlikely without changing the fundamental nature of the GC-FTIR interface.

Two alternative techniques have been developed to circumvent this limitation. The first is based on the principle of matrix isolation (Reedy et al. 1985), in which the effluent from the GC is mixed with argon. The mixture is then deposited as bands on a gold-plated drum maintained at approximately 12 K, and the reflectance spectrum of any one compound can be repeatedly scanned. With this method, GC-FTIR spectra may be obtained for samples in the subnanogram range. This technology is incorporated into the Cryolect® system from Mattson Instruments.

An analogous technique is used in the Tracer® system (Bio Rad Laboratories) introduced by Haefner et al. (1988). In this system, fractions eluting from GC are condensed as solids onto a sample deposition window cooled with liquid nitrogen (77 K). During a run the window assembly is moved slowly and GC effluent is frozen on the window in a continuous line. The infrared beam is then focused on the window, and the radiation transmitted through the sample is focused onto a small-area, high-sensitivity detector. The Tracer® system is far more sensitive than light pipe detectors (see below) primarily because the sample area is reduced 100-fold (0.8 mm²) in a light pipe as opposed to 0.008 mm² in the Tracer® interface). More importantly, at the end of a particular GC-FTIR run, unlike with a light pipe system, the window can be repositioned so that any desired compound is placed in the microsampling optical path and scanned repeatedly. Thus, the spectrum of any compound can be acquired after the run, at any desired resolution and measurement time for improved detection limits. An added advantage of both trapping systems is that the cold temperatures used minimize temperature-related band-broadening effects, so that IR absorbances are sharper than in standard IR spectra. Bands are particularly sharp with matrix isolation, where intermolecular interactions such as H bonding are virtually eliminated.

6.2.1. GC-FTIR Sensitivity: Light Pipe versus Trapping Interfaces

The sensitivity of the two interface types has been compared by analyzing a series of caffeine concentrations with a light pipe interface (Hewlett-Packard GC Infrared detector (IRD), a matrix isolation (MI) instrument (Cryolect® MI-FTIR, Mattson Instruments) and a cooled window instrument (Tracer® system, BioRad Laboratories) (Jackson et al. 1993). Standard solutions of caffeine in toluene were prepared at 500, 100, 50, 10, 1, 0.5, 0.1, 0.05, and 0.01 ng/μl.

For the GC-IRD, spectra were recorded by injecting 1 μl of sample using both split (ratio 38:1) and splitless modes. With splitless operation, spectra were obtained with 500, 100, and 50 ng of caffeine injected, whereas split injections generated spectra only from the 500 ng/μl sample, corresponding to 13 ng of caffeine on the column. The effective detection limit for caffeine was about 10 ng. On the Cryolect® and Tracer® systems, caffeine spectra were obtained from samples as small as 40 pg. These experiments demonstrated that the trapping systems are more sensitive than the light pipe interface by two to three orders of magnitude.

A dramatic example of the differences in sensitivity is shown in Figure 6.1, which shows a sample of the volatiles collected from a longhorn beetle, *Anaglyptus subfasciatus,* run on both light pipe and Tracer® instruments. With the light pipe interface, the IR Gram-Schmidt chromatogram showed essentially two peaks (Fig. 6.1), whereas the Tracer instrument detected numerous peaks, and was even more sensitive than a GC flame ionization detector (Leal et al. 1995). In addition, the spectrum for the minor component that appeared at 20.46 min in the vapor-phase measurement was of poor quality compared to that recorded with the trapping system (Fig. 6.2). Both spectra were generated from four scans at 8-cm⁻¹ resolution. However, the increased sensitivity of the Tracer® and Cryolect®

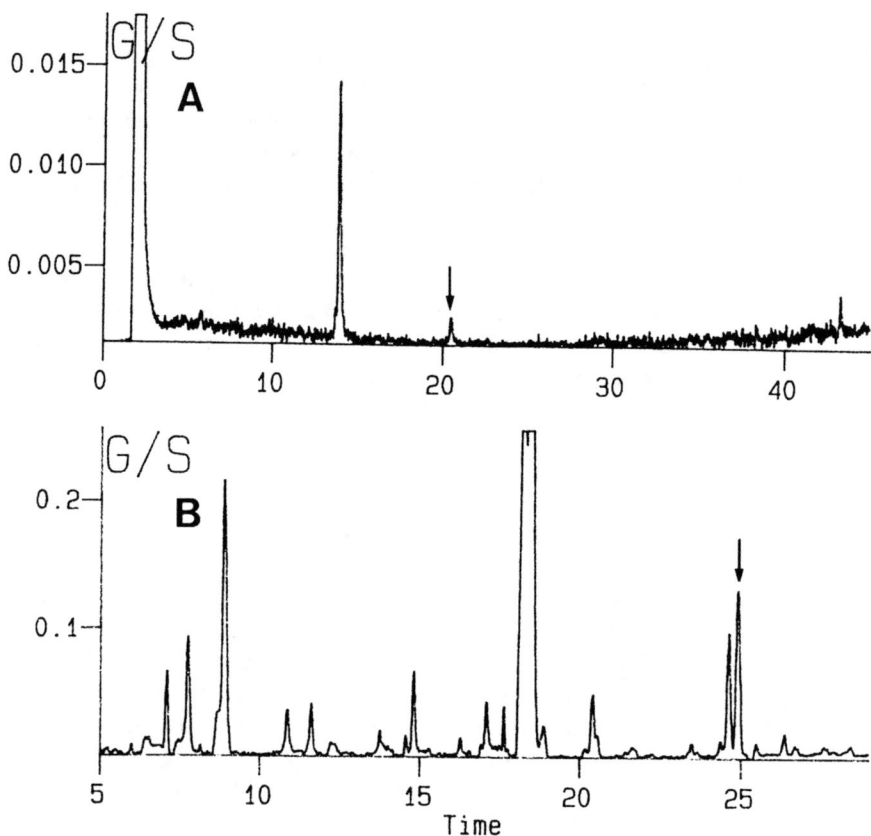

Figure 6.1. Gram-Schmidt chromatograms from splitless injections of an insect phero-mone sample on GC-FTIR systems equipped with a light pipe interface FTS-45RD/GC32 (**A**), or a trapping system FTS-40/Tracer (**B**). Arrows indicate the same component, as detected by the two different systems.

systems are not without disadvantages. First, the instrumentation is much more costly than a light pipe system, and second, the former systems require more sophisticated maintenance, making them less amenable for general-purpose, multiuser systems.

6.2.2. Optimizing Vapor-Phase GC-FTIR Conditions

For vapor-phase FTIR measurements both the GC and the light pipe interface must be optimized. A wide chemical dynamic range (maximum injectable amount of sample divided by the minimum amount of sample detected by the FTIR) is desired so that a range of sample concentrations can be measured. A narrow-bore column (0.25 mm or less) would give the best resolution and sensitivity, but

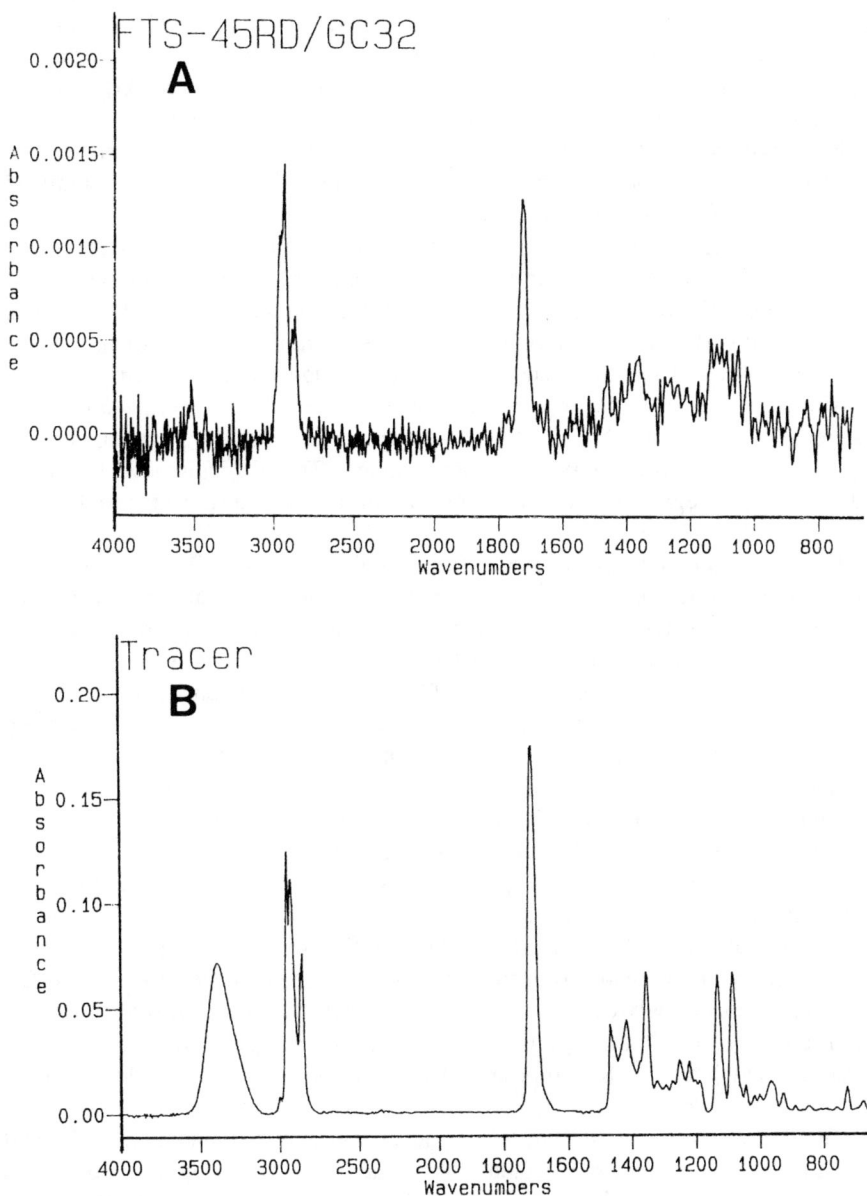

Figure 6.2.   FTIR spectra of 3-hydroxy-2-octanone, a minor component of the pheromone system of a longhorn beetle. With the light pipe interface (**A**) only the carbonyl and CH stretching bands appeared, whereas the Tracer system (**B**) generated a detailed spectrum with a low signal-to-noise ratio.

narrow-bore columns are easily overloaded. Consequently, GC-FTIR is typically performed using 0.32-mm i.d. capillary columns of 20–50 m length, with 0.25 micron film thickness. If necessary, sample capacity can be increased by using a thicker film column, but again at the cost of decreased resolution.

Sensitivity and GC resolution are optimized by selecting the appropriate chromatographic conditions. The broadening of the later GC peaks can be minimized by selecting efficient multistep oven temperature programs.

Furthermore, to avoid degradation of the GC resolution, the volume of the light pipe V_{cell} must be no larger than the volume of carrier gas between the half-height points $V_{1/2}$ of the narrowest peak in the chromatogram (Griffiths 1977). For optimal performance of a typical light pipe (1 mm i.d, 12 cm long, ~ 94 μl) a minimum carrier gas flow rate of 2 ml/min through the light pipe is required, giving a peak half-width of ~ 3 s. Because increase in the oven temperature decreases the flow rate, the flow rate should be set such that at higher temperatures the minimum required flow rate is maintained. This difficulty can be easily overcome by keeping the flow rate constant during the run with electronic pressure control.

In theory, the light pipe temperature should be 20–30°C higher than the maximum oven temperature. However, Gram-Schmidt chromatogram (GSC) intensity (i.e., sensitivity), interferogram intensity and spectral absorbance all decrease with temperature (Namba 1990), for different reasons. Decrease of the GSC was greatest; the intensity of the GSC at 350°C was only 20% of that at 50°C. The decrease of specific absorbances (e.g., 1763 cm^{-1}) was inversely proportional to the absolute temperature of the light pipe, caused by the decrease of the sample concentration due to expansion of the carrier gas. Finally, the decrease in the interferogram intensity is caused by the high unmodulated heat flux reaching the detector. This can drive the detector into a nonlinear response range, causing a decrease in response to the modulated light actually coming through the light pipe (Brown et al. 1985).

The user also should be reminded that the lifetime of the light pipe seals will be prolonged by operation at lower than maximum temperatures (normally 250°C). Condensation of analytes in the light pipe must be avoided, but sensitivity and light pipe lifetime should be a matter of concern as well. Generally, it is a good practice to maintain the light pipe at 150°C or so with carrier gas flowing through, even when not in use.

GC-FTIR is sensitive to both water and carbon dioxide contaminants derived from the sample or the environment. Water vapor bands appear as sharp signals above 3500 cm^{-1} and in the range 1460–1650 cm^{-1} in the spectrum (Fig. 6.3). This contamination may be caused by exposure of the detector to ambient air while refilling the detector Dewar with liquid nitrogen. Injected samples should be dry to avoid these interferences as well as damage to the KBr windows of the light pipe. Carbon dioxide is less problematic than water because it appears

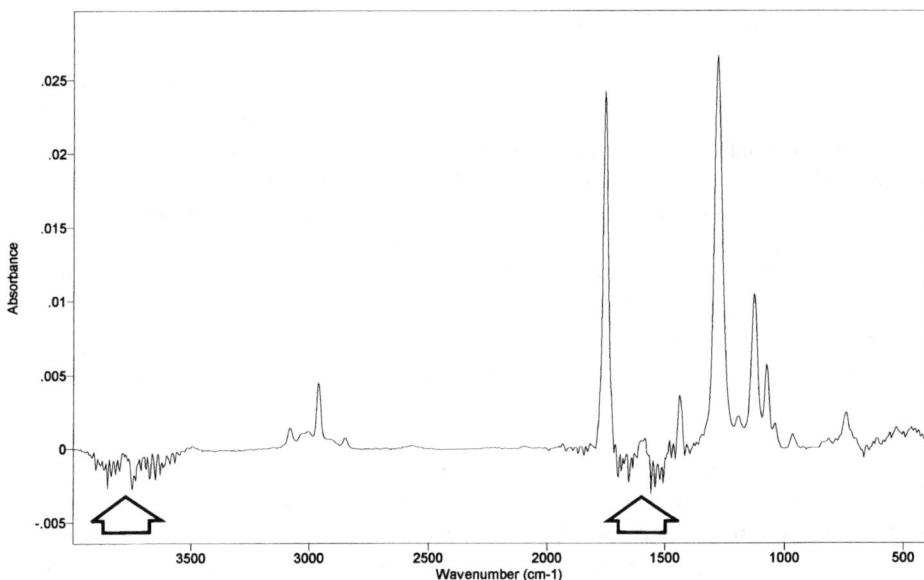

Figure 6.3. FTIR spectrum of methyl phthalate obtained in a light pipe interface system. Absorption corresponding to water vapor (*arrows*) appears as negative bands because the contamination was higher at the time of the background acquisition. The selection of a background region very close to the peak may minimize the intensity of the bands (as shown), but it would not eliminate the bands without losing some spectral features.

in a region of the spectrum that generally does not interfere with absorption bands of most organic compounds.

6.3. The Infrared Spectrum

The IR spectrum of a molecule is a composite of all the absorptions from vibrations of the molecule as a whole, and vibrations of individual groups of atoms such as functional groups. In particular, differences in bonds and bond strengths are reflected in the energy required to excite specific vibrational modes, leading to predictable shifts of absorption bands due to structural features. Thus, the positions of absorption bands not only reveal which functional groups may be present but also provide more detailed information, such as the ring sizes of cyclic compounds and whether functional groups are conjugated.

The spectrum can be divided into four parts for examination. First, the region from 4000 to 2400 cm^{-1} contains bands due to stretching frequencies of single bonds to hydrogen. The utility of these bands varies, depending on their strength and position. They are often diagnostic for functional groups such as alcohols, amines, amides and carboxylic acids. Some of the weaker C-H stretching bands are also useful, such as those from vinyl groups or aldehydes (see below).

The region from 2400 to 1900 cm^{-1} contains absorbances due to stretching vibrations of functional groups such as acetylenes and nitriles, or those with cumulated double bonds such as allenes, isocyanates, and isothiocyanates. Because this region is relatively uncluttered, even a weak absorption in this region is usually diagnostic for the presence of one of these functional groups.

The region from 1900 to 1500 cm^{-1} contains absorptions from C=C, C=O, C=N, and N=O stretches, and from N-H bending. The bands from various types of carbonyl groups (aldehydes, ketones, acids, esters, amides, etc.) are probably most useful because of their strength. The exact position of the band can also provide considerable information about the type of carbonyl, conjugation, and ring size.

Finally, the region from 1500 to 600 cm^{-1}, known as the "fingerprint" region, is characterized by large numbers of bands, due to molecular vibrations as a whole, to combination and overtone bands, and to a variety of stretching and bending vibrations. Because of the cluttered and complex nature of this region, assignments of particular bands within this region should be made with caution. This region is most useful from two aspects. First, the presence or absence of absorption bands can be used to corroborate (or deny) tentative assignments made with bands from other parts of the spectrum. Second, as implied by the name, the fingerprint region of the IR spectrum is essentially unique to a particular compound. Thus, this region can be used to confirm the identification of geometric isomers or diastereomers, for example, which may be difficult to conclusively determine by other methods. However, the IR spectra of enantiomers are identical. It must also be remembered that IR is essentially a qualitative technique, so that the strengths of absorbances cannot be used to determine anything other than a very general idea of the number of a particular type of functional group present. Consequently, the strengths of absorbances are usually only reported as strong (s), medium (m), weak (w), or variable (v); sometimes the abbreviation (br) is also used to indicate a broad peak.

A large body of empirical data has been accumulated concerning the absorption frequencies characteristic of various functional groups. This information has been summarized in correlation charts (Roeges 1994; Skoog 1985; Silverstein et al. 1991; Willard et al. 1981), which are invaluable for identification purposes because they enable intelligent guesses to be made as to what functional groups are present or absent in a molecule. It should be stressed that the absence of a particular band can be as useful or more useful than its presence, because the absence of the absorption band characteristic of a particular functional group is strong evidence that that functional group is not present in the molecule.

Characteristic group frequencies for condensed-phase infrared are readily available, and these are also applicable to some extent to spectra obtained by the trapping techniques (tracer & matrix isolation). Because condensed-phase infrared databases (e.g., Sadtler Index) have far more entries (~70,000) than vapor-phase libraries (9600 entries), computer-assisted interpretation of spectra of the former is more convenient and more productive.

6.4. Survey of Vapor-Phase Absorptions of Important Functional Groups

The group frequencies and general appearance of vapor-phase IR spectra differ markedly from those in the condensed-phase because the intermolecular effects such as H bonding, which broaden and enlarge peaks in condensed phase spectra are absent in the dilute vapor in the GC effluent. Thus, those familiar with interpretation of condensed phase spectra may find some adjustment is necessary. The differences have been previously addressed in a detailed group frequency publication for a wide variety of organic compounds (Nyquist 1984). To expeditiously apply GC-FTIR to the elucidation of pheromones and other semiochemicals, we have prepared our own user library (Leal et al. 1992). The following group frequency correlation data are based largely on an updated version of this database.

6.4.1. C=O Stretching

The strong carbonyl stretching band (Table 6.1) appears in vapor-phase spectra in the range 1655–1875 cm^{-1}. Carbonyl groups in aliphatic saturated ketones appear in the interval 1728–1740, and conjugation shifts the bands to lower frequencies. C=O stretching frequencies (abbreviated νC=O) for cycloalkanones decrease with increasing ring size: hexanones, 1736–1740 cm^{-1}; cyclopentanone, 1767 cm^{-1}. An α,β-unsaturation decreases the absorption frequency by 23 cm^{-1}.

The aldehyde νC=O bands appear in the region 1670–1744 cm^{-1}, with saturated aldehydes in the upper frequencies (1744 cm^{-1}). An epoxide adjacent to the aldehyde decreases the νC=O frequency by 8 cm^{-1}. α,β-Unsaturated aldehydes appear in the interval 1697–1713 cm^{-1}, a decrease of approx. 30 cm^{-1} compared to the saturated counterparts.

Aromatic aldehydes appear in the region 1713–1720 cm^{-1}, 28 cm^{-1} lower than the saturated aldehydes. Chelation gives rise to a 40-cm^{-1} decrease in frequency, to 1670–1678 cm^{-1}. An aldehyde C-H stretching band in the neighborhood of 2811 and 2710 cm^{-1}, along with a carbonyl band, is good evidence for the presence of an aldehyde.

Three bands are characteristics of the spectra of carboxylic acids, namely, OH, C=O, and C-O stretching bands. The νC=O bands appear in the region 1716–1778 cm^{-1}. Aliphatic acids absorb at 1778 cm^{-1}, while α,β-unsaturated ones appear ~27 cm^{-1} lower (1751–1759 cm^{-1}). The νC=C bands of α,β-unsaturated acids are also strong and appear at ~ 1656 cm^{-1}. Benzoic acids absorb at 15-cm^{-1} lower frequencies than the aliphatic counterparts, 1759–1770 cm^{-1}. Intramolecular hydrogen bonding due to an OH in the ortho position causes a decrease of 44

Table 6.1. Vapor-Phase Infrared Group Frequencies for Carbonyl Groups in Various Environments[a]

Functional group	Wavenumber (cm⁻¹)	Example
Ketone		
R-C(=O)R	1728–1740 (vs)	4-decanone (1728), acetone (1740), 2-decanone (1732), 2-undecanone (1732)
Six-member ring	1736–1740 (vs)	4-methylcyclohexanone (1736), 1,4-cyclohexanedione (1740), 1,3-cyclohexanedione (1736)
α,β-Unsaturated six-member ring	1693–1703 (vs)	2-cyclohexen-1-one (1709), 3-methyl-2-cyclohexen-1-one (1703), isopiperitenone (1693), carvone (1697)
Five member ring	1767 (vs)	cyclopentanone (1767)
α,β-Unsaturated five-member ring	1744 (vs)	2-cyclopenten-1-one (1744)
Quinones	1670–1678 (vs)	*p*-quinone (1678), methyl- and ethyl-quinone (1674), thymoquinone (1670), 2-hydroxy-1,4-naphthoquinone (1678)
Chelated quinone	1655 (vs)	plumbagin (1655)
Aldehydes		
R-C(=O)H	1744 (vs)	(Z)-11-16Ald (1744), citronellal (1744), hexanal (1744), β-acaridial (nonconjugated CHO, 1744)
R-CHO with an epoxide adjacent	1736 (vs)	2,3-epoxyneral (1736), 2,3-epoxygeranial (1736)
U-C(=O)H, U=α,β-unsaturation	1697–1713 (vs)	farnesals (1697), perillaldehyde (1709), neral and geranial (1700), (*E*)-2-hexenal and (*E*)-2-decenal (1713), α-acaridial (1712), β-acaridial (1712)
Ar-C(=O)H	1713–1720 (vs)	vanillin (1713), *p*-hydroxybenzaldehyde (1720), *o*-tolualdehyde (1716)
Chelated Ar-C(=O)H	1670–1678 (vs)	2-hydroxy-4-methylbenzaldehyde (1674), methyl 2-formyl-3-hydroxybenzoate (1670), 2-hydroxybenzaldehyde (1678)
Acid		
R-C(=O)OH	1778 (vs)	hexanoic, octanoic, nonanoic acid (1778)
U-C(=O)OH	1751–1759 (vs)	nerolic acid (1751), farnesoic acid (1752), (*E*)-2-hexenoic acid (1759)
Ar-C(=O)OH	1759–1770 (vs)	*m*-toluic acid (1763), *m*- and *o*-chlorobenzoic acid (1767 and 1770), *p*- and *m*-hydroxybenzoic acid (1759 and 1762)
Ar-C(=O)OH with an ortho OH	1716 (vs)	salicylic acid
Anhydride		
Five-member ring	1867–1875 (s) 1763–1813 (vs)	succinic anhydride (1875, 1813), methyl succinic anhydride (1875, 1813), dimethyl succinic anhydride (1763), maleic anhydride (1801), phthalic anhydride (1867, 1805)

Continued on next page

Table 6.1. *Continued*

Functional group	Wavenumber (cm⁻¹)	Example
Open-chain anhydride	1824 (m) 1763–1767 (m)	butyric anhydride (1824, 1767), isobutyric anhydride (1824, 1763)
Ester		
R-C(=O)OR	1755–1763 (vs)	methyl undecanoate (1759), ethyl oleate (1755), dimethyl maleate (1763), methyl palmitoleate (1760)
R-O-C(=O)H	1744–1747 (vs)	8-heptadecenyl formate (1747), 8,11-heptadecadienyl formate (1747), neryl formate (1744)
R-O-C(=O)CH₃	1760–1763 (s)	E7,Z9-dodecadienyl acetate (1763), Z11-16:Ac (1762), Z9-14:Ac (1762), citronellyl acetate (1760), (E)-2-octenyl acetate (1763)
Ar-C(=O)-OR	1740–1755 (vs)	methyl phthalate (1755), methyl 4-methylbenzoate (1743), methyl p-hydroxybenzoate (1743), hexyl 3-hydroxybenzoate (1740), methyl 3-formyl-4-hydroxybenzoate (1743)
Ar-C(=O)OR with an *ortho* OH	1697 (vs)	methyl salicylate
Lactone		
γ-lactone	1813 (vs)	γ-decalactone (1813), γ-octalactone (1813)
δ-lactone	1774 (vs)	δ-decalactone (1774)
ε-lactone	1770 (vs)	ε-caprolactone (1770)

^{a}Bands are qualitatively characterized as very strong (vs), strong (s), medium (m), or weak (w).

cm⁻¹ in the carbonyl band absorption frequency. Fumaric acid absorbs at 1767 cm⁻¹, but maleic acid seems to form an acid:anhydride equilibrium. A band at 1743 cm⁻¹, probably acid vC=O, is seen, along with the major carbonyl band at 1801 cm⁻¹ (from maleic anhydride).

Anhydrides exhibit two carbonyl stretching bands in the region 1867–1875 and 1763–1813 cm⁻¹. The occurrence of two bands is very clear in open anhydrides, but the lower frequency band is much more intense or seemingly the sole vC=O band in anhydrides in five-member rings.

Aliphatic esters show a carbonyl band in the interval 1755–1763 cm⁻¹, whereas benzoates are at 1740–1743 cm⁻¹. The effect of an *ortho*-positioned OH is the same as in benzoic acid, and a vC=O:HO band appear at 1697 cm⁻¹. Formates absorb in the interval 1744–1747 cm⁻¹. The vC=O band of acetates is normally the most intense of their spectra. Lactones exhibit carbonyl absorptions in the interval 1770–1813 cm⁻¹, except for macrolactones that display at ~ 1747 cm⁻¹. The five-membered ring lactones, japonilure and buibuilactone, sex pheromone constituents of many scarab species (Leal 1997) give a strong carbonyl band at 1813 cm⁻¹.

6.4.2. OH Stretching

The OH stretching bands (Table 6.2) of aliphatic carboxylic acids appear in the range 3572–3584 cm^{-1}, whereas benzoic and α,β-unsaturated acids are seen in the interval 3572–3587 cm^{-1}.

Phenolic OH stretches appear in the range 3645–3653 cm^{-1}, except for phenols *ortho*-substituted with CHO or COOH, which display in the interval 3201–3271 cm^{-1}. The greatest effect is due to an aldehyde group (~448 cm^{-1}) and the smallest to a –COOH group (374 cm^{-1}). Interactions with an *ortho* methoxy group also decrease the νOH frequencies, although to a lesser extent (<100 cm^{-1}).

Alcohol OH stretches are seen in the previously reported ranges (Nyquist 1984) of primary (3670–3680 cm^{-1}), secondary (3658–3670 cm^{-1}), and tertiary (3640–3648 cm^{-1}), but allylic alcohols display in the interval 3648–3668 cm^{-1}.

6.4.3. C-O Stretching

The C-O stretching frequencies (Table 6.3) in alcohols are not easily predictable, but it can be stated that allylic alcohols exhibit in the range 990–1018 cm^{-1}, which distinguishes them from other alcohols, even when the νOH band is not seen.

In phenols, the νC-O bands appear in the interval 1138–1269 cm^{-1}, but usually the OH bending absorption (1072–1161 cm^{-1}) has a stronger intensity.

The C-O stretching frequencies of ethers appear in the region 1131–1141 cm^{-1}

Table 6.2. *Vapor-Phase OH Stretching Frequencies*[a]

Functional group	Wavenumber (cm^{-1})	Example
Acid	3572–3587 (m)	hexanoic, nonanoic acid (3576), fumaric acid (3584), *m*-toluic acid (3587), *m*- and *o*-chlorobenzoic acid (3584 and 3572), *p*- and *m*-hydroxy-benzoic acid (both at 3587), cinnamic acid (3583), salicylic acid (3580), nerolic acid (3580)
Alcohol/phenol	3591–3668 (w)	(Z)-8-dodecenol (3659), (E,E)-9,11-tetradecadienol (3668), 2-decanol (3567), 2,4-pentanediol (3653, 3591), cinnamyl alcohol (3637), nerol (3652), perillyl alcohol (3653), 5-isopropyl-3-methylphenol (3653), *p*- and *m*-hydroxybenzoic acid (3645 and 3652)
o-Methoxyphenol	3579–3595 (m)	2-methyl-4-methylphenol (3595), vanillin (3579)
Phenol with a carbonyl group in the *ortho* position	3201–3271 (m)	salicylic acid (3271), methyl salicylate (3259), 2-hydroxy-benzaldehyde (3201)

[a]Bands are qualitatively characterized as very strong (vs), strong (s), medium (m), or weak (w).

Table 6.3. *Vapor-Phase C-O Stretching Frequencies[a]*

Functional group	Wavenumber (cm⁻¹)	Example
Alcohol	1002–1130 (w)	(*E,E*)-9,11-tetradecadienol (1049), (*Z*)-8-dodecenol (1049), (*E,E*)-8,10-dodecadienol (1049), nerol (1002), perillyl alcohol (1007), 2,3-epoxynerol (1034), linalool (1107), 2-decanol (1130), 2,4-pentanediol (1130), cinnamyl alcohol (1018), citronellol (1034)
Phenol	1138–1269 (m)	α-naphthol (1138), 2-methoxy-4-methylphenol (1240; δOH 1149), 5-isopropyl-3-methylphenol (1269; δOH 1157), *p*-hydroxybenzaldehyde (1256, δOH 1157), vanillin (1184, δOH 1157), salicylic acid (1184, δOH 1072), *p*- and *m*-hydroxybenzoic acid (1265, δOH 1080 and 1260, δOH 1080), methyl *p*-hydroxybenzoate (1192, δOH 1107), methyl salicylate (1254, δOH 1161)
Ether	1203–1275 (s)	vinyl butyl ether (1203), butyl ether (1131), ethyl ether (1141), vanillin (1275), 2-methoxy-4-methylphenol (1273)
Acid	1103–1184 (m)	alkanoic acid (1136, 1102), (*E*)-2-hexenoic acid (1138, δOH 1362), *m*-toluic acid (1165, δOH 1358), *m*- and *o*-chlorobenzoic acid (1184, δOH 1362 and 1184, δOH 1340), salicylic acid (1149, δOH 1362), benzoic acid (1180, δOH 1346), *p*- and *m*-hydroxybenzoic acid (1161, δOH 1366 and 1169, δOH 1334), fumaric acid (1122, δOH 1369), cinnamic acid (1119, δOH 1362), nerolic acid (1107)
Anhydride	1018–1257 (s) 891–1165 (s)	butyric anhydride (1030), isobutyric anhydride (1018), succinic anhydride (1053, 906), dimethyl succinic anhydride (1215, 1165), maleic anhydride (1230, 891), phthalic anhydride (1257, 910)
Ester C-C(=O)-O	1173–1307 (s)	methyl undecanoate (three bands pattern with the most intense at 1173), dimethyl malate (three bands, 1261, 1215, and 1176), diethyl methylmalonate (four bands
O-C-C	1107–1222 (m)	from 1245 to 1157), methyl 3-formyl-4-hydroxybenzoate (1277, 1215), methyl *p*-hydroxybenzoate (1273, 1165), methyl phthalate (1284, 1130), methyl 3-hydroxybenzoate (1288, 1222), methyl salicylate (1307, 1215), ethyl oleate (1242, 1176)
Formate	1161–1169 (vs)	neryl formate (1161), 8-heptadecenyl formate (1169)
Acetate C-C(=O)-O O-C-C	1231–1235 (vs) 1022–1053 (w)	(*Z,E*)-9,12-tetradecadienyl acetate (1235, 1046), (*Z*)-11-hexadecenyl acetate (1235, 1042), (*Z*)-9-tetradecenyl acetate (1235, 1041), (*Z,E*)-7,9-dodecadienyl acetate (1231, 1045), perillyl acetate (1231, 1022), citronellyl acetate (1234, 1053)
Lactone	1156–1230 (m)	japonilure (1156, 1029), γ-decalactone (1164, 1038), γ-octalactone (1161, 1026), ε-caprolactone (1165, 1076), δ-decalactone (1230, 1140)

[a]Bands are qualitatively characterized as very strong (vs), strong (s), medium (m), or weak (w).

for R-O-R whereas for anisoles (Ar-O-R) they appear at 1273–1275 cm^{-1}; vinyl butyl ether appears in between the two regions.

The vC-O bands of carboxylic acids appear in the range 1134–1184 cm^{-1}. α,β-Unsaturated acids, such as nerolic and farnesoic acids are seen at ~ 1107 cm^{-1}. A band of medium intensity due to δOH (OH bending) appears in the interval 1334–1369 cm^{-1}.

Two C-O-C stretching bands appear in the spectrum of anhydrides in the regions 891–1165 and 1018–1275 cm^{-1}; in the open-chain anhydrides only one band appears, which is the strongest in the spectrum.

Esters exhibit two vC-O bands due to R-C(=O)-O and O-R (or O-Ar) stretchings. In the benzoate esters they appear in the regions 1273–1307 and 1165–1222 cm^{-1}, respectively. Aliphatic esters display the two bands in the interval 1157–1261 cm^{-1}. The most intense band in the spectrum of formates is the vC-O band in the range 1161–1169 cm^{-1}. Likewise, in acetates this band appears in the interval 1231–1235 cm^{-1} along with a vO-C-C band in the range 1022–1053 cm^{-1}. Lactones exhibit two bands in the intervals 1156–1230 and 1026–1140 cm^{-1}.

6.4.4. Determination of Double-Bond Configuration and Position

Infrared absorption at 967–970 cm^{-1} (condensed phase) has been used to determine the presence of 1,2-disubstituted E double bonds in organic compounds, but the potential of using another band, the =CH stretching of Z double bonds, for the assignment of double-bond configuration has been neglected until recently. The stretching vibrations of sp^2-hybridized CH bonds give weak but useful bands above 3000 cm^{-1}. The importance of these bands has been used in the assignment of Z-unsaturated esters (Doumenq et al. 1989) as well as in the characterization of an unsaturated lactone pheromone (Leal 1990). The effects of proximity to functional groups and position in the carbon chain on band positions in monoenes, and both conjugated and nonconjugated dienes and polyenes, has been the subject of several excellent recent studies (Attygalle 1994; Attygalle et al. 1994, 1995; Svatos & Attygalle 1997). These studies have shown that IR spectra, when interpreted carefully, can provide large amounts of information about double-bond placement and configuration. An analogous study of disubstituted piperidines and pyrrolidines (Garrafo et al. 1994) has also been reported.

The IR spectra of (Z)-9-dodecenyl acetate and (E)-4-tridecenyl acetate are provided as examples (Fig. 6.4). They differ basically in the occurrence of the =CH stretching in the former at 3012 cm^{-1} and the out-of-plane C-H bending in the latter at 967 cm^{-1}. Both bands appear in the spectrum of (Z,E)-9,12-tetradecadienyl acetate (=CH stretch, 3020 cm^{-1}; C-H bending, 963 cm^{-1}). The CH bending bands appear in general in the region 960–980 cm^{-1}. In conjugated dienes, IR spectra can be very useful in determining the configuration (Coffelt et al. 1979; Attygalle 1994). (Z,Z) Isomers have no absorption in the region 945–990 cm^{-1}, (Z,E) and

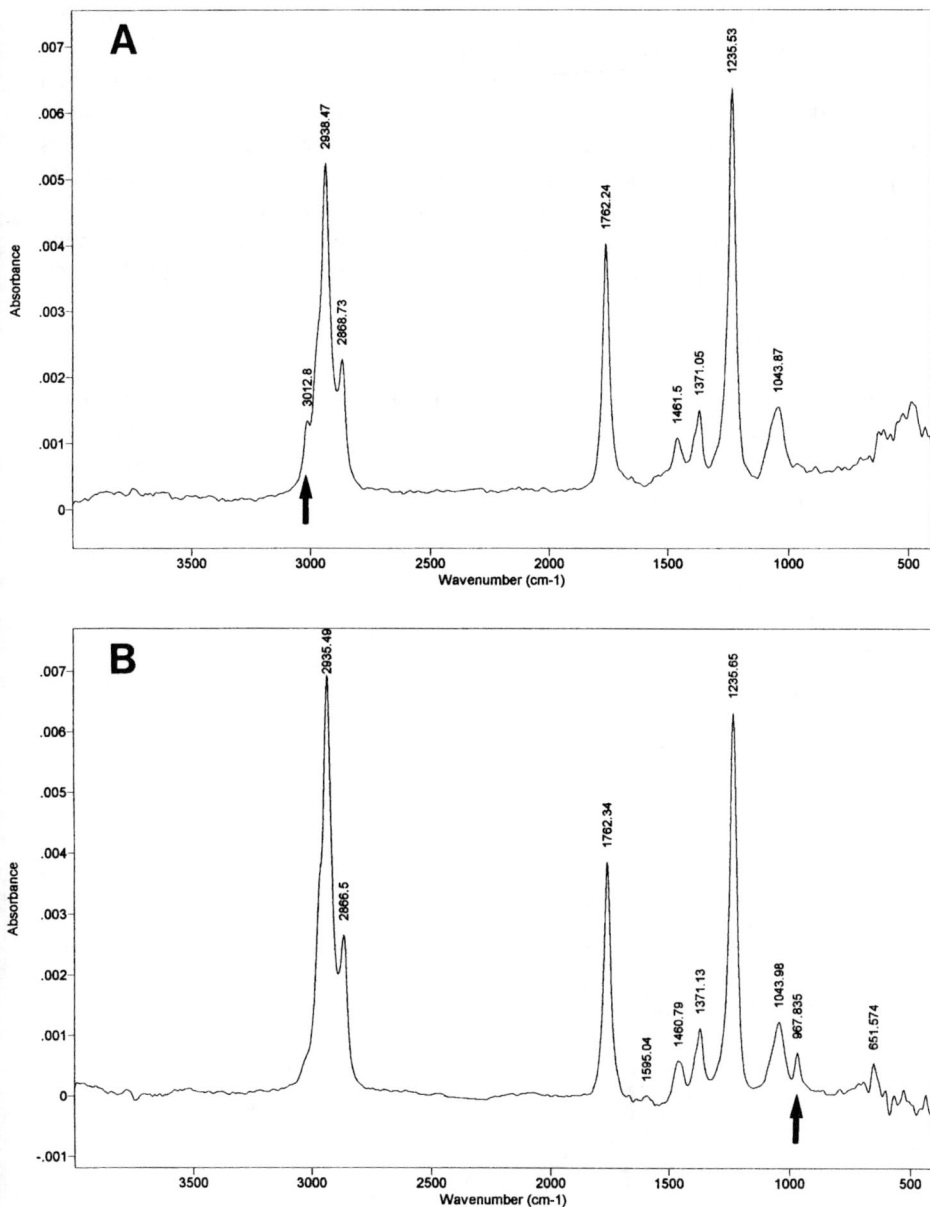

Figure 6.4. Vapor-phase FTIR spectra of (**A**) (*Z*)-9-dodecenyl acetate and (**B**) (*E*)-4-tridecenyl acetate, showing (*arrows*) the weak but characteristic olefinic C-H stretch band for *Z*9-12:Ac and the out of plane C-H bending band of *E*4-13:Ac.

(*E,Z*) isomers have a pair of bands at about 945 and 979 cm^{-1}, and (*E,E*) isomers show a single band at about 980–986 cm^{-1}.

Normally the =CH stretch of compounds in *trans* configurations is not seen in the IR spectra because it is buried under aliphatic CH stretching bands. However, it has been observed that for alkenyl acetates =CH stretching bands appear at 3011–3013 cm^{-1}. In monounsaturated acetates having a double-bond one carbon in from the far end of the chain (i.e., CH$_3$CH=CH) the band was shifted to 3021–3022 cm^{-1}, while (*Z*)-2- and (*Z*)-3-alkenyl acetates showed the stretching absorption at 3028–3029 cm^{-1} and 3017–3018 cm^{-1}, respectively (Atty-galle et al. 1994).

A band diagnostic for the CH stretch of terminal C=C double bonds occurs in the neighborhood of 3080 cm^{-1}. These vinyl groups also give rise to a C=C stretching vibration in the vicinity of 1645 cm^{-1}. Both bands appear in the spectrum of (*R,Z*)-7,15-hexadacadien-4-olide (Fig. 6.5), a sex pheromone of a scarab beetle (Leal et al. 1996). We also have used these bands in the identification of (*E*)-2,7-octadienyl acetate (Fig. 6.6), a sex pheromone component of a lygaeid bug (Aldrich et al., 1997).

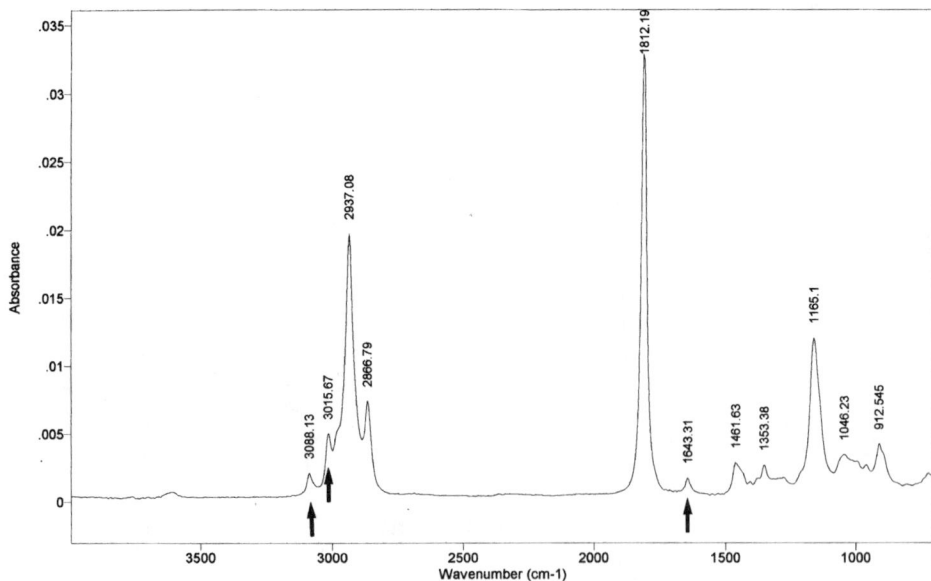

Figure 6.5. Vapor-phase FTIR spectrum of (*R,Z*)-7,15-hexadecadien-4-olide, sex phero-mone of a scarab beetle. The band at 3015 cm^{-1} (arrow) is due to a double bond in the *cis* configuration whereas the occurrence of a terminal double bond can be inferred from the bands at 1643 and 3088 cm^{-1} (arrows).

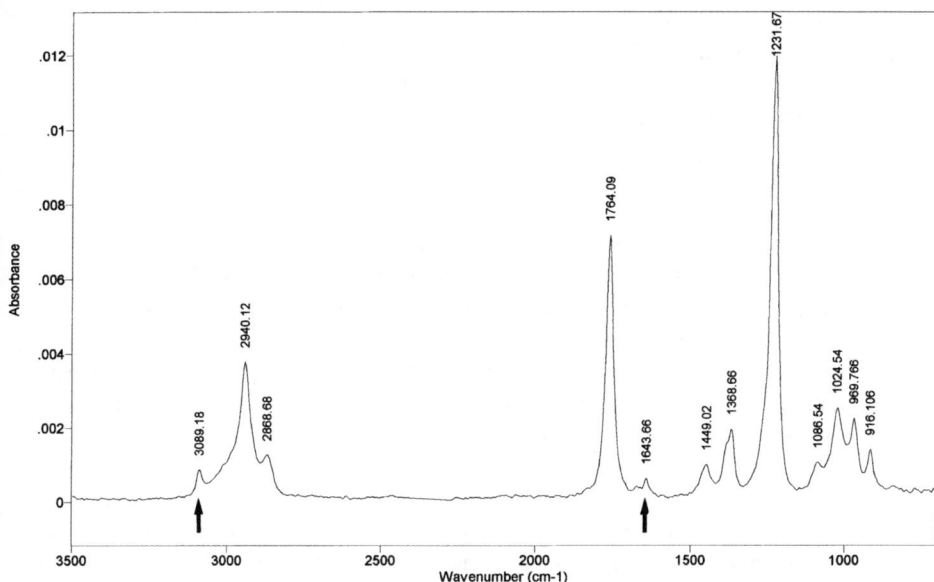

Figure 6.6. Vapor-phase IR spectrum of a lygaeid bug sex pheromone. The arrowed bands at 1645 and 3089 cm^{-1} are characteristic of a terminal double bond. The C=O stretching band of the α,β-unsaturated ester appears at 1764 cm^{-1}.

6.5. Ultraviolet Spectroscopy

UV spectroscopy generally refers to the measurement of electronic transitions induced by absorption of electromagnetic radiation of wavelengths 200–380 nm (the near ultraviolet). Electronic transitions also occur at wavelengths shorter than 200 nm (the vacuum ultraviolet), but this region of the spectrum is not readily accessible with standard instruments. Longer wavelengths, in the visible regions (380–800 nm), are easily accessible instrumentally but are of less importance in the structural elucidation of most classes of organic compounds, because the majority of organic compounds are colorless.

The intensity of radiation absorbed by a substance is a function of the concentration of the substance in the light path and the thickness of the sample. To avoid the complication of losses due to reflection and scattering, I_0 is defined as the radiant power transmitted through a blank and I is the radiation transmitted through the sample. The transmittance T is defined as the ratio of I to I_0:

$$T = I/I_0.$$

Absorbance A is defined as the logarithm of the reciprocal of transmittance:

$$A = \log(1/T) = -\log(I/I_0) = \log(I_0/I).$$

Absorbance is proportional to concentration and path length, according to the Beer-Lambert law (often referred to simply as Beer's law). For purposes of characterization of a chromophore, intensity of absorption is expressed in terms of molar absorbtivity ε given by

$$\varepsilon = (MA)/(Cl), \text{ or rearranged, } A = (\varepsilon Cl)/M,$$

where M is the molecular weight, C is the concentration (in grams per liter), and l is the path length through the sample in centimeters. UV absorptions are usually reported in terms of the extinction coefficient(s) and the wavelength(s) (λ_{max}) at which maximum absorption occurs.

Dilute solutions may be adequate for UV spectroscopy because molar absorptivity values frequently exceed 10,000. UV is nondestructive, but for its application, the target compound must be pure. Small contaminants with high molar absorptivity can be very misleading, because the UV spectrum obtained from a sample in solution is a composite of all materials in the sample (including the solvent). To circumvent this problem, UV spectra may be obtained on individual compounds separated by high pressure liquid chromatography (HPLC), using a diode array detector (see later and chapter 2).

To avoid poor resolution and difficulty in spectral interpretation, a solvent should not be employed for measurements near or below its ultraviolet cutoff, the wavelength at which absorbance for the solvent alone approaches one absorbance unit. Hexane, methanol, and ethanol have cutoffs at 210 nm (for a complete list see Willard et al. 1981). UV grade 95% ethanol is readily available, but absolute ethanol may cause problems due to traces of benzene.

6.6. Information from UV Spectra

UV spectroscopy is the least useful of the spectroscopic methods commonly used in the identification of new compounds, for several reasons. First, only a subset of organic compounds, usually those with aromatic groups or conjugated double bonds (chromophores) that contain valence electrons with relatively low excitation energies, absorb in the readily accessible UV range. When measured at high resolution, spectra of UV-active compounds contain considerable fine structure because electronic transitions are influenced by rotational and vibrational modes. However, in practice, UV spectra are generally recorded at low resolution, so this detail (which is not useful for structural identifications) is lost.

Second, UV spectra are relatively featureless so that the information that can be gleaned from spectra is limited. Detailed explanations of the characteristic UV absorptions for organic compounds, and of empirical rules for predicting the positions of absorption maxima, can be found in any standard text on identification of organic compounds (e.g., Silverstein et al. 1991).

Third, the spectrum may be a composite from two or more isolated chromophores in different parts of the molecule. Thus, in the absence of other information about the structure, interpretation of UV spectra usually is not possible.

However, a few general points may provide some hints as to the types of structural units present, as follows:

1. Compounds that are transparent to UV (210–400 nm) have no conjugated double or triple bonds or aromatic rings.

2. Compounds with extended chromophores (i.e., multiple conjugated double or triple bonds or aromatic rings) have large extinction coefficients (>10,000). Extension of conjugation in a carbon chain is always associated with a pronounced shift toward longer wavelengths and usually with an increase in the intensity of absorption. In very long chromophores, the absorption will extend into the visible region, and the compound will be colored.

3. Simple conjugated systems such as dienes and α,β-unsaturated carbonyls have extinction coefficients in the range of ~10,000–20,000, with λ_{max} in the region between 215 and 300 nm, depending on factors such as attached substituents and ring strain. As the conjugation is increased, the intensity of the absorption increases, and λ_{max} moves to longer wavelengths.

4. Low intensity absorptions ($\varepsilon < 100$) in the region 270–350 nm are probably due to a forbidden $n-\pi^*$ transition in aldehydes and ketones. For other carbonyl compounds, this weak transition usually occurs at wavelengths too short to be observed.

5. Absorptions with extinction coefficients in intermediate ranges ($\varepsilon = 1000$–10,000) are usually due to single aromatic chromophores. The absorption may be strongly influenced by substituents (auxochromes), which may have little or no absorption themselves but which strongly influence the intensity and position of the absorption of the chromophore to which they are attached. Auxochromes that are pH dependent (e.g., amino and phenolic OH groups) are particularly useful, because the absorption will change with pH (Field et al. 1995). For example, phenolic compounds that are colorless under acidic or neutral conditions are often colored under basic conditions.

6. Cross-conjugated systems, in which the conjugation cannot be traced out in one continuous line, behave as separate chromophores.

Overall, UV spectra may be most useful for confirming the presence of structural fragments tentatively assigned on the basis of information from other types of spectra. Compilations of spectra (e.g., the Sadtler Indexes, Sadtler Research Laboratories, 1996) may be useful as a source of spectra of model systems for comparison with the spectrum of an unknown. However, it must be remembered that substituents have a very marked influence on both the position and intensity of absorptions, and so the chromophore in a model system should match that of the proposed target structure as closely as possible.

6.7. Spectra from UV Detectors Used with HPLC

Unlike GC, HPLC has no detector that is as versatile, universal in applicability, and reliable as the flame ionization detector. UV detectors are commonly used in HPLC, despite the inherent limitation that the compounds of interest must contain a chromophore in order to be visible to the detector. Of these, diode array detectors, which can obtain and store UV spectra on every compound eluting in an HPLC run, are most useful (Miller et al. 1982). Furthermore, a peak purity algorithm can be used to check that the spectrum seen is indeed due to a single compound, and is not a composite from two partially resolved compounds. These detectors are discussed in detail in chapter 2.

6.8. Acknowledgments

The writing of this chapter was supported in part by a special coordination fund for promoting science and technology (SCF) in the basic research core system by the Science and Technology Agency, Japan. I thank Bio Rad Japan for the comparative vapor-phase infrared measurements. I am indebted to T. Suzuki, S. Matsuyama (Tsukuba University), T. Nakano (Bio-Rad Japan), D. Ibarra, H. Hubert, P. Zarbin, S-C. Xu, and H. Miyazawa (NISES) for their critical review of a draft version of the manuscript.

6.9. References

Aldrich, J.R., W.S. Leal, R. Nishida & A.P. Khrimian. 1997. Semiochemistry of aposematic seed bugs. Entomol. Exp. et Appl. **84**:127–135.

Attygalle, A. 1994. Gas phase infrared spectroscopy in characterization of unsaturated natural products. Pure Appl. Chem. **66**:2323–2326.

Attygalle, A.B., A. Svatos, C. Willcox & S. Voerman. 1994. Gas-phase infrared spectroscopy determination of double bond configuration of monounsaturated compounds. Anal. Chem. **66**:1696–1703.

Attygalle, A.B., A. Svatos, C. Willcox & S. Voerman. 1995. Gas-phase infrared spectroscopy determination of double bond configuration of some polyunsaturated pheromones and related compounds. Anal. Chem. **67**:1558–1567.

Brown, R.S., J.R. Cooper & C.L. Wilkins. 1985. Light pipe temperature and other factors affecting signal in gas chromatography/Fourier transform infrared spectrometry. Anal. Chem. **57**:2275–2279.

Coffelt, J.A., K.W. Vick, P.E. Sonnet & R.E. Doolittle. 1979. Isolation, identification, and synthesis of a female sex pheromone of the navel orangeworm, *Amyelois transitella* (Lepidoptera: Pyralidae). J. Chem. Ecol. **5**:955–966.

Doumenq, P., M. Guiliano & G. Mille. 1989. GC/FTIR potential for structural analysis of marine origin complex mixtures. J. Environ. Anal. Chem. **37**:235–244.

Field, L.D., S. Sternhell & J.R. Kalman. 1995. Organic Structures from Spectra, 2nd ed. John Wiley & Sons, Chichester.

Garrafo, H.M., L.D. Simon, J.W. Daly & T.F. Spande. 1994. *Cis*- and *trans*-configurations of α,α'-disubstituted piperidines and pyrrolidines by GC-FTIR; application to deca-hydroquinoline stereochemistry. Tetrahedron **50**:11329–11338.

Griffiths, P.R. 1977. Optimized sampling in the gas chromatography-infrared spectroscopy interface. Appl. Spectrosc. **31**:284–288.

Griffiths, P.R., S.L. Pentoney Jr., A. Giorgetti & K.H. Shafer. 1986. The hyphenation of chromatography and FT-IR spectrometry. Anal. Chem. **58**:1349A–1366A.

Haefner, A.M., K.L. Norton, P.R. Griffiths, S. Bourne & R. Curbelo. 1988. Interfaced gas chromatography and Fourier transform infrared transmission spectrometry by eluate trapping at 77°K. Anal. Chem. **60**:2441–2444.

Jackson, P., G. Dent, D. Carter, D.J. Schofield, J.M. Chalmers, T. Visser & M. Vredenbregt. 1993. Investigation of high sensitivity GC-FTIR as an analytical tool for structural identification. J. High Res. Chrom. **16**:515–521.

Leal, W.S. 1990. (*R,Z*)-5-(–)-(Oct-1-enyl)oxacyclopentan-2-one, the sex pheromone of the scarab beetle *Anomala cuprea.* Naturwissenchaften **78**:521–523.

Leal, W.S. 1997. Evolution of sex pheromone communication in plant-feeding scarab beetles. *In:* Pheromone Research: New Directions, eds. R.T. Carde & A.K. Minks, pp. 505–513, Chapman and Hall, New York.

Leal, W.S., Y. Kuwahara, S. Matsuyama, T. Suzuki & T. Ozawa. 1992. GC-FTIR potential for structure elucidation. J. Braz. Chem. Soc. **3**:25–29.

Leal, W.S., X. Shi, K. Nakamuta, M. Ono & J. Meinwald. 1995. Structure, stereochemistry, and thermal isomerization of the male sex pheromone of the longhorn beetle, *Anaglyptus subfasciatus.* Proc. Natl. Acad. Sci. USA **92**:1038–1042.

Leal, W.S., S. Kuwahara, M. Ono & S. Kubota. 1996. (R,Z)-7,15-Hexadecadien-4-olide, sex pheromone of the yellowish elongate chafer, *Heptophylla picea.* Biorg. Med. Chem. **4**:315–321.

Miller, J.C., S.A. George & B.G. Willis. 1982. Multichannel detection in high-performance liquid chromatography. Science **218**:241–246.

Namba, R. 1990. Industrial application of GC/FT-IR pp. 470–518. *In:* Practical Fourier Transform Infrared Spectroscopy & Laboratory Chemical Analysis, eds. J.R. Ferraro and K. Krishnan, Academic Press, San Diego.

Nyquist, R.A. 1984. The Interpretation of Vapor-Phase Infrared Spectra. Sadtler Research Laboratories, Philadelphia.

Reedy, G.T., D.G. Ettinger & J.F. Schneider. 1985. High-resolution gas chromatography/ matrix isolation infrared spectrometry. Anal. Chem. **57**:1602–1609.

Roeges, N.P.G. 1994. Guide to the Complete Interpretation of Infrared Spectra of Organic Structures. John Wiley & Sons, New York.

Sadtler Research Laboratories. 1996. The Sadtler Standard Spectra. Sadtler Division, Bio-Rad Laboratories, Philadelphia.

Silverstein, R.M., G.C. Bassler & T.C. Morrill. 1991. Spectrometric Identification of Organic Compounds, 5th ed. John Wiley & Sons, New York.

Skoog, D.A. 1985. Principles of Instrumental Analysis, 3rd ed., Saunders, Philadelphia.

Svatos, A. and A.B. Attygalle. 1997. Characterization of vinyl-substituted carbon-carbon double bonds by GC/FT-IR analysis. Analyt. Chem. **69**:1827–1836.

Willard, H.H., L.L. Merritt, Jr., J.A. Dean & F.A. Settle, Jr. 1981. Instrumental Methods of Analysis, 6th ed. Wadswoth, Belmont, CA.

7

Microchemical Techniques

Athula B. Attygalle

7.1. Introduction

Semiochemicals are usually found in complex mixtures with other compounds that do not play a role in chemical communication. To characterize semiochemicals, these mixtures first must be separated into their individual components by a chromatographic technique. Preferably, the separation should be conducted without any prior chemical modification. Unfortunately, many samples are not directly amenable to direct chromatographic analysis. Derivatization often enables analysis, and sometimes even better characterization, of such samples. In particular, microchemical reactions can be useful for

1. characterizing functional groups,

2. obtaining more helpful spectra,

3. improving separations,

4. imparting volatility and thermal stability for gas chromatographic (GC) analysis,

5. reducing excessive volatility,

6. determining enantiomeric composition, and

7. enhancing detectability.

A large number of microchemical reactions are now available, and a comprehensive coverage is not the intention of this chapter. This chapter concentrates on useful techniques generally applicable to semiochemical characterization. Although most derivatization methods described in the literature are for milligram quantities of material (Knapp 1979; Blau & King 1977; Ma & Ladas 1976), with a little modification most of these procedures can be used at nanogram levels.

7.2. Derivatization and Degradation Methods (Microreactions)

A key step in the process of identification of a semiochemical is to determine which functional groups are present. Sometimes the electron-impact (EI) mass spectrum of the compound may show features suggesting the presence of certain functional groups (Table 7.1). When the molecular ion of an EI-mass spectrum or the pseudomolecular ion of a chemical ionization (CI) mass spectrum is recognizable, the accurate mass of this ion can be obtained by high-resolution mass spectrometry. The sample size required depends on the mass spectrometric sample introduction method. An amount of about 50–100 ng is usually sufficient for GC-high-resolution mass spectrometry. For relatively nonvolatile samples, direct sample introduction or fast atom bombardment methods are used, however,

Table 7.1. Common Electron-Impact Fragment Ions Diagnostic of Functional Groups

Functional Group	Characteristic Ions in EI Mass Spectrum (*m/z*)	Remarks
Primary alcohols	31 [$H_2C=O^+H$]	A molecular ion is weak or absent. Usually a peak for M-18 is observed.
Methyl carbinols	45 [$CH_3-HC=O^+H$]	A molecular ion is weak or absent. Usually a peak for M-18 is observed.
Aldehydes	29 [CHO^+], 44	A molecular ion is weak. Usually a peak for M-18 is observed.
Formates	47 [$HCO(OH_2)^+$]	A molecular ion is absent. Usually a peak for M-46 is observed.
Acetates	43 [CH_3CO^+], 61 [$CH_3CO(OH_2)^+$]	A molecular ion is absent. Usually a peak for M-60 is observed.
Propionates	57 [$C_2H_5CO^+$], 75 [$C_2H_5CO(OH_2)^+$]	A molecular ion is absent. Usually a peak for M-74 is observed.
Butyrates	71 [$C_3H_7CO^+$], 89 [$C_3H_7CO(OH_2)^+$]	A molecular ion is absent. Usually a peak for M-88 is observed.
Long-chain methyl esters	59 [$^+COOCH_3$], 74[a] [$H_2C=C(OH)OCH_3^{+\cdot}$]	In addition to the molecular ion, a peak for M-32 is observed.
Long-chain ethyl esters	73 [$^+COOC_2H_5$], 88[a] [$H_2C=C(OH)OC_2H_5^{+\cdot}$]	
Methyl ketones	43 [CH_3CO^+], 58[a] [$H_2C=C(OH)CH_3^{+\cdot}$]	In addition to the molecular ion, a peak for M-15 is observed.
Ethyl ketones	57 [$C_2H_5CO^+$], 72[a] [$H_2C=C(OH)C_2H_5^{+\cdot}$]	In addition to the molecular ion, a peak for M-29 is observed.
Propyl ketones	71 [$C_3H_7CO^+$], 86[a] [$H_2C=C(OH)C_3H_7^{+\cdot}$]	In addition to the molecular ion, a peak for M-43 is observed.
Primary amine	30 [$H_2C=N^+H_2$]	
Primary amide	59[a] [$H_2C=C(OH)NH_2^{+\cdot}$]	

[a]McLafferty rearrangement ion; a hydrogen atom must be present on the carbon atom γ to the carbonyl group.

the amount required is substantially more. Once the accurate mass of the molecular ion is known, the molecular formula and the number of double-bond equivalents can be computed unambiguously. Sometimes the molecular formula alone suggests the functional groups to be expected. Other spectroscopic methods such as infrared (IR) and nuclear magnetic resonance (NMR) can also provide additional information about functional groups. The indicated functional groups are then verified by microchemical reactions.

Microscale reactions use different methodology than conventional methods utilizing measuring cylinders, separatory funnels, and reflux condensers. However, basics of microchemistry are easy to master. With a little practice, most microreactions can be performed with relative ease and with simple apparatus. However, experiments must be organized to keep the number of manipulations to an absolute minimum to minimize contamination and loss of the sample. A microchemist must learn to recognize common contaminants such as phthalates, silicones, polyaromatic hydrocarbons; stabilizers such as butylated hydroxytoluene (BHT = 2,6-di-*t*-butyl-4-methylphenol); and other artifacts immediately by their mass spectra (Middleditch 1989). Solvents must be the best purity available and glass-distilled before use. Plastic and rubber seals must be avoided because they often introduce plasticizers and other impurities. When necessary, Teflon® tape and seals must be used instead of paraffin tape (Parafilm®), which introduces hydrocarbon impurities.

Some microreactions are instantaneous, and can be conducted simply by mixing the reagents in a microliter syringe (Hoff & Feit 1964). However, most microreactions require more time. They are conveniently conducted in sealed melting-point tubes (2 mm i.d.) or containers made from Pasteur pipettes (Fig. 7.1A) with a microtorch (Supelco; Radioshack). Using these containers, microreactions of 5- 50-μl scale can be performed with ease. Alternatively, for those who are not comfortable with sealing tubes with a torch, a variety of screw-capped borosilicate glass vials with V-shaped bottoms are commercially available (Reacti-vials®, Pierce; V-vials, Wheaton Scientific; Fig. 7.1B). The Teflon®-lined silicone rubber septa used with these vials provide a relatively tight seal, and usually the contents can be heated to about 180°C if necessary without any significant sample loss. However, at elevated temperatures, there is a risk that the septum could burst and the sample will be lost. The sample vial should be placed in a screw-capped metal container before heating to reduce the danger from a vial bursting. Foil-lined caps must be avoided, since the adhesives used in these caps introduce contaminants. It is always good practice to run a blank reaction at the desired temperature and analyze the contents under the same conditions used for the sample. This will also indicate whether the system will withstand the high reaction temperature.

To measure and transfer known amounts of liquids, microliter syringes with fine needles are recommended. Syringes with fused silica needles (the type used for on-column GC injections) are particularly useful for microliter samples.

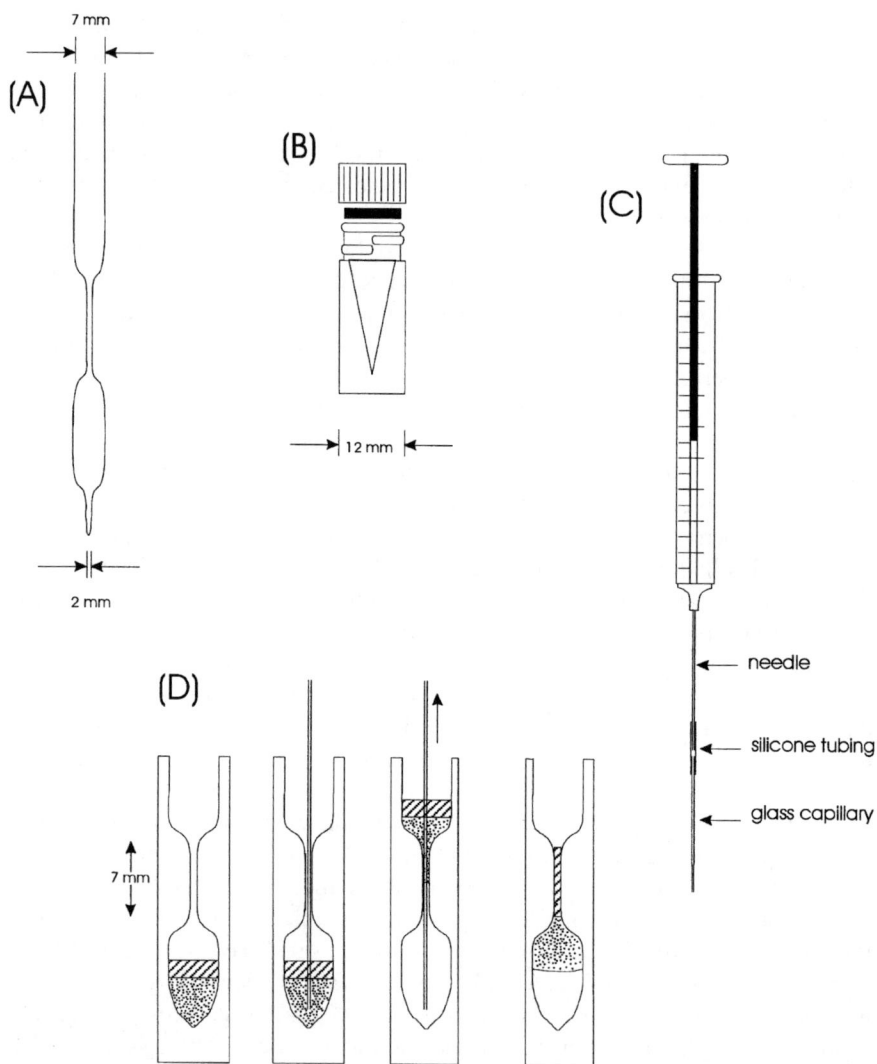

Figure 7.1. Micromanipulations. (**A**) A microreaction tube made from a Pasteur pipette; the narrow neck is sealed using a microtorch after adding the reagents to the bottom chamber. (**B**) A commercially available conical-bottomed vial. (*C*) A 10-μl syringe with a glass capillary attached to the needle. (*D*) The use of a Keele vial.

However, using microliter syringes to transfer liquids often leads to cross-contamination. Cleaning a syringe after measuring liquids is time-consuming. A simple way to avoid this problem is to attach a piece of glass or fused silica capillary to the syringe needle, with a short piece of silicone rubber tubing (Fig. 7.1C). With a little practice it is possible to draw liquids only into the capillary, which can be discarded after use, without contaminating the silicone rubber piece. As an alternative, graduated glass capillaries can be used (Fisher Scientific). The micropipette controller marketed by Brand (Wertheim/Main, Germany) is an ideal device for transferring a few microliters of liquid. Another option is to insert a glass capillary tube snugly into an appropriately cut disposable polypropylene tip used with Pipetteman® or Eppendorf type micropipettors. The transfer of exact quantities of solids is even more difficult. It is difficult to weigh quantities less than 1 mg accurately. It is best to weigh about 10 mg accurately, dissolve in a known quantity of solvent, and transfer the appropriate volume of the solution.

Heating of a reaction mixture is done in an oven (a GC oven is ideal) since the temperature can be controlled precisely. Alternatively, a metal block heater with holes for the sample tubes can be used; commercial models provide sample holders with different size holes. However, temperature control is less precise than with a GC oven.

Stirring a microscale reaction mixture efficiently is nearly impossible. When using screw-cap vials, swirling or vortexing, mixing with a syringe, or bubbling a slow stream of nitrogen are the best alternatives. If the reaction volume is at least 50 µl, a small magnetic stir bar, or a vaned stirrer, placed inside a Reactivial® can be used. When a reaction is conducted in a sealed tube, the vessel can either be sonicated in an ultrasonic bath or vortexed briefly.

Some microreactions require considerable workup and/or cleanup before analysis. Solid-phase extraction procedures are useful if polarity of the product is significantly different from the reagents (see chapter 1). Often a product must be extracted from an aqueous mixture of spent reagents. The device known as the "Keele vial" (Wheaton Scientific Products) is useful for separating small amounts of immiscible liquids (Attygalle & Morgan 1986). Using this device, derivatization and extraction can be performed in the same container, and the product can be extracted into a small volume (5–10 µl) of solvent (Fig. 7.1D). When a "Keele vial" is used for extraction, the required liquid layer is drawn into the narrow capillary neck from which an aliquot can be withdrawn conveniently. This is achieved by drawing air from the bottom chamber with a syringe, or by adding more of the heavier liquid to the bottom chamber (Attygalle & Morgan 1986).

Many microchemical derivatizations require the evaporation of the solvent or excess reagents on completion of the reaction, usually with a slow stream of clean nitrogen or argon.

Filtration of a few microliters of a mixture is not straightforward. Centrifugation and removal of the supernatant with a microliter syringe is recommended. For larger samples, a syringe filter of the type used for removal of particles from high-pressure liquid chromatography (HPLC) solvents can be used. To minimize contamination, polymeric membranes must be avoided and inorganic membranes based on alumina matrices should be used (ANOPORE™, Phenomenex). In particular, the tube-tip syringe filters are ideal for the microfiltration of aqueous- or organic-based samples of small volumes. The sample hold up volume can be as small as 7.5 µl. In practice, the filter should be prewashed with several volumes of solvent and the filtrate analyzed for contaminants before the sample is filtered.

7.2.1. Reactions of Alkenes and Alkynes

Alkenes and alkynes are hydrocarbons bearing carbon-carbon double bonds and triple bonds, respectively. These compounds may be linear or branched and may contain rings.

7.2.1.1. Reactions to Determine Number of Carbon-Carbon Double-Bonds (Hydrogenation)

Hydrogenation of a compound followed by GC-MS analysis of the product(s) is a reliable procedure for determining the number of nonaromatic double bonds or triple bonds in a molecule. Characterization of branched alkenes is also facilitated by hydrogenation. The mass spectrum of a methyl-branched unsaturated compound may not suggest the position of the methyl group, whereas the saturated analog cleaves preferentially at branch points, yielding diagnostic fragments.

Typically, reduction can be accomplished by mixing a small sample in hexane or ether (50 µl) and ~ 0.5 mg of 10% Pd on activated charcoal catalyst in a stirred, conical-bottomed glass vial (Pierce Chemical Co.). A balloon filled with hydrogen is attached to the vial and kept in place for about 8–10 h. The reduced sample is filtered and a small aliquot of the supernatant is analyzed by GC-MS. For example, in the structure elucidation of epilachnene (**1**), a defensive allomone of Mexican bean beetle pupae (Attygalle et al. 1993a), hydrogenation changed the molecular weight by two mass units. The molecular formula of epilachnene obtained from accurate mass measurements was $C_{16}H_{29}NO_2$. The major product (**2**) from hydrogenation showed a molecular ion at m/z 269 and established that epilachnene is a monocyclic compound bearing one carbon-carbon double bond.

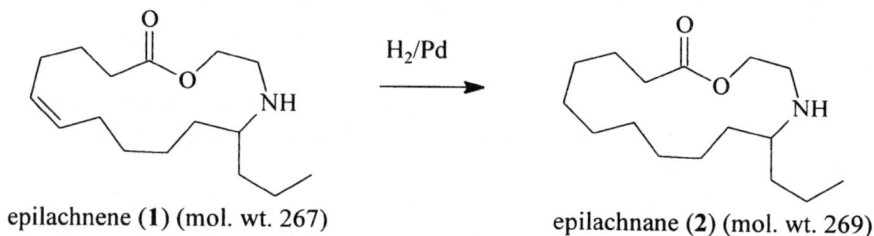

epilachnene (**1**) (mol. wt. 267)   epilachnane (**2**) (mol. wt. 269)

Alternatively, hydrogenation can be performed on samples as small as 5–10 ng, by a reaction gas chromatographic technique (Beroza & Sarmiento 1966). In this procedure, a small amount of catalyst is placed in the insert liner of the injection port, or in a precolumn, and hydrogen is used as the carrier gas. The reduction of carbon-carbon double and triple bonds takes place almost instantaneously.

The catalyst is prepared by evaporating an aqueous solution (150 ml) of palladium chloride (25 mg) and sodium hydroxide (11.2 mg) in contact with a silica support such as Chromosorb™ W (100–120 mesh, 1.5 g) in a rotary evaporator. Once the material is free flowing, it is further dried in an oven at 150°C for 12 h. The material is then packed between two silanized glass wool plugs in a glass or a metal tube and hydrogen (40 ml/min) is passed over it for 60 min at 200°C. This activated catalyst can be used as a precolumn packing, placed between two silanized glass wool plugs, for packed columns, or used in the insert liner of the injection port for capillary columns.

Using this reaction gas chromatogrpahic procedure, the homofarnesene isomer (3) from the Dufour's gland of *Myrmica* ants was shown to bear four double bonds (Attygalle & Morgan 1982). Furthermore, the mass spectrum of the homofarnesane (4) obtained indicated the presence of an ethyl group at C-7 position of homofarnesene (3).

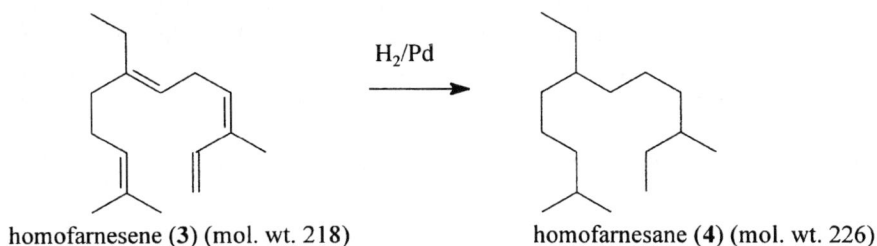

homofarnesene (3) (mol. wt. 218)   homofarnesane (4) (mol. wt. 226)

Hydrogenation was the key step in the determination of absolute configuration of the ant alarm pheromone, manicone (5) [(4E)-4,6-dimethyl-octen-3-one]. The reduction of the double bond of natural manicone resulted in two diastereoisomers, which gave two peaks on gas chromatographic analysis on a chiral stationary phase. Under similar conditions, racemic synthetic manicone gave four peaks. A comparison of retention times with those obtained from a synthetic sample established the absolute configuration of natural manicone as S (Bestmann et al. 1987).

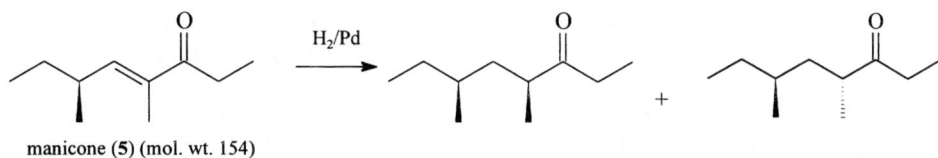

manicone (5) (mol. wt. 154)

Similarly, hydrogenation of neocembrene (**6**), a queen-specific chemical of Pharaoh's ant, gave a mixture of products with similar mass spectra. Molecular ions of all these spectra appeared at m/z 280. In this way, neocembrene was shown to be a monocyclic compound with four double bonds (Edwards & Chambers 1984) (note that uptake of four hydrogen atoms also can be due to a triple bond, instead of two double bonds).

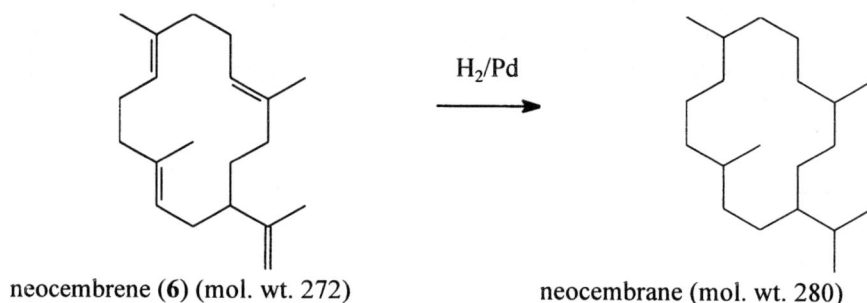

$$H_2/Pd$$

neocembrene (**6**) (mol. wt. 272)   neocembrane (mol. wt. 280)

7.2.1.2. Partial Reduction of Carbon-Carbon Double Bonds

For polyunsaturated compounds, partial hydrogenation, preferably to the corresponding monoenes may be valuable for locating the double bonds since there are several efficient methods to determine double-bond position and configuration of monoenes. An efficient way to carry out this random reduction is by the use of diimide (**7**) (Corey et al. 1961). The reagent diimide (N_2H_2) is formed in situ by the oxidation of hydrazine by hydrogen peroxide. The *syn* addition of hydrogen atoms to the double bond takes place via a cyclic transition state.

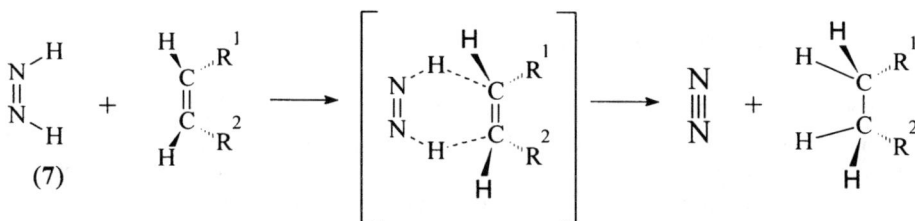

Typically the polyene is dissolved in a few microliters of ethanol and mixed with a solution of NH_2NH_2 (10 μl, 10% vol/vol hydrazine hydrate in ethanol) (warning: hydrazine is highly toxic) and H_2O_2 (10 μl, 0.6% in ethanol). The reaction mixture is heated at 60°C for a predetermined period of time. Control of reaction time is critical to avoid overreduction. The time required depends entirely on the chemistry and the amount of the polyene being used. For optimal results, the conditions must be worked out using a known polyene. The cooled reaction mixture is quenched with dilute HCl (20 μl, 10%) and the mixture is extracted with hexane (3×15 µl). The combined hexane layers are concentrated and analyzed by GC-mass spectrometry (GC-MS). For example, (Z,E)-9,11-tetradecadienyl acetate [(Z,E)-9,11-TDDA] gave (Z)-9-tetradecenyl acetate and (E)-11-tetradecenyl acetate plus the saturated product, tetradecyl acetate, and the starting material (Fig. 7.2). (Note that different double bonds are reduced at different rates. In the case of (Z,E)-9,11-TDDA, the E-11 double bond reacts faster than the Z-9 bond; see Fig. 7.2). Similarly, (4Z,7Z,10Z)-4,7,10-tetradecatri-enyl acetate gave three expected monoenes together with the three dienes and the completely reduced product (Attygalle et al. 1996).

Many applications of this procedure are found in the literature. Yamaoka et al. (1987) employed this procedure to establish the structure of the trail pheromone of the termite *Reticulitermes speratus* as (3Z,6Z,8E)-dodecatrien-1-ol. Similarly, Nesbitt et al. (1973) used the method to reduce a terminal double bond, which facilitated determination of the configuration of the double bond at C-9 of (9E)-9,11-dodecadienyl acetate, a pheromone component of the red boll worm moth.

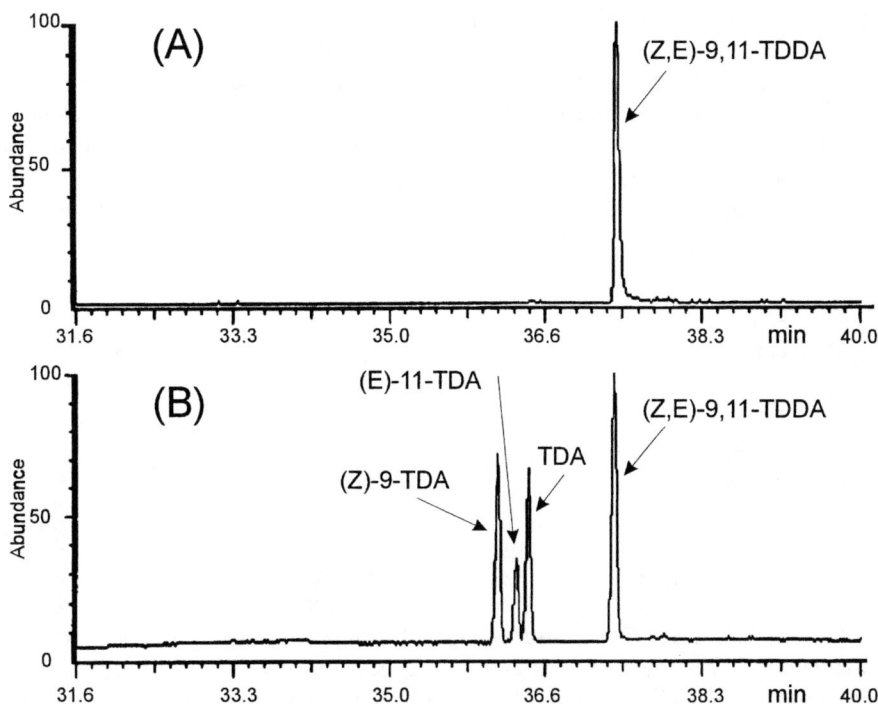

Figure 7.2. Gas chromatogram of (Z,E)-9,11-tetradecadienyl acetate [(Z,E)-9,11-TDDA] before reduction (**A**), and after partial reduction (**B**) [(Z)-9-TDA = (Z)-9-tetradecenyl acetate, (E)-11-TDA = (E)-11-tetradecenyl acetate, and TDA = tetradecyl acetate]. [DB-5 coated 0.25 mm × 30 m fused silica column. Oven temperature was held at 60°C for 4 min and programmed at 4°C/min, to 220°C].

Yamamoto et al. (1991a) used this procedure in conjunction with the dimethyldi-sulfide addition reaction (section 7.2.1.6.3) to locate the double bonds in heneico-sapentaenoic acid. Recently, this random reduction procedure was used in the identification of (3E,8Z,11Z)-3,8,11-tetradecatrienyl acetate, the major sex phero-mone component of the tomato pest *Scrobipalpuloides absoluta* (Attygalle et al. 1995a), for which less than 100 ng of material was available.

7.2.1.3. Bromination of Carbon-Carbon Double Bonds

An easy method for differentiating gas chromatographic peaks corresponding to saturated and unsaturated compounds is valuable to microchemists. Bromination shifts all the peaks due to unsaturated compounds, leaving those of saturated compounds unchanged (Attygalle & Morgan 1984b). Pyridinium tribromide (Ald-rich) dissolved in dichloromethane (15 mg/ml) is a more useful bromination reagent than bromine itself (Burk et al. 1986). The brominating agent is added in slight excess to the alkenes dissolved in dichloromethane, and after about 2 h the excess reagent is removed by washing with aqueous $NaHSO_3$ solution. The organic layer is then separated and analyzed by GC.

Bromination follows a stereospecific trans addition mechanism. Hence the reaction can be used to elucidate the geometry of low-molecular-weight alkenes (Burk et al. 1986). Although the pair of enantiomers produced from each alkene is not resolved by achiral GC phases, the diastereomers from *cis* and *trans* alkenes are well separated. Usually the adduct from a *trans* alkene elutes before that of the corresponding *cis* isomer.

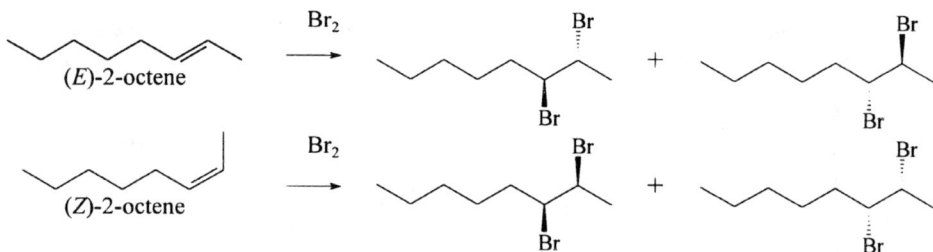

7.2.1.4. Isomerization of Carbon-Carbon Double Bonds

Depending on the spatial arrangement of groups attached to a carbon-carbon double bond, two geometric isomers, designated as *cis* (Z) and *trans* (E), are possible for each double bond. When only one geometric isomer is available, either as a synthetic standard or from a natural extract, and a complete mixture of isomers is required for GC retention time evaluations, an equilibrium mixture of geometric isomers can be produced by subjecting the double bond to an isomerization reaction. Heating an alkene sample to about 90°C with thiophenol (Henrick et al. 1975; Schwartz et al. 1986) or thiobenzoic acid (Tanaka et al.

1985) for about 24 h results in isomerization. Ultraviolet (UV) light, or the radical initiator 2,2′-azobisisobutyronitrile (AIBN; Schwartz et al. 1986) may accelerate the isomerization. Excess reagent is removed by extracting the reaction mixture with dilute NaOH. The isomerization reaction is particularly useful for the determination of the configuration of a conjugated diene system, since gas chromatographic elution order of the four isomers of a conjugated system is generally predictable. For example, when (8*E*,10*E*)-8,10-dodecadienol (**8**) (50 mg) was heated with thiophenol (1 mg) at 100°C for 1.5 h, a mixture containing (*E,E*)-(61%) (**9**), (*E,Z*)-(20%) (**10**), (*Z,E*)-(14%) (**11**), and (*Z,Z*)-(5%) (**12**) isomers was obtained (McDonough et al. 1993).

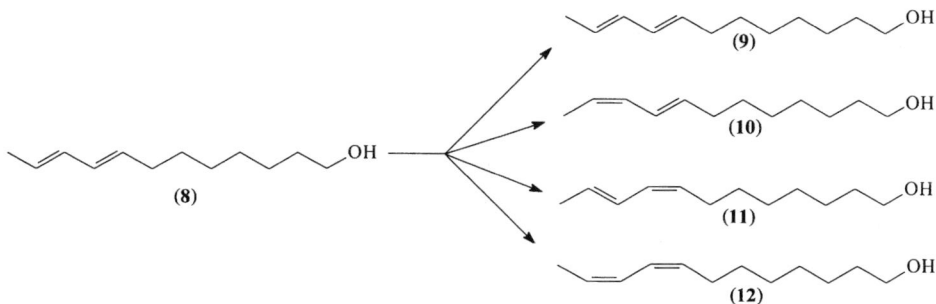

Similarly, Hall et al. (1980) isomerized (10*E*,12*E*)-10,12-hexadecadienal by heating with 0.5% thiophenol for 1 h at 100°C.

Isomerization of double bonds can be achieved also by heating with aqueous nitrous acid (Sonnet 1974). This treatment does not cause migration of the double bond. The review by Sonnet (1980) is an excellent source for information on other isomerization reactions.

7.2.1.5. Reactions to Determine the Carbon Skeleton (Hydrogenolysis)

The simplest derivative of an organic compound is its parent alkane or cycloalkane, and complete hydrogenolysis to the hydrocarbon skeleton has been employed in the structure elucidation of many natural products. Hydrogenolysis can be achieved by heating the sample with a catalyst such as palladium or platinum, in a hydrogen atmosphere at a temperature of at least 300°C. This temperature is much higher than that normally used for hydrogenation (Beroza & Sarmiento 1963). Alternatively, hydrogenolysis can be performed by a reaction gas chromatographic procedure called carbon-skeleton chromatography. This procedure is very similar to that described for hydrogenation by reaction gas chromatography (section 7.2.1.1). However, for hydrogenolysis, the catalyst is kept at much higher temperature. Hydrogenolysis saturates multiple bonds and replaces functional groups containing halogen, oxygen, sulfur, and nitrogen by hydrogen (even carboxyl or ester groups are converted to methyl groups). The resulting product is predominantly the parent alkane, sometimes accompanied by a few of its lower homologues. GC-MS analysis of the products enables determination of the length, branching, and cyclic nature of the carbon skeleton of the parent compound. When milder conditions are used (less catalyst and lower temperatures), sometimes both hydrogenation and hydrogenolysis products and some starting material are found in the resulting mixture.

For example, methyl linolenate (**13**) gave heptadecane and octadecane as hydrogenolysis products in addition to the hydrogenation product methyl stearate (Fig. 7.3).

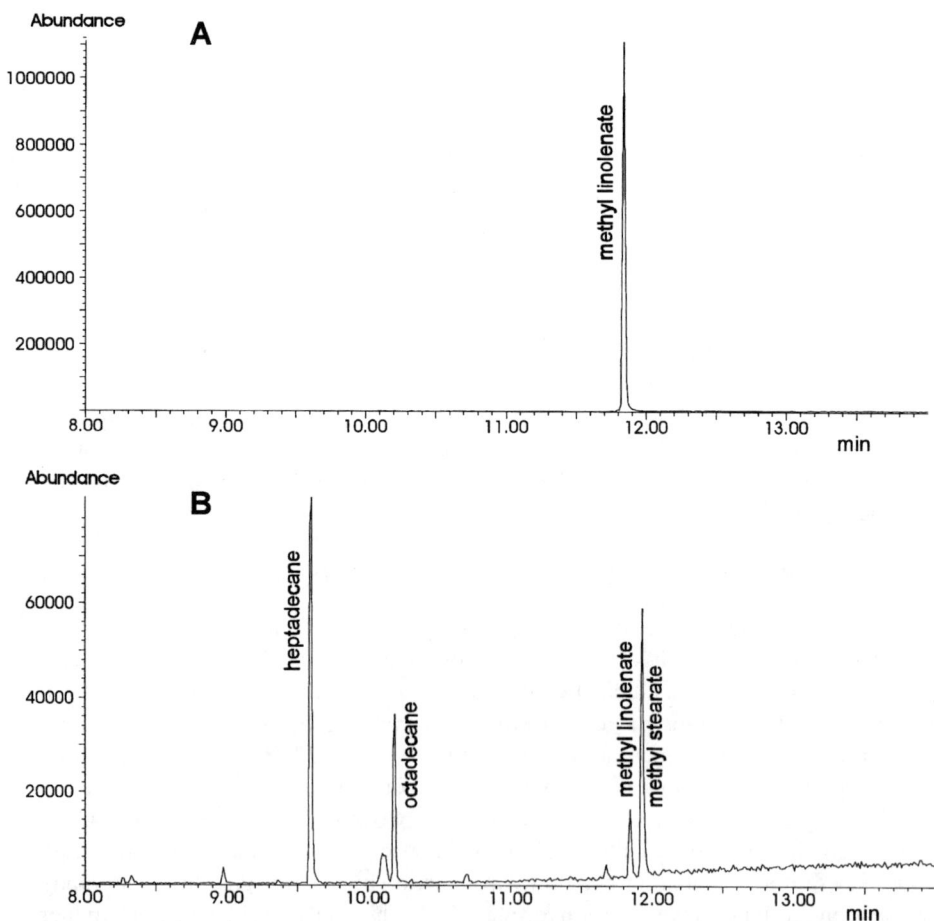

Figure 7.3. Hydrogenolysis of methyl linolenate. (**A**) A gas chromatogram of methyl linolenate sample before hydrogenolysis. (**B**) After hydrogenolysis. (DB-5 coated 0.25 mm × 30 m fused silica column. Oven temperature was held at 60°C for one min and programmed at 20°C/min, to 280°C and held for 15 min).

methyl linolenate (mol. wt. 292) (**13**)

Hydrogenolysis

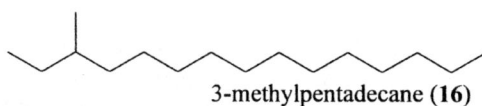

heptadecane (mol. wt. 240)   octadecane (mol. wt. 254)   methyl stearate (mol. wt. 298)

For the identification of the position of the methyl group in 10-methyl-2-tridecanone (**14**), the sex pheromone of southern corn root worm, Guss et al. (1983) used the carbon-skeleton procedure to convert the pheromone to 4-methyltridecane (**15**).

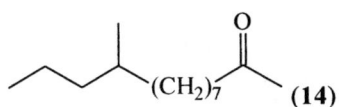

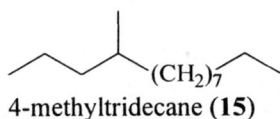

4-methyltridecane (**15**)

In a similar structure determination, Cavill et al. (1980) obtained 3-methylpentadecane (**16**) from 13-(1-methylpropyl)tridecanolide (**17**) isolated from the Argentine ant.

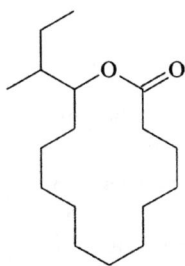

hydrogenolysis

3-methylpentadecane (**16**)

(**17**)

By hydrogenolysis, cyclopentanoid (iridoid) compounds such as anisomorphal (dolichodial, **18**) and iridodial (**19**) can be reduced to their basic cyclopentanes by removing the aldehyde groups (Cavill et al. 1976).

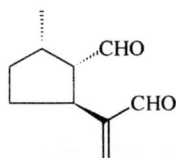

anisomorphal
[dolichodial, (**18**)]

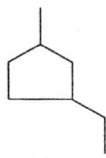

1-ethyl-3-methylcyclopentane (**20**)

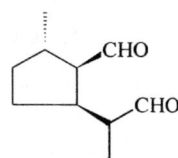

iridodial (**19**)

Even steroids are reduced to the parent hydrocarbons by the hydrogenolysis procedure (Adhikary & Harkness 1969); progesterone (pregn-4-ene-3,20-dione, **21**) gave 5-α- and 5-β-pregnane (**22**) in addition to two peaks for 20-oxo-pregnane.

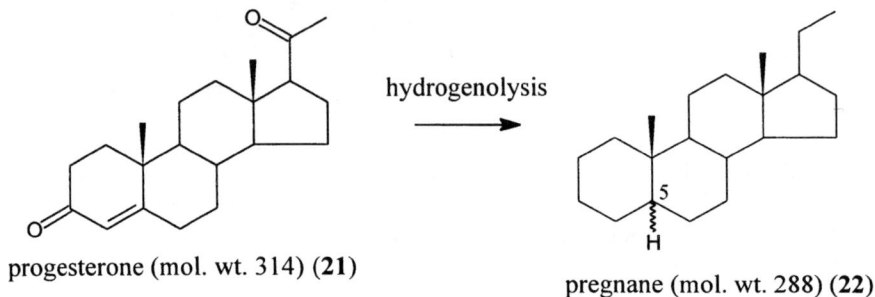

progesterone (mol. wt. 314) (**21**)

pregnane (mol. wt. 288) (**22**)

Brownlee and Silverstein (1968) used the hydrogenolysis procedure directly with GC fractions without the intervention of a solvent. In one application, nonane was the major hydrocarbon produced from the bicyclic ketal brevicomin (**23**), the sex attractant of Western pine beetle.

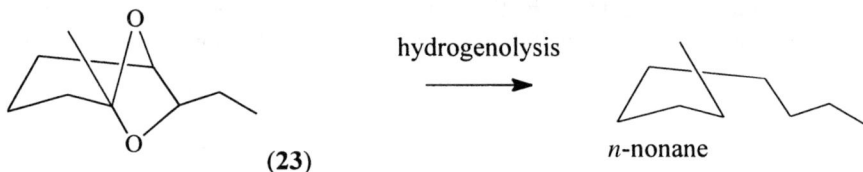

(**23**)

n-nonane

Bierl-Leonhardt et al. (1981a) described a much simpler microscale hydrogeno-lysis procedure. In this method, the substances trapped by micropreparative gas chromatography are heated to 250°C with 5% platinum on Al_2O_3 and $LiAlH_4$ in a sealed tube. The products are solvent extracted (hexane) and examined by GC-MS. Apparently, $LiAlH_4$ acts as the hydrogen source for this hydrogenation. For example, the pheromone of Comstock mealybug (2,6-dimethyl-1,5-heptadien-3-ol acetate, **24**), gave 2,6-dimethylheptane (**25**) indicating the positions of the methyl groups in the parent molecule (Bierl-Leonhardt et al. 1980a).

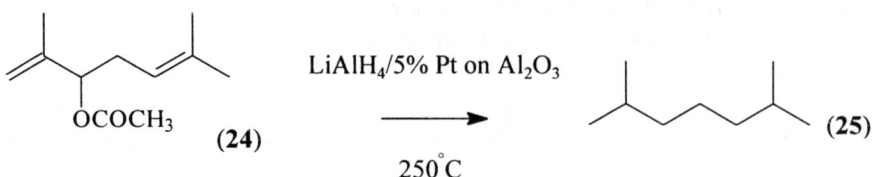

Similarly, in the identification of the sex pheromone of the citrus mealybug, *Planococcus citri,* the alcohol (**26**) obtained from the hydrolysis of the pheromone (an acetate) was subjected to hydrogenolysis. This procedure removed the -CH$_2$OH group and saturated the carbon-carbon double bond to yield 1,1-dimethyl-2-isopropylcyclobutane. In fact, the same cyclobutane derivative was obtained when an authentic sample of 1-methyl-2-isopropenylcyclobutaneethanol (grandisol, a pheromone component of the boll weevil *Anthonomus grandis* (**28**) was subjected to the same reaction (Bierl-Leonhardt et al. 1981).

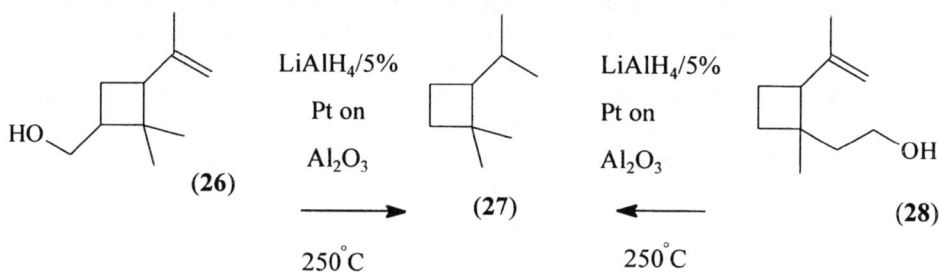

Hydrogenolysis is useful in the identification of alkaloids. For example, 1-(2-hydroxylethyl)-2-(12-aminotridecyl) pyrrolidine (**29**) was reduced to its basic carbon skeleton by high-temperature hydrogenolysis with Pt and LiAlH$_4$. The most abundant hydrocarbon in the resulting mixture was identified as *n*-heptadecane, indicating an uninterrupted chain of 17 carbon atoms in the parent alkaloid (Attygalle et al. 1993b).

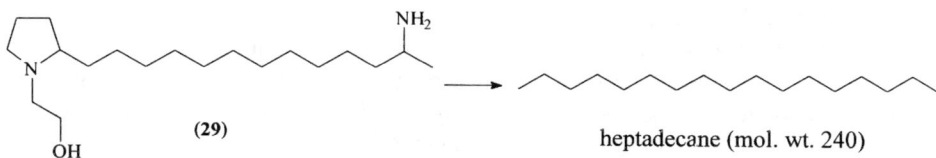

heptadecane (mol. wt. 240)

7.2.1.6. Reactions to Determine Double-Bond Positions and Configurations

Determining the position and configuration of double bonds in microamounts of unsaturated compounds has always been a formidable challenge. Although NMR methods can be used for compounds containing up to three double bonds (Rossi et al. 1982), in practice, the amount of sample required precludes the use of NMR for many samples of chemical ecological interest. Undoubtedly mass spectrometry (Hogge & Millar 1987; Schmitz & Klein 1986) and gas-phase infrared spectrometry (Attygalle et al. 1994, 1995b) are the preferred techniques since only a few nanograms of material are sufficient for a complete determination.

The determination of the position of the double bond of an unsaturated compound from electron-impact mass spectra is an arduous task. However, if the spectra of all possible isomers of a monounsaturated system are available, it is possible to make a reliable estimate by comparing the relative intensities of ions in each ion cluster of standard spectra with those of the unknown compound. The reliability of the method can be improved when combined with GC retention time data.

Horiike et al. (1981; Horiike & Hirano 1987) first described this empirical approach with double-bond isomers of tetradecenyl acetate. Later they extended this method to monounsaturated alcohols (Horiike & Hirano 1984; Horiike et al. 1986, 1990). Two other groups have verified and extended the method for other monounsaturated compounds (Leonhardt et al. 1985; Lanne et al. 1985). The convenience of this method is that it requires no derivatization. Therefore, the sensitivity reaches the detection limit of the mass spectrometer. The limitation is that the method requires sets of synthetic samples of all double-bond isomers, therefore it becomes impracticable for polyunsaturated and branched compounds.

Furthermore, mass spectra of a compound obtained under different conditions and with different instruments are slightly different from each other. When very small differences in ion intensities are important, it becomes necessary to obtain spectra under identical conditions. Fortunately, this is not so critical because the ratios of the intensities of some ions show a moderately linear correlation to the position of the double bond (Leonhardt et al. 1985). For linear monounsaturated compounds, this method can indicate a double-bond position with an accuracy of at least ±1. However, it is risky, at best, to claim determination of double-bond positions of an unknown by comparison of its mass spectrum with published or computer library spectra. Unquestionably, the reliability of the method increases if the spectrum of the unknown and those of the standards are obtained under identical conditions.

Chemical-ionization mass spectrometry of certain compounds enables the location of double bonds. The recommended reagent gas for this purpose is nitric oxide, although other gases such as methyl vinyl ether and methylamine have also been used (Budzikiewicz 1988; Malosse & Einhorn 1990). However, some of these gases are corrosive and detrimental to mass spectrometers, therefore they are rarely used in routine analysis.

For conjugated dienes, isobutane CI spectra have been used to locate the conjugated system in aldehydes, alcohols, formates, acetates, and hydrocarbons (Einhorn et al. 1985; Doolittle et al. 1985). For this purpose $(CH_3)_3C$-Cl is considered even better than isobutane as the reagent gas for locating conjugated double bonds in long-chain compounds (Einhorn et al. 1987). Fabrias et al. (1989) have used this procedure in the identification of (11Z,13Z)-11,13-hexadecadienol.

Under electron-impact conditions the spectra from E- and Z-isomers of mononunsaturated long-chain compounds are virtually identical. However, chemical-ionization mass spectra offer some promise to distinguish between E- and Z-isomers. Vékey et al. (1988) differentiated E- and Z-isomers of monounsaturated acetates by isobutane CI-MS.

More generally, chemists resort to derivatization and degradation methods to locate double bonds. The double bond can be cleaved and the products identified.

Alternatively, the compound is converted into a derivative that yields characteristic fragment ions revealing the double-bond position when subjected to GC-MS analysis.

7.2.1.6.1 Ozonolysis of Carbon-Carbon Double Bonds

Ozonolysis is a simple technique, which has found wide applicability for the location of carbon-carbon double bonds (Davison & Dutton 1966; Nickell & Privett 1966). Development of ozonolysis as a reliable microchemical technique was the result of extensive investigations by Beroza & Bierl (1966, 1967). Ozone adds to the double bond to form a cyclic ozonide (**30**), which can be reduced with a reagent such as triphenylphosphine (Ph$_3$P), or thermally cleaved, to yield carbonyl compounds. Aldehydes are formed if olefinic hydrogen atoms are present at the double bond. Ketones are generated from disubstituted olefinic carbons. These carbonyl compounds are readily identified by GC-MS as they are, or after derivatization (section 7.2.4.3). However, to obtain reliable results, the unsaturated compound must be relatively pure. When a mixture is ozonolyzed, it is difficult to ascertain which carbonyl fragment originated from which unsaturated compound.

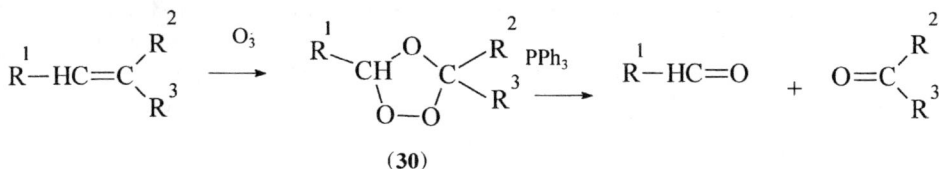

(30)

A relatively inexpensive microgenerator for ozone can be assembled from simple components (Beroza & Bierl 1969; Attygalle & Morgan 1988). Essentially, a slow stream of oxygen (1–2 ml/min) is passed through a high-voltage spark generated by a Tesla coil (e.g., hand-held vacuum tester). The formation of ozone can be monitored by the change of color of starch-iodide indicator paper dipped in dilute sulfuric acid.

Typically, the substance to be ozonized is dissolved in a solvent such as hexane, carbon disulfide, dichloromethane, carbon tetrachloride, or pentyl acetate. About 10–50 µl of this solution is placed in a cone-shaped glass vial cooled in an ice bath (to prevent rapid solvent evaporation), and a slow stream of ozone is bubbled via a fused silica capillary (0.2 mm i.d.) for ~ 20–30 s. Using a slow flow of ozone, reactions can be performed even with solution volumes as small as 5 µl. The reaction mixture can be analyzed directly by GC-MS. Customarily, a few microliters of a solution of Ph$_3$P in dichloromethane is added to reduce the ozonides to carbonyl compounds. Excess Ph$_3$P, and the product Ph$_3$P=O, can be removed by percolating through a short silica column, otherwise significant GC peaks are observed for these compounds later in the gas chromatogram. However, when only trace amounts of carbonyl compounds are expected, the crude mixture can be analyzed directly by gas chromatography.

Solvents often contain traces of unsaturated compounds. Therefore it is advisable to subject the solvent to ozonolysis and analyze the products by GC-MS before the compound under investigation is checked. Moreover, trace amounts of small carbonyl compounds produced may be obscured by the solvent. In fact, it is possible to conduct ozonolysis under solventless conditions. The compound is trapped in a glass capillary by micropreparative GC, and after passing ozone the capillary is sealed and crushed in the injection port of the gas chromatograph. The ozonides are cleaved thermally by the heat of the injection port and chromatographic peaks for the carbonyl compounds can be observed directly (Cronin & Gilbert 1973; Attygalle & Morgan 1983).

Traditionally, ozonolysis has been used extensively in natural product chemistry. However, due to the development of more efficient techniques, ozonolysis is rarely used these days with simple monoenes. For some applications, however, the technique is still useful. For example, ozonolysis was employed elegantly in the chirality determination of the ant trail pheromone, faranal (31) (Kobayashi et al. 1980). One of the ozonolysis products from faranal is a dimethyl dialdehyde. The gas chromatographic retention time on an achiral column of this dialdehyde was identical to that obtained from cis-1,2-dimethyl-4-cyclohexene (32). On the other hand, the retention time of the diastereomeric dimethyl dialdehyde derived from trans(±)-1,2-dimethyl-4-cyclohexene (33) was different from that from the natural product. In this way, it was demonstrated that the absolute configuration of faranal must be either 3R,4S, or 3S,4R.

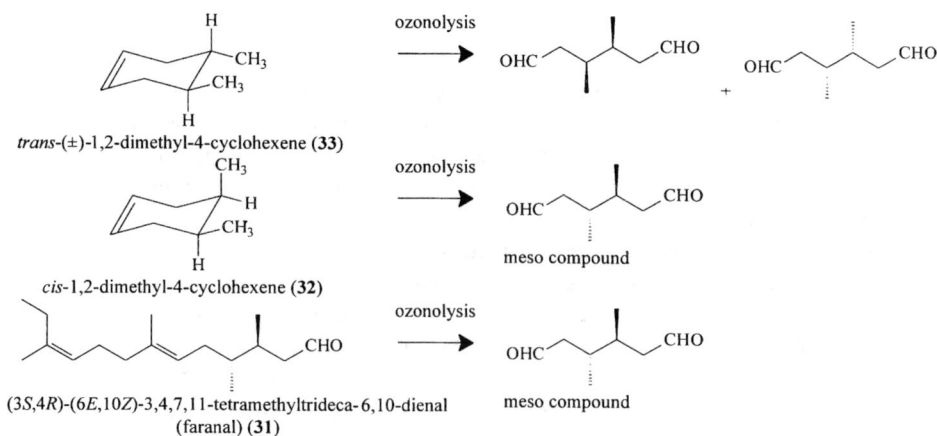

trans-(±)-1,2-dimethyl-4-cyclohexene (33)

cis-1,2-dimethyl-4-cyclohexene (32)

(3S,4R)-(6E,10Z)-3,4,7,11-tetramethyltrideca-6,10-dienal
(faranal) (31)

meso compound

Under controlled conditions polyunsaturated compounds can be subjected to incomplete ozonolysis (Attygalle et al. 1989). In fact, it is possible to ozonize a single double bond at a time of any polyunsaturated compound. However, the optimal conditions must be worked out with a known polyene before the ozonolysis of an uncharacterized compound is attempted. For example, (4*E*,6*E*,10*Z*)-4,6,10-hexadecatrienyl acetate (**34**), a pheromone component of the cocoa pod borer moth, was subjected to partial ozonolysis and all the six expected aldehydes were detected after derivatizing the mixture with *N,N*-dimethylhydrazine (section 7.2.4.3).

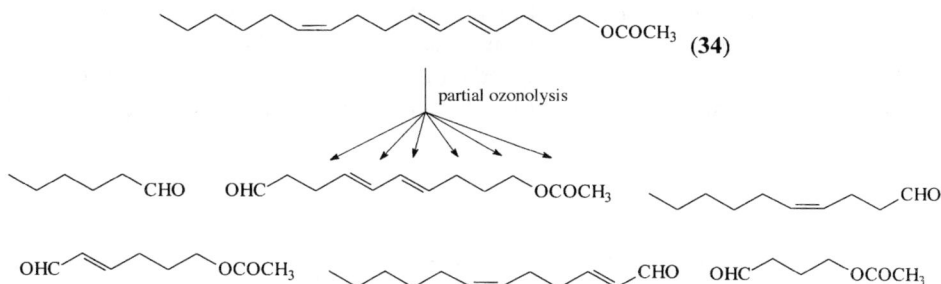

Acetylenic bonds are more resistant than olefinic bonds to ozonolysis. As a result, partial ozonolysis products can be obtained from compounds bearing both types of bonds. In the identification of (13*Z*)-13-hexadecen-11-ynyl acetate (**35**), a pheromone of the pine processionary moth, a product of molecular weight 252 (**36**) was obtained by microozonolysis (Guerrero et al. 1981).

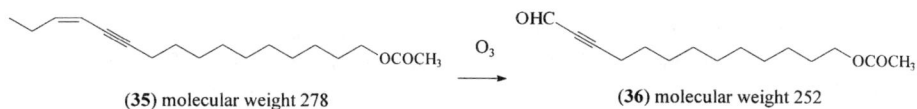

(35) molecular weight 278   (36) molecular weight 252

7.2.1.6.2. Other Oxidative Cleavage Reactions of Carbon-Carbon Double Bonds

Traditionally, oxidative cleavage of double bonds to yield carboxylic acids has been used often in natural product structure elucidation. Nowadays, these methods are less popular since efficient nondestructive methods that require far less material are available. However, occasionally oxidative cleavage becomes useful for polyunsaturated compounds. To obtain reliable results at least a few micrograms of pure material must be isolated. The oxidative cleavage of a polyunsaturated compound produces a mixture of mono- and dicarboxylic acids, which can be converted to methyl esters (section 7.2.5.2) and characterized by GC-MS. Although several reagents, such as periodate and hypochlorite, are available for this oxidation, ruthenium tetroxide (RuO_4) appears to be one of the rapid and more efficient oxidizing agents (Carlsen et al. 1981). Apparently, the solvent system $CCl_4/H_2O/CH_3CN$ (2:2:3) is very important for efficient execution of this oxidation. One limitation is the possibility of side reactions, such as oxidation of certain functional groups (Carlsen et al. 1981).

Typically, the material to be oxidized is dissolved in CCl_4/CH_3CN (1:1, 50 μl) and mixed with a saturated aqueous solution of sodium metaperiodate $(NaIO_4)(100$ μl) containing RuO_2 (~10 μg) (note that ruthenium dioxide hydrate can be substituted by ruthenium trichloride hydrate). The mixture is shaken or sonicated for 1.5 h at room temperature, and extracted with dichloromethane. The organic layer is separated and filtered through a short column of Florisil®. The resulting extract containing carboxylic acids can be derivatized and analyzed by GC-MS.

For example, Murata et al. (1991) conducted oxidative cleavage for the character-ization of the ketodiene, (6Z,26Z)-6,26-pentatriacontadien-2-one (**37**) an integu-mental lipid of the Guam brown tree snake, and obtained eicosadioic acid (**38**) and nonanoic acid (**39**) among the products.

A modification of the oxidative cleavage procedure is called the Lemieux reaction. This reaction, involving periodate and catalytic amounts of potassium permanganate, cleaves double bonds yielding ketones, aldehydes or, more usually, carboxylic acids formed by further oxidation of the aldehydes (Lemieux & von Rudloff 1955). Braconnier et al. (1985) used this reaction for the determination of the position of the double bond in (Z)-1,17-diamino-9-octadecene, a defensive alkaloid of ladybird beetles (Braconnier et al. 1985). The diacetyl derivative of this compound was cleaved, and the products were converted to methyl esters and identified as methyl 8-acetylaminononanoate and methyl 9-acetylaminonon-anoate. Similarly, Cimino et al. (1991) used this reaction to clarify the absolute stereochemistry of haminol-A (**40**), an alarm pheromone of a Mediterranean opisthobranch. The natural alcohol was first derivatized with (R)-(−)-1-(1-naph-thyl)ethyl isocyanate (**41**) and the product (**42**) was oxidized by the Lemieux procedure. The acid obtained (**43**) was methylated, and an NMR study of the product (**44**) established the absolute configuration.

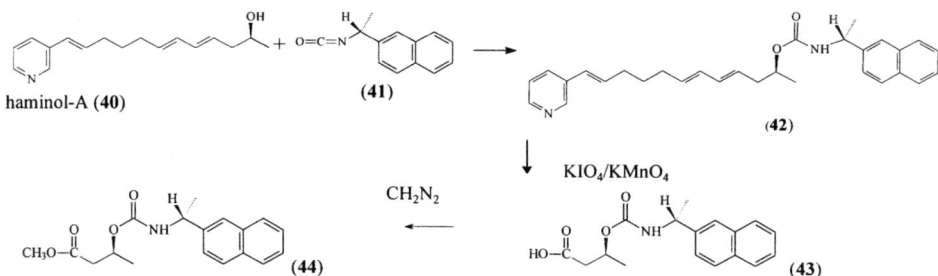

7.2.1.6.3. Methylthiolation of Carbon-Carbon Double Bonds

Undoubtedly, this is the most efficient and recommended method for the location of double bonds in carbon chains, particularly for monounsaturated compounds. The iodine-catalyzed addition of dimethyl disulfide (DMDS) followed by GC-MS analysis of the adducts was first described by Francis and Veland (1981) for monounsaturated hydrocarbons and fatty-acid methyl esters (Francis 1981). The method was later improved for nanogram-scale derivatizations and applied to pheromones (Buser et al. 1983; Attygalle & Morgan 1986). The method has been successfully employed for many natural products including alkenes (Billen et al. 1986; Lanne et al. 1988; Scribe et al. 1990), methyl alkenoates (Dunkelblum et al. 1985; Scribe et al. 1988; Yamamoto et al. 1991b), and alkenals (Leonhardt & DeVilbiss 1985; Ho et al. 1996). The method has been further extended to locate the double bond in a methyl ester of an unsaturated ketoacid (Schulz et al. 1988). Ayasse et al. (1990) have shown the feasibility of applying the DMDS method even for unsaturated macrocyclic lactones. Mason et al. (1990) have used the procedure in the structure elucidation of long-chain monounsaturated ketones as large as (Z)-26-pentatricontenone.

The DMDS derivatives show excellent gas chromatographic properties. For example, the isomers (Z)-6-pentadecene and (Z)-7-pentadecene, which were difficult to resolve by gas chromatography, were well separated once the corresponding DMDS derivatives were formed (Billen et al. 1986).

The addition of DMDS to double bonds proceeds smoothly at ambient temperatures with iodine or boron trifluoride catalysis. The reaction is slow but yields are nearly quantitative if sufficient time is allowed (24 h). For monounsaturated compounds, the reaction is stereospecific. (Z)- and (E)-Isomers yield *threo-* and *erythro*-isomers, respectively (**45, 46**), as a result of *trans* addition (Caserio et al. 1985).

H⸱⸱⸱ ⸝C=C⸝ ⸱⸱⸱H
R¹   R²
cis-alkene

DMDS/I₂
⟶

CH₃S
H⸱⸱⸱ | ⸱⸱⸱H
⸝C—C⸝
R¹ SCH₃ R²

threo-adduct (45)

H⸱⸱⸱ ⸝C=C⸝ ⸱⸱⸱R²
R¹   H
trans-alkene

DMDS/I₂
⟶

CH₃S R²
H⸱⸱⸱ | ⸱⸱⸱
⸝C—C⸝
R¹ SCH₃ H

erythro-adduct (46)

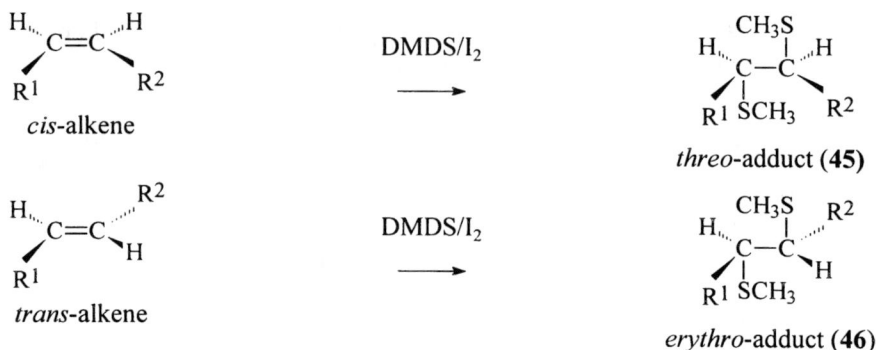

The *erythro*- and *threo*-isomers usually show nearly identical EI-mass spectra, however, they are well resolved by capillary gas chromatography. The *threo*-isomers usually elute first. Therefore the geometry of the unknown can be determined by comparing the retention times with those obtained from the derivatives of standards. In the mass spectrometer, the DMDS adducts cleave preferentially across the carbon-carbon bond between the two carbon atoms bearing the CH_3S substituents. For example, the spectrum of the DMDS adduct of (Z)-9-tricosene shows two intense ions at m/z 173 and 243 representing $CH_3(CH_2)_7CH=S^+CH_3$ and $CH_3(CH_2)_{12}CH=S^+CH_3$ fragments, respectively (Fig. 7.4A).

$$CH_3—(CH_2)_m—\overset{\overset{CH_3S}{|}}{HC}—\underset{\underset{SCH_3}{|}}{CH}—(CH_2)_n—R \quad \xrightarrow{EI\text{-}MS} \quad CH_3—(CH_2)_m—\overset{\overset{CH_3S^+}{||}}{HC} \;+\; \underset{\underset{^+SCH_3}{||}}{CH}—(CH_2)_n—R$$

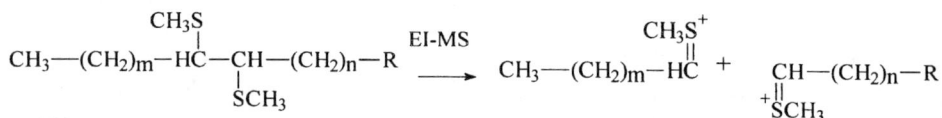

Similarly, the adduct of (Z)-11-octadecenyl acetate gave a molecular ion at m/z 404 and fragments at m/z 259 and 145 (Fig. 7.4B). The peak at m/z 199 occurs from loss of acetic acid from the m/z 259 fragment, a typical fragmentation pattern shown by adducts of acetates. In addition, losses of 47 mass units (CH_3S-) from the molecular ion, and 48 mass units (CH_3SH) from other ions are also observed.

EI mass spectra of DMDS adducts are very informative. Spectra of derivatives from compounds such as long-chain alcohols and acetates (Fig. 7.4) which do not show molecular ions under EI conditions, show significant molecular ions after derivatization. The DMDS procedure is suitable for samples as small as 1–50 ng, or even with subnanogram quantities by searching the acquired GC-MS data for the expected molecular and fragment ions (there are several names for this procedure; mass chromatography, ion chromatography, mass fragmentog-

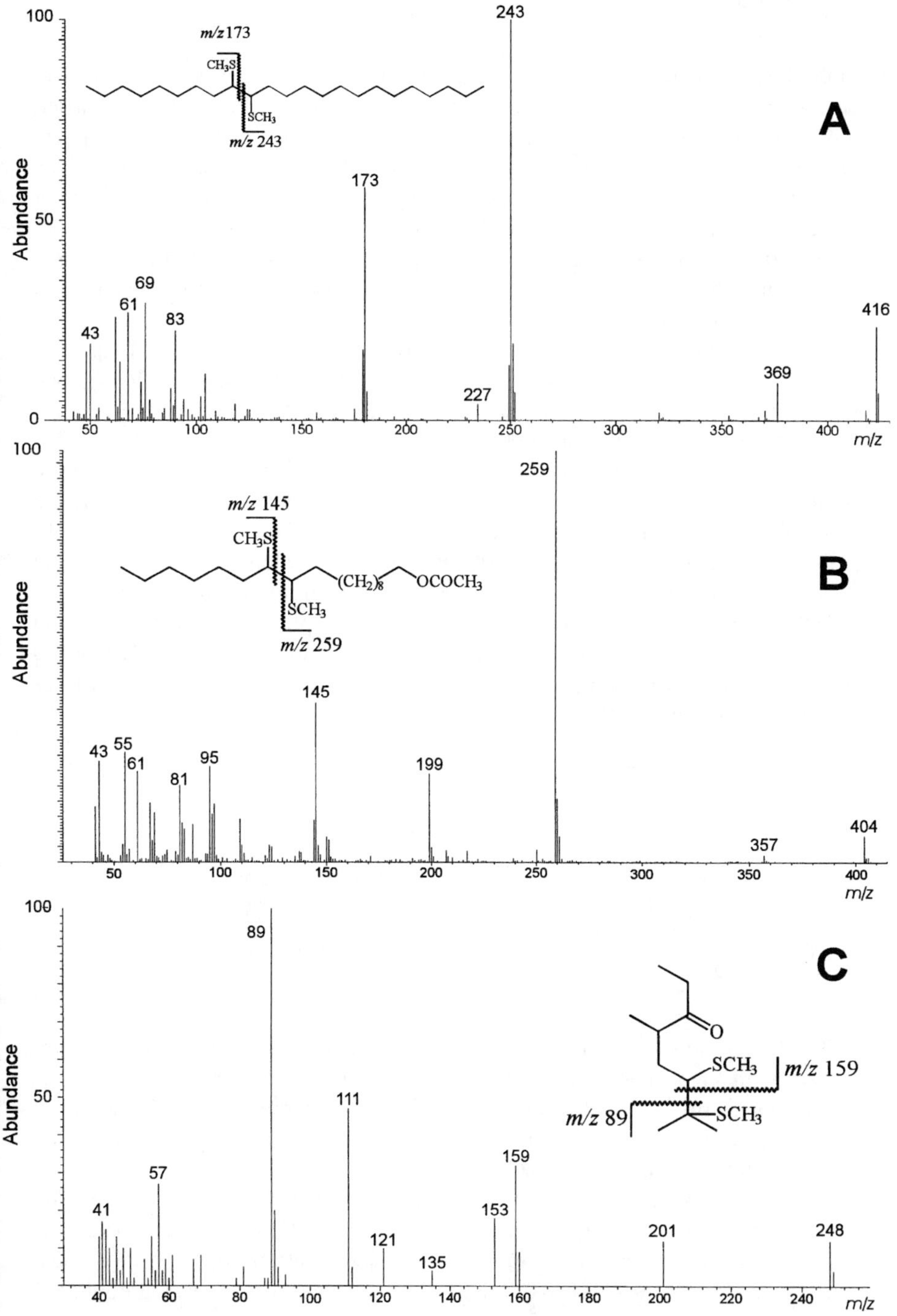

Figure 7.4. Electron-impact mass spectra (70 eV) of DMDS adducts of (Z)-9-tricosene (**A**), (Z)-11-octadecenyl acetate (**B**), and 4,7-dimethyl-6-octen-3-one (**C**).

raphy). The reaction rate is faster in carbon disulfide than in solvents such as hexane or heptane. If higher temperatures are used (80°C) the reaction is faster, however, unwanted side products containing only one methylthio group are known to appear. Typically, a few nanograms of the compound to be derivatized is dissolved in a solvent such as hexane (10 μl), and a solution of iodine in ether (5%, 10 μl) and dimethyl disulfide (1 μl) are added. The mixture is kept at room temperature for 8–10 h and the iodine is decolorized with a minimum volume of aqueous sodium thiosulfate (5%). The adduct is extracted into hexane and examined by GC-MS.

With aldehydes and ketones the location of the double bond is not straightforward since the nominal mass of a carbonyl group is the same as that of two methylene groups. For example, mass spectra of the DMDS adducts of 5-dodecenal (**47**) and 7-dodecenal (**48**) both show two intense ions at m/z 131 and 159. Therefore it is not possible to recognize which fragment bears the aldehyde group by low-resolution EI-mass spectrometry. Of course, if synthetic standards are available the GC retention times can be compared. On the other hand, high-resolution GC-MS provides an easy solution to this problem. If a high-resolution mass spectrometer is not available, the carbonyl group can be derivatized or reduced before the DMDS derivatization. For example, in the characterization of (Z)-26-pentatricontenone, after adding DMDS Mason et al. (1990) derivatized the keto group with O-methylhydroxylamine. The final derivative gave a mass spectrum with high intensity diagnostic ions.

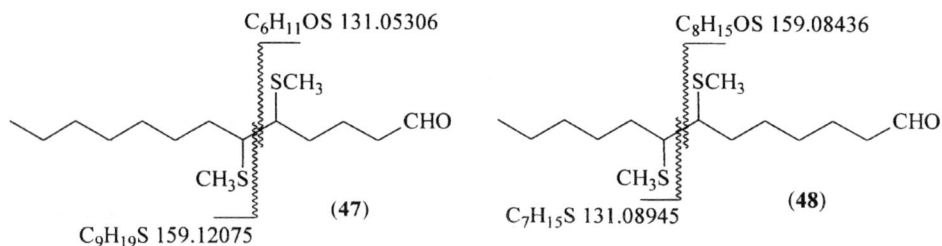

The DMDS procedure has been applied to terpenoid and other branched compounds with a reasonable degree of success (Attygalle et al. 1993c). For example the mass spectrum of the derivative (**49**) from 4,7-dimethyl-6-octen-3-one shows two peaks at m/z 89 and 159 indicating the presence of an isopropylidene moiety in the parent molecule (Fig. 7.4C).

Carbon-carbon bonds conjugated with a carboxymethyl or carbonyl group do not give the desired adducts (Leonhardt & DeVilbiss 1985). However, sometimes this feature can be convenient. For example, neral (**50**) and geranial (**51**) yield only monoadducts, the spectra of which indicate the presence of an isopropylidene moiety in the parent compounds (Attygalle et al. 1993c).

(50)   DMDS/I₂ →   *m/z* 89 (100%)

(51)   DMDS/I₂ →   *m/z* 89 (100%)

Polyunsaturated Compounds

Although not as straightforward as for monounsaturated compounds, the DMDS method has been extended to straight-chain polyunsaturated compounds. However, there are some limitations (Vincenti et al. 1987; Takano et al. 1989). For the purpose of easy understanding of mass spectra of polyene adducts, it is convenient to group polyenes into three groups.

(1) Polyenes with Double Bonds Separated by at Least Four Methylene Groups

When the double bonds are separated by at least four methylene groups, the method is readily applicable to polyenes since the fragmentation pathways of the adducts are similar to those described for monoene adducts. For example, the DMDS diadduct (molecular weight 704) (**52**) obtained from (9Z,19Z)-9,19-heptatriacontadiene gives fragments at *m/z* 173 (100%) and 299 (48%), which indicate clearly the double bonds at positions 9 and 19 of the parent diene. Although peaks expected for *m/z* 531 and 405 fragments ions bearing three CH₃S-groups are not observed, significant signals are observed at *m/z* 483 (5%) and 437 (40%), and at *m/z* 357 (18%) and 311 (47%), which originate from a loss of 48 (elements of methanthiol) and 94 (two methanthiol units) mass units from the *m/z* 531 and 405 fragments, respectively (Carlson et al. 1989). Similarly, 2,13-octadecadienyl acetate was identified from the mass spectrum of the *bis*-DMDS derivative (Tonini et al. 1986). The double bond at the 2 position was much more resistant to derivatization than that at the 13 position; even after 3 days of reaction, the mixture contained 70% of the monoadduct. The method has been extended to alkatrienes such as 1,7,13-pentacosatriene; however, as the unsaturation increases the mass spectra of the adducts become more complex.

m/z 173(100%)   m/z 299(48%)

CH₃S   CH₃S

C₈H₁₇   C₁₇H₃₅

SCH₃   SCH₃

m/z 531(0%)   m/z 405(0%)   (52)

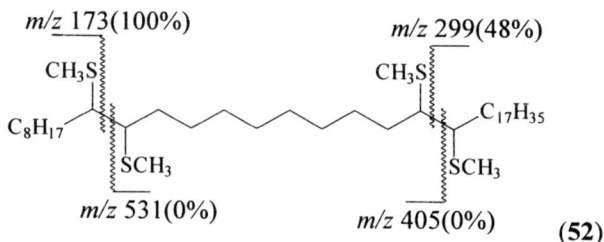

Murata et al. (1991) applied the method to (6Z,26Z)-pentriacontadien-2-one and other similar ketodienes. Although the positions of the double bonds could not be assessed unambiguously by this method alone, because they are carbonyl compounds, a reduction of the carbonyl group, or high-resolution GC-MS, would have provided definitive evidence. Christie et al. (1989) employed the method to locate the double bonds in methyl 13-cyclopent-2-enyl-tridec-4-enoate.

(2) Compounds with Double Bonds Separated by Zero to Three Methylene Groups

When two CH=CH type carbon-carbon double bonds are not separated by a methylene group, or they are separated only by one, two, or three methylene groups, the DMDS addition takes place via a different reaction pathway compared to that observed for isolated monoenes for which a single stereospecific product is always obtained. Evidently, both double bonds participate in the reaction, and depending on the number of intermediate methylene groups four-, five-, or six-membered cyclic thioethers are formed, and due to the poor stereoselectivity of the reaction a complex mixture of diastereomers results (Vincenti et al. 1987). For example, derivatization of (7Z,11Z)-7,11-hexadecadienyl acetate gave eight products representing the eight expected diastereoisomers. The mass spectra of these stereoisomers are, however, very similar and sufficiently informative to locate the two double bonds for most examples. In addition to the desired adducts, polyunsaturated compounds sometimes produce small amounts of other cyclic sulfur compounds as well.

(2.1) Conjugated Dienes

When the two CH=CH type bonds are conjugated, the DMDS derivatization reaction forms a tetrasubstituted tetrahydrothiophene (**53**) bearing methylthio groups at the 3 and 4 positions of the ring.

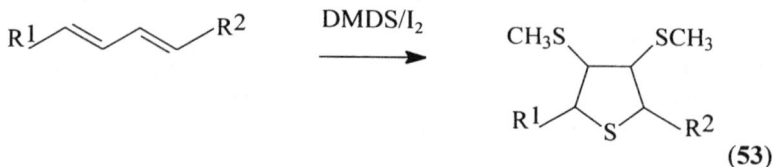

R^1 ⟍⟋⟍⟋ R^2 $\xrightarrow{\text{DMDS/I}_2}$ CH₃S ⟍ ⟋ SCH₃

R^1 ⟍ S ⟍ R^2

(**53**)

Unfortunately, the EI-mass spectrum obtained from the DMDS adduct of conjugated diene acetates may not show diagnostic fragments that would assist in locating the double bonds of the parent compound (Vincenti et al. 1987). However, general mass spectral characteristics at least indicate that the diene is conjugated. For example, the spectrum of the adduct of (3*E*,5*Z*)-3,5-tetradecadienyl acetate (**54**) is dominated by a peak at *m/z* 283 (loss of CH_3S- and CH_3SH from the molecular ion), and a peak at *m/z* 223 (base peak resulting from the loss of acetic acid from *m/z* 283 ion).

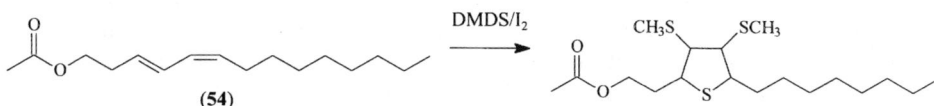

In contrast to the spectra of the acetate adducts, those obtained from adducts of conjugated alkadienes exhibit significant fragment ions, which are useful for the location of the diene system. Takano et al. (1989) demonstrated this fact by examining spectra of six positional isomers of conjugated pentadecadienes.

(2.2) Methylene-Interrupted Dienes

Dienes with the two double bonds separated by a single methylene group react with DMDS to give four-membered cyclic thioethers (thietanes (**55**). The mass spectra of these derivatives show peaks diagnostically useful for the location of the double bond system.

For example, Ho et al. (1996) used this procedure to locate the double bonds in (11*E*,14*E*)-11,14-octadecadienal, a sex pheromone component of the female tea cluster caterpillar. The mass spectrum of the DMDS adduct showed three key fragments at *m/z* 215 (26–35%), 175 (20%), and 103 (19–38%). As described above, since this compound is an aldehyde it is not possible to assign the position of the double bonds on these data alone. Therefore, Ho et al. (1996) reduced the aldehyde to the corresponding alcohol (**56**), repeated the DMDS procedure, and located the diene system successfully.

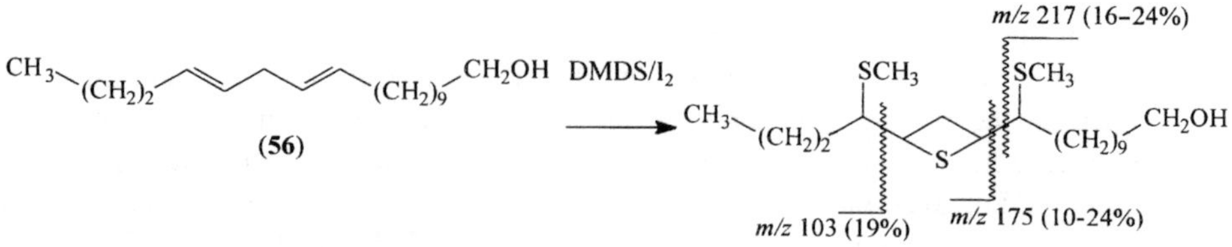

m/z 217 (16–24%)

CH$_3$~(CH$_2$)$_2$ ⟍⟍⟍⟍ (CH$_2$)$_9$⟍CH$_2$OH DMDS/I$_2$ ⟶ CH$_3$~(CH$_2$)$_2$ SCH$_3$ S SCH$_3$ (CH$_2$)$_9$⟍CH$_2$OH

(56)

m/z 103 (19%)   m/z 175 (10–24%)

Unfortunately, the DMDS reaction with dienes interrupted by a single methyl-ene group does not yield thietanes as the sole product. Under different DMDS and iodine concentrations several products can be expected. In fact, under controlled conditions (1 μmol of diene with 0.3 ml of DMDS and 4 mg of iodine at 35°C for 30 min), Yamamoto et al. (1991b) were able to avoid the cyclization and obtain a mixture of methyl 12,13-*bis*(methylthio)octadec-9-enoate (**56A**) and methyl 9,10-*bis*(methylthio)octadec-12-enoate (**56B**) from methyl linoleate (**56C**). The mass spectra of these adducts are easier to interpret than those of thietanes.

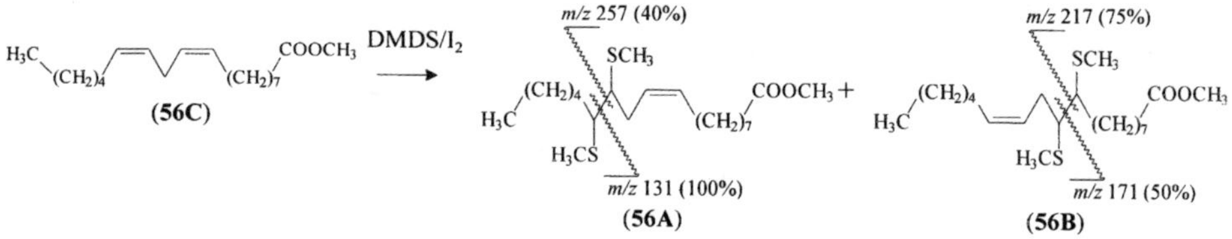

m/z 257 (40%)   m/z 217 (75%)

H$_3$C~(CH$_2$)$_4$ ⟍⟍⟍ (CH$_2$)$_7$⟍COOCH$_3$ DMDS/I$_2$ ⟶   (CH$_2$)$_4$ SCH$_3$ COOCH$_3$ + (CH$_2$)$_4$ SCH$_3$ COOCH$_3$
H$_3$C (CH$_2$)$_7$ H$_3$C (CH$_2$)$_7$
H$_3$CS   H$_3$CS

(56C)   m/z 131 (100%)   m/z 171 (50%)

(56A)   (56B)

In fact, when methyl linoleate is allowed to react with DMDS/I$_2$ for 24 h, in addition to thietanes, other cyclic products are also observed (Carballeira et al. 1994).

(2.3) Dienes Interrupted by Two Methylene Groups

Dienes in which the two double bonds are interrupted by two methylene groups give five-membered cyclic thioethers (**57**).

R^1⟍ ⟍⟍⟍⟍⟍⟍ R^2 DMDS/I$_2$ ⟶   R^1 ⟍⟋S⟍⟋ R^2
                                        CH$_3$S   SCH$_3$

(57)

For example, the spectrum of the derivative from (5Z,9Z)-5,9-octadecadienyl acetate (**58**) shows diagnostic ions at m/z 259 and 261 for the fragments arising from the loss of each side chain of the thiolane ring. In addition, an ion at m/z 173 for the CH$_3$-(CH$_2$)$_7$-CH=S$^+$CH$_3$ is also very informative. These fragments, together with those observed at m/z 434 (5%, M$^+$), 213 (61%, 261-CH$_3$SH),

211 (100%, 259-CH$_3$SH) and 153 (60%, 261-CH$_3$SH-CH$_3$COOH), conclusively establish the position of the two double bonds in the carbon chain (Vincenti et al. 1987).

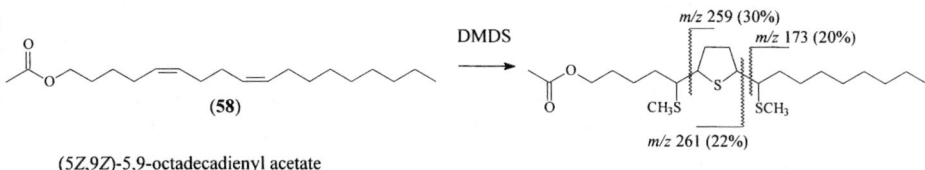

(58)

(5Z,9Z)-5,9-octadecadienyl acetate

(2.4) Dienes Interrupted by Three Methylene Groups

Dienes in which the two double bonds are interrupted by three methylene groups give six-membered cyclic thioethers **(59)**. The mass spectra of these derivatives provide definitive data for the location of the double bond system.

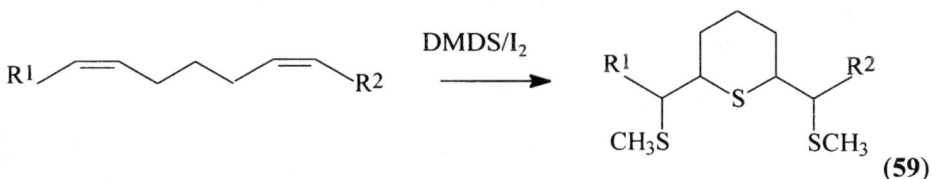

(59)

The spectrum of the derivative obtained from (3Z,8Z)-3,8-octadecadienyl acetate **(60)** shows diagnostic ions at m/z 287 and 247 for the two fragments arising from the loss of each side chain of the tetrahydrothiopyran ring. In addition, a signal is observed at m/z 187 for the CH$_3$-(CH$_2$)$_8$-CH=S$^+$CH$_3$ ion arising from the alkyl-end of the molecule. Further evidence for the conclusive proof is provided by the peaks observed at m/z 434 (6% M$^{•+}$), 239 (43%, 287-CH$_3$SH), 199 (20%, 247-CH$_3$SH) and 139 (66%, 247-CH$_3$SH-CH$_3$COOH) (Vincenti et al. 1987).

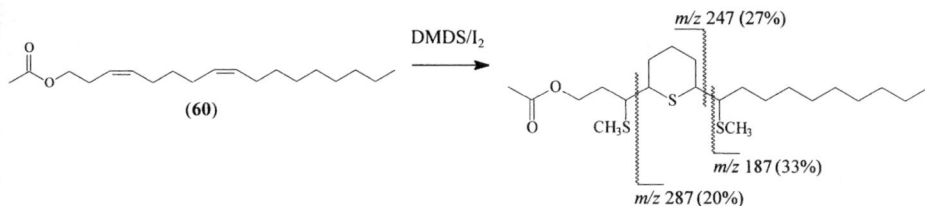

(60)

(2.5) Triply Unsaturated Compounds

The mass spectra of DMDS derivatives from triply unsaturated compounds bearing double bonds separated by less than three methylene groups are even more complicated. Although useful information can be derived, considerable expertise is required to interpret such data. For example, the three isomers (3E,8Z,11Z)-3,8,11-tetradecatrienyl acetate (**61**), (3E,8Z,12Z)-3,8,12-tetradecatrienyl acetate (**62**), and (3E,8Z)-3,8,13-tetradecatrienyl acetate (**63**) can be differentiated on the basis of the spectra of their DMDS derivatives (Griepink et al. 1996). However, most fragments are common to all three spectra, and the few peaks which are unique to each spectrum are of very low intensity (<2%). The signal at m/z 75 (which can be considered as characteristic for the ω-2,6 system) is the base peak of the spectrum of the derivative of (3E,8Z,12Z)-3,8,12-tetradecatrienyl acetate. Although not as intense, this fragment is present in the spectra of the other two derivatives as well. Similarly, a base peak can be expected at m/z 61 in the spectra of derivatives of ω-1,6 compounds. However, the m/z 61 signal is present in virtually any DMDS spectrum, sometimes even as the base peak (Vincenti et al. 1987). Although the DMDS method may not determine positions of the double bonds separated by less than three methylene units in a polyene convincingly, the spectra of derivatives provide a better fingerprint to compare an unknown compound with authentic material than a comparison of the spectra of underivatized compounds.

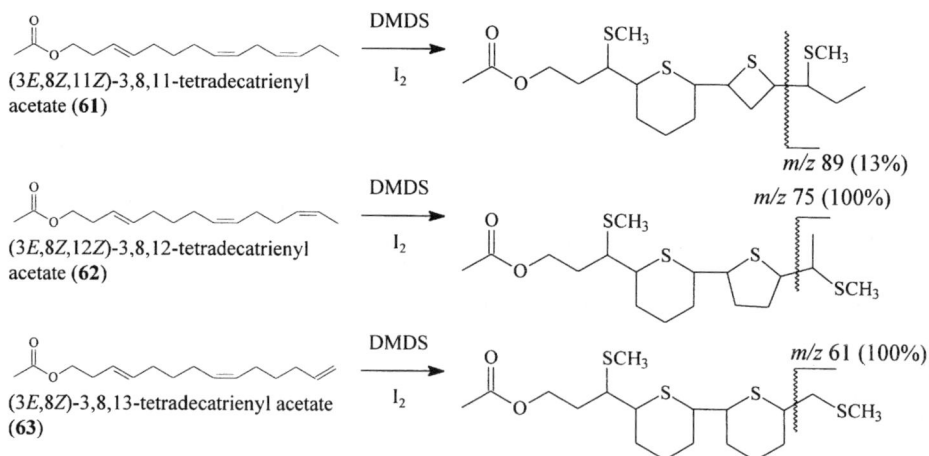

7.2.1.6.4. Methoxymercuration-Demercuration of Carbon-Carbon
            Double Bonds

A methoxymercuration-demercuration procedure, first described by Abley et al. (1970), was one of the most popular methods for double-bond location before the DMDS method took precedence. However, for certain special applications the formation of these methoxy derivatives is still useful.

$$R^1—HC{=}CH—R^2 \quad \xrightarrow{Hg(OAc)_2} \quad$$

$$\begin{array}{c} HgOAc \\ | \\ R^1—HC\overset{+}{-}CH—R^2 \end{array}$$

$$\Big| \ CH_3OH$$

Left branch:

$$\begin{array}{c} CH_3O \\ | \\ R^1—HC—CH—R^2 \\ | \\ HgOAc \end{array}$$

$$\Big\downarrow \ NaBH_4$$

$$\begin{array}{c} R^1—HC—CH_2—R^2 \\ | \\ CH_3O \end{array}$$

Right branch:

$$\begin{array}{c} OCH_3 \\ | \\ R^1—HC—CH—R^2 \\ | \\ AcOHg \end{array}$$

$$\Big\downarrow \ NaBH_4$$

$$\begin{array}{c} OCH_3 \\ | \\ R^1—H_2C—CH—R^2 \end{array}$$

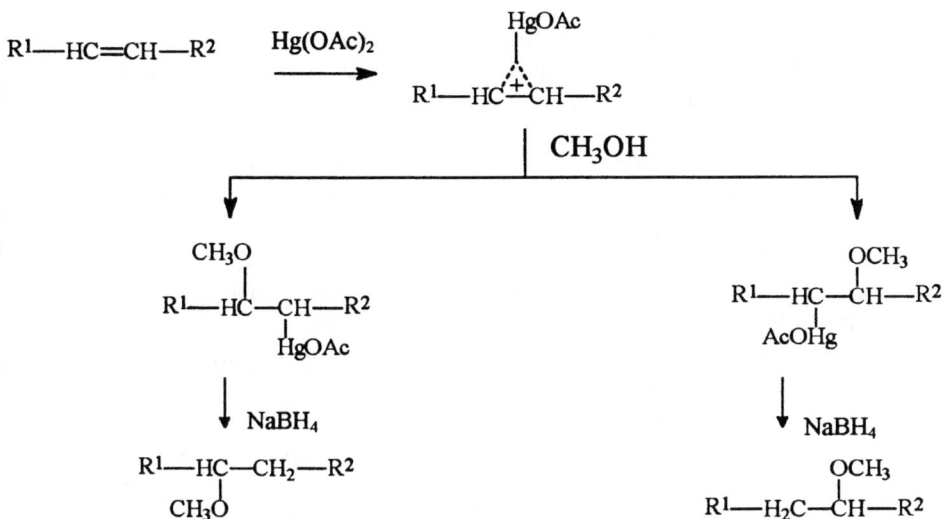

For example, this procedure is particularly useful for locating terminal double bonds. For terminal double bonds, the addition of mercuric acetate follows a regioselective path to yield the expected Markovnikov type adduct (i.e., the methoxy group becomes attached to the more substituted alkene carbon). Therefore only one adduct is formed, and the methoxy group is attached to the penultimate carbon atom of the chain. The mass spectra of these adducts show a very intense ion at m/z 59, which is often the base peak of the spectrum, indicating that the parent compound contained a terminal double bond.

However, when the double bond is in the middle of a carbon chain the reaction, as expected, is not selective. The reaction usually produces two coeluting regioisomers for each double bond, and as a result a composite mass spectrum is obtained. Under electron-impact conditions, the derivatives fragment on either side of the methoxy group to yield four intense fragment ions. From the masses of these ions, the original double-bond position can be deduced (Baker et al. 1982; Blomquist et al. 1980; Vostrowsky & Michaelis 1981; Vostrowsky et al. 1981).

The technique is not directly applicable to unsaturated amines; the free amino group must be acetylated (section 7.2.7.2) before addition of mercuric acetate (Jones et al. 1982). For example the mass spectrum of the derivative obtained from 2-(5-hexenyl)-5-(8-nonenyl)pyrrolidine (**64**), a defensive alkaloid from *Monomorium* ants, has a base peak at m/z 59 indicating the terminal unsaturation.

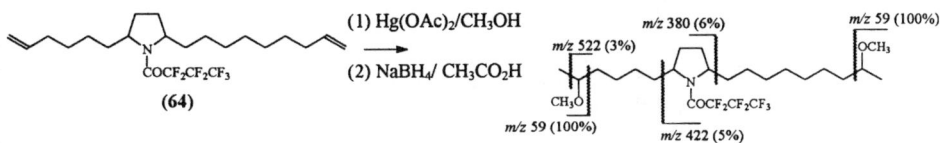

(1) Hg(OAc)$_2$/CH$_3$OH
(2) NaBH$_4$/ CH$_3$CO$_2$H

m/z 522 (3%); m/z 380 (6%); m/z 59 (100%)
m/z 59 (100%); m/z 422 (5%)

(**64**) COCF$_2$CF$_3$

A number of 2,6-dialkenylpiperidines with terminal double bonds have been identified from ants, and the trifluoroacetyl derivatives have been subjected to methoxymercuration-demercuration to assign the positions of the double bonds (Jones et al. 1990).

The mass spectra of derivatives of dienes are more informative and easier to interpret than those from the DMDS technique. For example, the spectrum obtained from the mixed dimethoxy derivatives of (6Z,9Z)-6,9-heptacosadiene showed significant fragment ions indicative of the exact location of the diene system (Howard et al. 1978; Blomquist et al. 1980). Under controlled conditions, it is possible to obtain monomethoxy derivatives from dienes (Nelson et al. 1984), therefore, the method is generally useful for polyunsaturated compounds (Vostrowsky et al. 1981).

Typically, the compound to be derivatized is dissolved in methanol (100 µl) and mixed with finely powdered anhydrous mercuric acetate (50 µg). The mixture is kept overnight in the dark with occasional shaking. The reaction is completed by adding finely powdered sodium borohydride until no more reaction is visible. The mixture is acidified with a drop of glacial acetic acid, and then partitioned between ether and water.

7.2.1.6.5. Epoxidation of Carbon-Carbon Double Bonds

There are several reasons why the formation of epoxide derivatives from compounds bearing CH=CH type double bonds becomes important to a microanalytical chemist. Although a large number of *cis/trans* isomers of the CH=CH system can be resolved by gas chromatography on polar stationary phases such as SP 2340 (a cyanopropylsilicone) (Heath et al. 1980), the separation of isomers of nonpolar compounds such as unfunctionalized alkenes is sometimes difficult. In contrast, the epoxide derivatives of *cis/trans* isomers are easy to resolve even on nonpolar GC phases. Usually, *cis* epoxides show longer retention times than the *trans* isomers (McDonough & George 1970).

For example, the *cis* and *trans* isomers of 9-tetradecenyl acetate were not resolved on a nonpolar stationary phase such as DB-5 (Fig. 7.5A), whereas the epoxides obtained from this mixture were easily resolved (Fig. 7.5B). In addition, the mass spectra of epoxides indicate the position of the former double bond (Aplin & Coles 1967). For example, the EI mass spectrum of the epoxide of 9-tetradecenyl acetate (**65**), although a molecular ion is absent, shows a diagnostic ion at m/z 213 (Fig. 7.5C), which allows the location of the double bond. The signal at m/z 153 results from a loss of elements of acetic acid from the m/z 213 ion.

m/z **213**

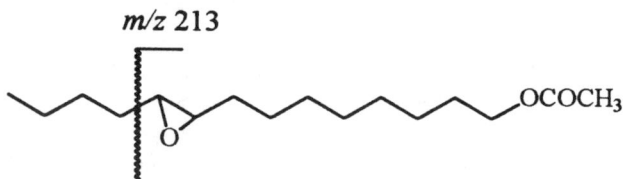

(**65**)

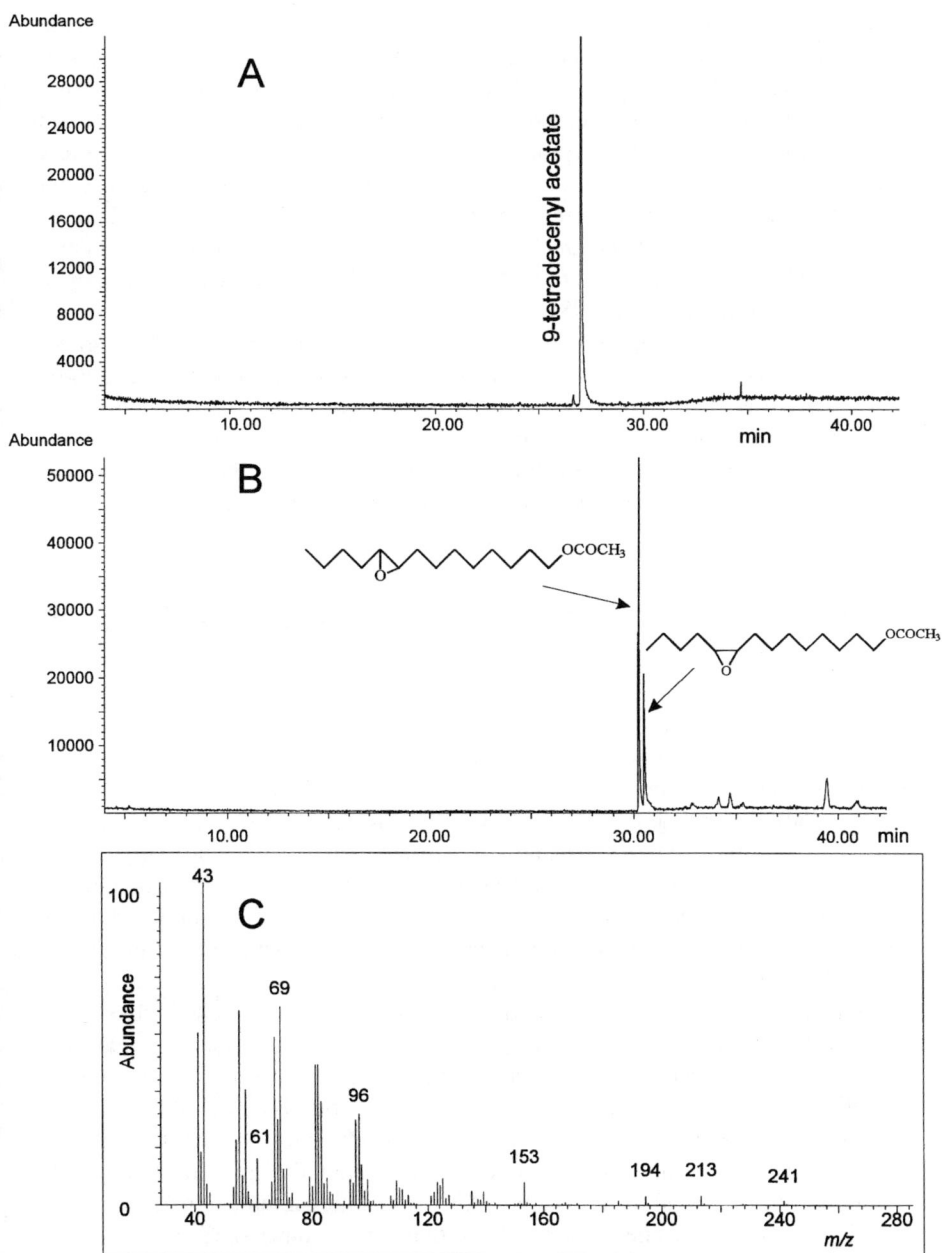

Figure 7.5. Gas chromatograms obtained from a 1:3 mixture of *cis* and *trans* 9-tetra-decenyl acetate before (**A**) and after epoxidation of the mixture (**B**). EI-mass spectrum (70 eV) of the epoxide of (*E*)-9-tetradecenyl acetate (**C**). (GC conditions as in Fig. 7.2).

The double bond positions of (Z)-5-tetradecenyl acetate, a pheromone component of *Scotia exclamationis* (Bestmann et al. 1980), and (Z)-1,17-diamino-9-octadecene, a defensive alkaloid of ladybird beetles (Braconnier et al. 1985), were determined from the mass spectra of the respective epoxide derivatives. Although the EI mass spectra of simple epoxides exhibit sufficient diagnostic ions to allow the location of the oxirane ring in the carbon chain, the diagnostic ions are not very intense. More useful information on more complicated epoxides is obtained by isobutane positive-ion chemical-ionization mass spectrometry (CI-MS) (Tumlinson et al. 1974; Tumlinson & Heath 1976), or CH_4/NO negative-ion chemical ionization (Bouchoux et al. 1987).

Alkenes and other unsaturated compounds are readily epoxidized with *m*-chloroperbenzoic acid (Schwartz & Blumbergs 1964; McDonough & George 1970; Bierl-Leonhardt et al. 1980b). Although any peracid can be used, *m*-chloroperbenzoic acid (*m*-CPBA, **66**) is the most commonly used reagent for this derivatization. The mass spectra of the epoxides may exhibit fragments from cleavages on either side of the epoxide (Bierl-Leonhardt et al. 1980b).

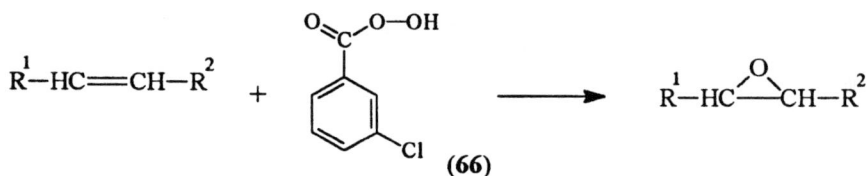

$$\overset{1}{R}-HC=CH-\overset{2}{R} \quad + \quad \text{(66)} \quad \longrightarrow \quad \overset{1}{R}-HC\overset{O}{\diagdown}CH-\overset{2}{R}$$

Typically, the compound to be derivatized is dissolved in a few microliters of dichloromethane and an equal volume of *m*-CPBA in dichloromethane is added (~ 2 mg/ml). The mixture is allowed to stand at room temperature for about 1 h, and aliquots of the mixture can be analyzed directly by GC-MS when only minute amounts of olefins are being derivatized. However, it is preferable to remove excess *m*-CPBA by washing with a few microliters of dilute sodium carbonate. With a little experience, this reaction can be performed successfully on samples as small as 10 ng (Attygalle & Morgan 1988).

In addition to monoenes, epoxidation has been applied to locate double bonds in polyunsaturated compounds (Hogge et al. 1985). The analyte is partially epoxidized to yield a mixture containing primarily monoepoxides. The remaining double bonds are hydrogenated and the resulting mixture is analyzed by GC-MS.

Compounds bearing CH=CH type double bonds and a terminal functional group, such as an acetate or a hydroxy, are not known to give interfering side reactions when allowed to react with *m*-CPBA at ambient temperatures and mild conditions. However, one must be aware that several other oxidation reactions may take place, particularly with branched compounds, or those containing carbonyl groups or activated aromatic rings. If the reaction mixture is heated for a prolonged period of time, even compounds bearing no double bonds react with *m*-CPBA. In branched compounds, hydroxyl groups are preferentially introduced into tertiary positions. In some cases hydroxyl groups are introduced into second-ary positions also (Asakawa et al. 1986). The yields of these hydroxylation reactions are usually less than 50%. Also phenols may be oxidized into quinones, and aromatic rings containing OMe groups are hydroxylated at *o*- and *p*-positions when heated with *m*-CPBA. Furthermore, under rigorous conditions, secondary alcohols are oxidized to ketones and cyclic secondary alcohols may yield lactones (Tori 1988). Reaction of peracids with ketones inserts an oxygen atom at the carbonyl carbon to yield esters (Baeyer-Villiger reaction).

7.2.1.6.6. Tetracyanoethylene Addition to Locate Trans-Trans Conjugated Double Bonds

Conjugated carbon-carbon double bonds of R^1-CH=CH-CH=CH-R^2 type are often encountered in sex pheromones of female Lepidoptera and other samples of chemical ecological interest, and determination of the exact stereochemistry is essential. Gas-phase infrared spectra obtained by gas chromatography-Fourier transform infrared (GC-FTIR) instruments provide diagnostic information (Atty-galle 1994; Svatoš & Attygalle 1997; Attygalle et al. 1995b). However, few laboratories have access to GC-FTIR equipment. Derivatization methods that provide unequivocal data about the stereochemistry are therefore important.

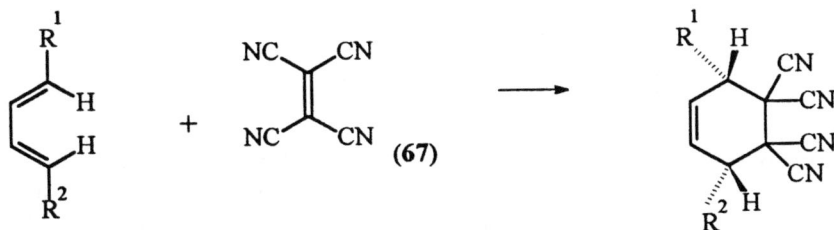

A reliable method to find out whether a R^1-CH=CH-CH=CH-R^2 system has the *trans-trans* configuration is the tetracyanoethylene (TCNE, **67**) derivatization procedure. TCNE adds selectively only to the *trans-trans* isomers of conjugated compounds of R′-CH=CH-CH=CH-R″ type, and the cycloadducts obtained from this Diels-Alder reaction are sufficiently thermally stable to be examined by gas chromatography. The high selectivity of this reaction has been utilized for selectively removing *trans-trans* isomers formed as impurities during the synthe-sis of the other three corresponding isomers of a conjugated system (Hall et al. 1975; Nesbitt et al. 1973; McDonough et al. 1993). The potential of TCNE addition as a microanalytical technique for structure determination has not been exploited thoroughly although Hall et al. (1980) and Nesbitt et al. (1975) have used the reaction for a few pheromone identifications. In the identification of the sex pheromone of spiny bollworm, Hall et al. (1980) established the (*E,E*)-conjugated system in (10*E*,12*E*)-10,12-hexadecadienal by this reaction.

Undoubtedly, the TCNE addition is an excellent method for identifying GC peaks corresponding to (E,E)-conjugated compounds in a complex chromatogram. For the formation of TCNE derivatives, dienes are dissolved in tetrahydrofuran (THF) (10 ng/µl). Typically, 10 µl of this solution is mixed with an equal volume of TCNE (5% in THF). The mixture is kept at room temperature for about 3 h and an aliquot can be analyzed by GC-MS (note that TCNE is a highly toxic irritant). Excess TCNE elutes very early in the chromatogram and does not interfere with the analysis.

For example, when a mixture of all four geometric isomers of 9,11-tetradecadie-nyl acetate was subjected to this derivatization, the GC peak for the *trans-trans* isomer disappeared and a new adduct peak appeared later in the chromatogram (Fig. 7.6). Furthermore, the EI-mass spectra of TCNE adducts provide information about the location of the conjugated system, although the molecular ions are usually weak or absent. For example, the spectrum of the TCNE adduct (**68**) of $(9E,11E)$-9,11-tetradecadienyl acetate, shows a peak at m/z 291 (Fig. 7.6), from a loss of acetic acid from the unobserved ion m/z 351 (M^+-29).

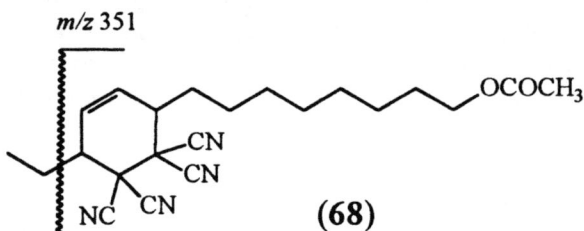

This reaction was employed to locate the position of the conjugated system in the sex pheromone of *Samia cynthia ricini*, the eri-silk worm moth (Bestmann et al. 1989). The major component of this pheromone system is $(4E,6E,11Z)$-4,6,11-hexadecatrienyl acetate (**69**). The EI-mass spectrum of this compound showed a molecular ion at m/z 278, the base peak at m/z 43, and small but significant ion at m/z 61 suggesting a hexadecatrienyl acetate. Furthermore, the relatively long gas chromatographic retention time, compared to hexadecyl acetate and any nonconjugated hexadecatrienyl acetate, indicated that a conjugated double bond system might be present. The presence of the m/z 305 fragment in the mass spectrum of the TCNE adduct established the $(4E,6E)$-conjugated system.

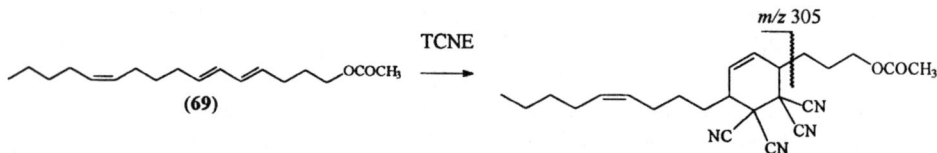

Furthermore, TCNE selectively adds only to certain branched conjugated compounds. This fact is useful for the elucidation of certain terpene structures. Einhorn

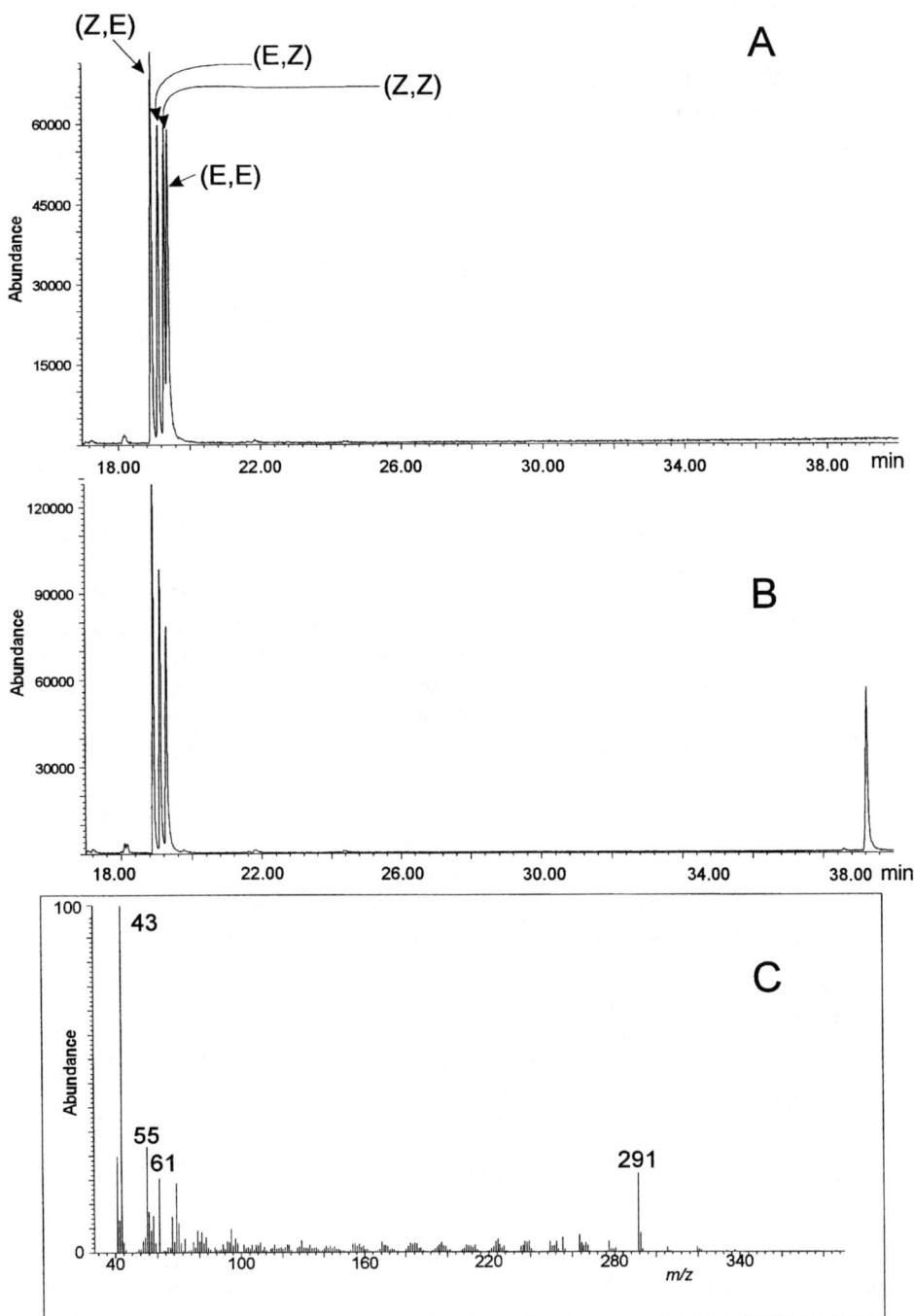

Figure 7.6. Gas chromatograms obtained from a mixture of (*Z,E*)-, (*E,Z*)-, (*Z,Z*)-, and (*E,E*)-9,11-tetradecadienyl acetate before (**A**) and after reaction with TCNE (**B**). The EI-mass spectrum (70 eV) of the TCNE adduct of (*E,E*)-9,11-tetradecadienyl acetate (**C**). (DB-5 coated 0.25 mm × 30 m fused silica column. Oven temperature was held at 60°C for 1 min and programmed at 10°C/min to 170°C, then 3°C/min to 250°C and held at 250°C for 10 min).

et al. (1990) recently used the TCNE procedure for confirming an (E,E) structure of a pine scale pheromone (70). The (Z,E) isomer showed no reaction with TCNE at 25°C.

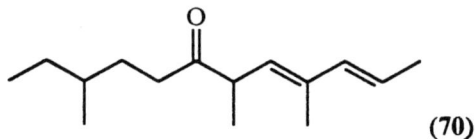

(70)

7.2.1.6.7. Reactions to Locate Conjugated Double Bonds

The addition of 4-methyl-1,2,4-triazoline-3,5-dione (MTAD) to conjugated dienes followed by GC-MS analysis of the cycloadducts is an excellent method for locating conjugated double-bonds (Young et al. 1990). MTAD adducts are more volatile than those of 4-phenyl-1,2,4-triazoline-3,5-dione (PTAD, 70), the reagent originally used for this purpose (Young et al. 1987). This derivatization procedure is particularly helpful because, unlike TCNE, both MTAD and PTAD react with all four isomers of a R-CH=CH-CH=CH-R^1 system. The reaction is instantaneous and takes place at ambient temperature. The reagent is dissolved in dichloromethane (1%) and added to a solution of dienes in dichloromethane until a slight pink color persists. After about 10 min, the reaction mixture is analyzed by GC-MS without any purification. In fact, for nanogram amounts of sample, the reaction can be performed directly in the barrel of the GC syringe. The mass spectra of the adducts obtained by GC-MS exhibit prominent molecular ions and fragment ions diagnostic of the position of the conjugated system in the parent diene. The GC retention indices of the adducts are rather high, since the adducts obtained are rather involatile, however, the method is applicable even for compounds with 20 carbon atoms (Attygalle, unpublished).

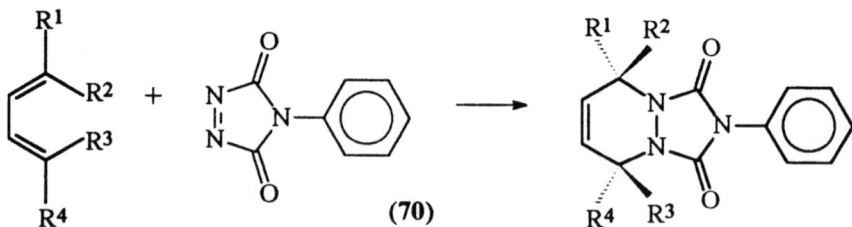

(70)

PTAD is also an especially mild oxidizing agent (room temperature in toluene) particularly good for oxidizing, or preparing, compounds sensitive to acid, base, or heat. The reagent oxidizes alcohols to the corresponding aldehydes or ketones (Cookson et al. 1966). This fact must be remembered if the conjugated compound contains primary or secondary OH groups.

Figure 7.7A shows a gas chromatogram obtained from a mixture of hexadecyl acetate and (*Z,E*), (*E,Z*) (*Z,Z*) and (*E,E*) isomers of 11,13-hexadecadienyl acetate. After the reaction, the diene peaks disappear completely while new peaks appear later in the chromatogram. Under the conditions used, the cyclo adducts of (*Z,E*), (*E,Z*), (*Z,Z*)-11,13-hexadecadienyl acetate coelute, while that of the (*E,E*) isomer is well resolved. The spectra of all four adducts are very similar. The spectrum of the adduct of (*Z,Z*)-11,13-hexadecadienyl acetate (**71**) is shown in Figure 7.7. The peaks at *m/z* 194 and 364 represent the respective losses of the side chains linked to carbon atoms α to the nitrogen atoms of the six-membered ring.

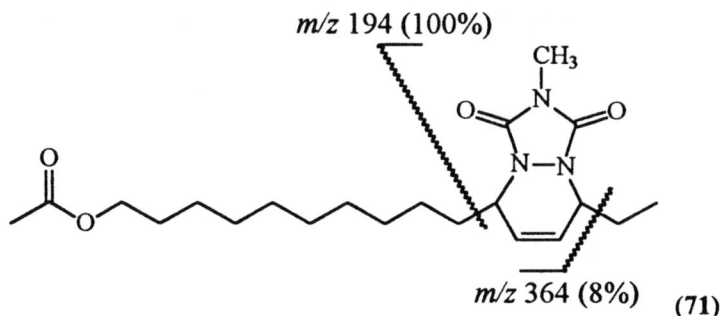

m/z 194 (100%)

m/z 364 (8%)   (**71**)

7.2.1.6.8. Other Reactions of Alkenes

A large number of reactions of alkenes that are useful to derive analytical information have been reviewed (Hogge & Millar 1987; Attygalle & Morgan 1988). Many of these methods have not found wide applicability, however, or have lost their popularity due to the advent of more efficient techniques. For example, the osmium tetroxide (OsO_4) method to convert olefins to glycol type derivatives followed by conversion to methyl or silyl ethers, or acetonides, and determination by GC-MS for the location of the double bond in a carbon chain, is no longer widely used. However, this method might be useful still for highly unsaturated compounds.

7.2.2. *Reactions of Alcohols and Phenols*

Alcohols and phenols are polar compounds and sometimes show poor gas chromatographic properties. In addition, mass spectra of alcohols rarely show molecular ions. To derive more structural information from mass spectra, and to reduce polarity of hydroxy compounds, alcohols are often derivatized before GC-MS analysis. A large number of procedures are available for this purpose (Knapp 1979). A few selected techniques are discussed in this section.

7.2.2.1. *Silylation of Hydroxy Groups*

Silylation both increases the volatility and thermal stability and improves gas chromatographic peak shapes of alcohols and phenols. Peak asymmetry of polar

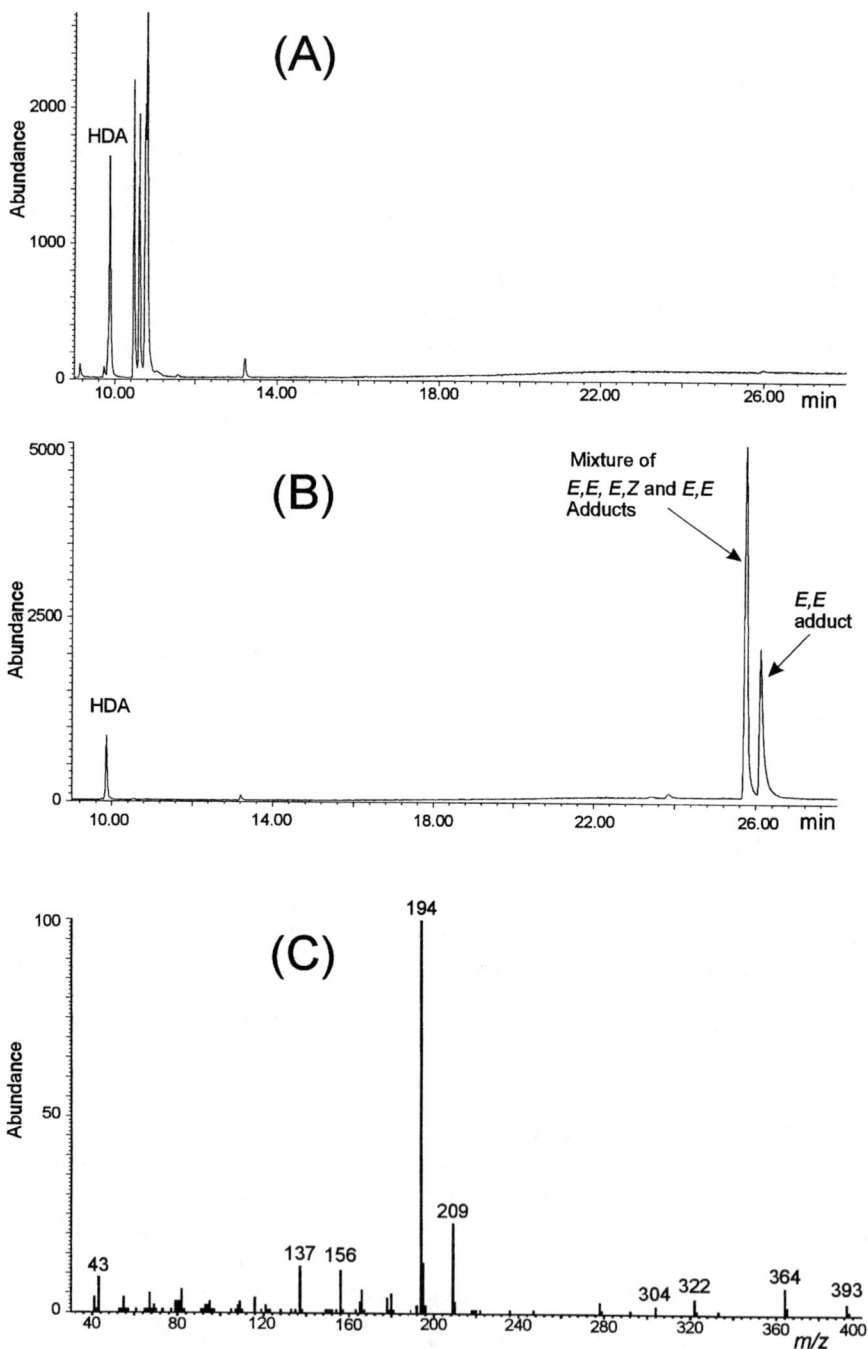

Figure 7.7. Gas chromatograms obtained from a mixture of hexadecyl acetate (HDA) and four isomers of 11,13-hexadecenyl acetate (*Z,E, E,Z, Z,Z,* and *E,E*) before (**A**), and after derivatization with MTAD (**B**) (DB-5 coated 0.25 mm × 30 m fused silica column. Oven temperature was held at 170°C for 1 min and programmed at 5°C/min to 270°C, and held at 270°C for 25 min). Electron-impact mass spectrum of the cycloadduct obtained from MTAD derivatization of (11*E*,13*E*)-11,13-hexadecadienyl acetate (**C**).

compounds is usually due to adsorption effects particularly with older columns with more active sites. For trace samples, the peaks can disappear altogether. When tailing or disappearance of peaks is observed, it may be necessary to derivatize polar samples, (or replace the column). When a compound contains hydroxyl groups, e.g., sugars, derivatization is essential to impart higher volatility to the compound. The most common derivatives are trimethylsilyl ethers (TMS). However, a large number of silylating reagents have been described for the preparation of many other ethers bearing alkyl groups other than methyl (Pierce 1968; van Look et al. 1995; Evershed 1993). Furthermore, in mass spectrometry, derivatives with larger alkyl groups often provide spectra with more abundant diagnostic ions.

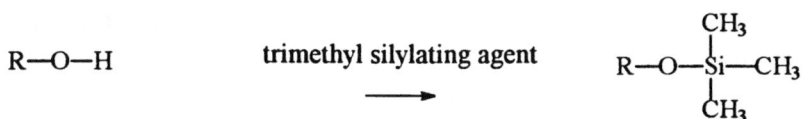

For the preparation of trimethylsilyl derivatives, many derivatization reagents are available (Pierce 1968; van Look et al. 1995). Convenient reagents for this purpose are *N,O*-bis(trimethylsilyl)acetamide (BSA, **72**), *N*-methyl-*N*-trimethyl-silyltrifluoroacetamide (MSTFA, **73**), and *N,O*-bis(trimethylsilyl)trifluoroaceta-mide (BSTFA, **74**). Alcohols are derivatized by simply warming (a solvent is not required) with the reagent. A purification step is not necessary, and the resulting mixture can be analyzed directly by GC-MS since the reagents and the byproducts are volatile.

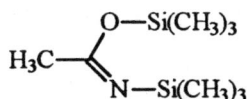 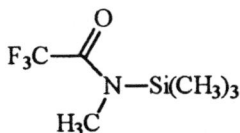 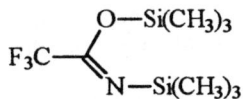

BSA (72)   **MSTFA (73)**   **BSTFA (74)**

For primary alcohols, the amount of additional structural information obtained from the mass spectrum of an alkylsilyl ether is rather limited. However, for secondary alcohols this is a useful technique for location of the hydroxyl group in a carbon chain. Schildknecht et al. (1983) have used trimethylsilylation (MSTFA) for the GC-MS identification of β-hydroxy carboxylic acids found as defensive allomones of a water beetle. MSTFA is a powerful silylating agent that is able to derivatize even carboxylic groups. The derivatization essentially requires mixing of the reagent with the compound to be derivatized and warming the mixture to about 80°C for 15–90 min. If necessary, the byproducts of the reaction, *N*-methyltrifluoroacetamide, and the excess reagent can be removed by a gentle stream of nitrogen. The residue is then dissolved in dichloromethane and analyzed by GC-MS.

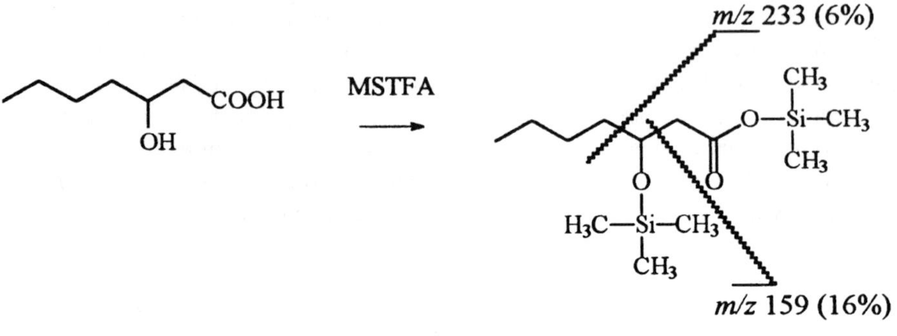

m/z 233 (6%)

m/z 159 (16%) (**75**)

In the electron-impact mass spectra of O-TMS derivatives of long-chain compounds, a molecular ion is usually absent. For example, in the mass spectrum of the disilyl derivative of 3-hydroxyheptanoic acid (**75**) even the signal for the M^+-15 ion at *m/z* 275, from which the molecular weight can be deduced, is of low intensity (0.9%). However, other diagnostic ions that enable the location of the OH group are clearly visible. The signals at *m/z* 233 (6.0%) and 159 (16%) indicate clearly that an OH group is present at the 3 position of the parent acid.

The silylation technique was used to identify the dihydroxy fatty acid obtained from the methanolysis of $N[15(\beta$-glucopyranosyl)oxy-8-hydroxy]-taurine (**76**), the oviposition-deterring pheromone of the European cherry fruit fly (Hurter et al. 1987). From the mass spectroscopic peaks at *m/z* 117, 245, and 303 of the silyl derivative (**77**), the underivatized compound was characterized as methyl 8,15-dihydroxypalmitate.

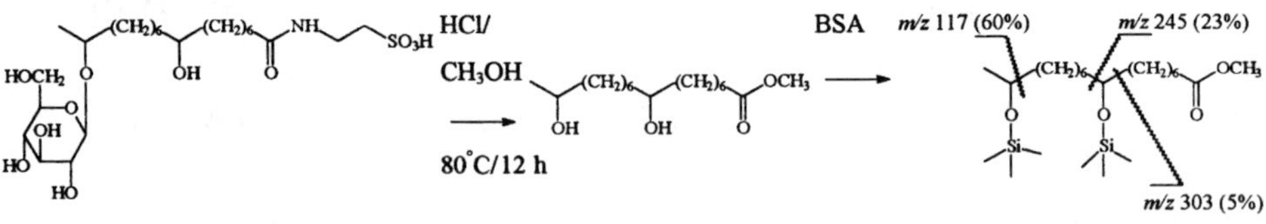

(**76**)   (**77**)

Silylation is often used in the characterization of sugars by GC-MS. For example, the sugar moiety of lurlene (**78**), the glycosidic sex pheromone of *Clamydomonas allensworthii*, was identified by a silylation technique. The pheromone was hydrolyzed with 2-M HCl, and after the mixture was extracted with ether, the aqueous phase was dried. The residue was derivatized with MSTFA (Starr et al. 1995; Jaenicke & Marner 1995), and by a comparison of GC-MS data obtained from this derivative with those from derivatized authentic sugars the pentose of lurlene was identified as D-xylose (**79**).

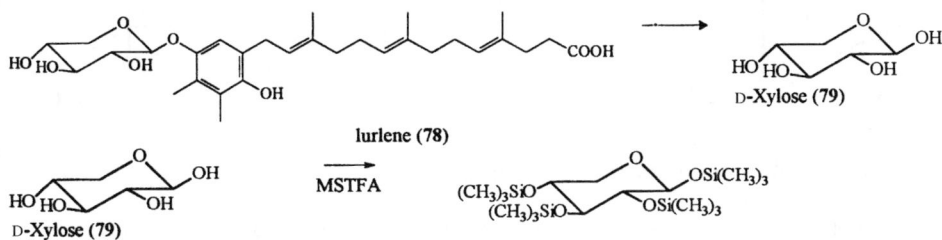

lurlene (**78**)

D-Xylose (**79**)

For pheromone identifications, *t*-butyldimethylchlorosilane (TBDMSCl, **80**) has been used as the silylating agent on several occasions. Typically, for silylation of hydroxyl groups with TBDMSCl, the alcohol in triethylamine containing 3% DMAP (4-dimethylaminopyridine) as catalyst is mixed with TBDMSCl and the mixture is kept at room temperature for 8 h. The mixture is evaporated to dryness and the residue is taken up in hexane for GC-MS analysis.

In the identification of the pheromone of Comstock mealybug, Bierl-Leonhardt et al. (1980a) converted the alcohol (**81**) obtained from the pheromone 2,6-dimethyl-1,5-heptadien-3-ol acetate to its *t*-butyldimethylsilyl ether. The mass spectrum of the derivative (**82**) showed characteristic ions at m/z 185 and 213, resulting from cleavages α to the siloxy group, indicating that the compound is a 3-alcohol and not a 4-alcohol.

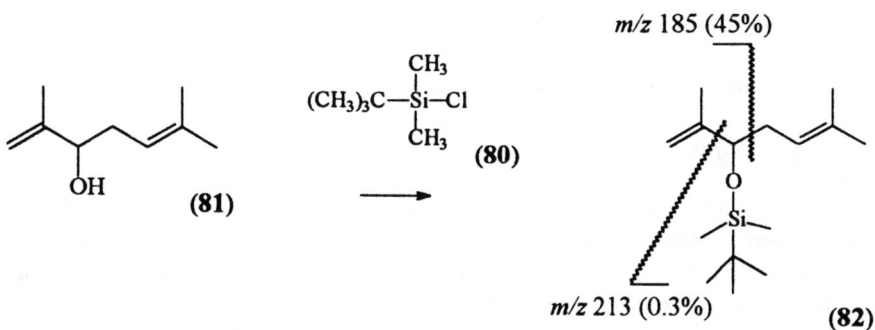

m/z 185 (45%)

m/z 213 (0.3%)

(**82**)

Similarly, in the identification of the sex pheromone of the citrus mealybug, *Planococcus citri* (Risso), the alcohol (**83**) obtained from the hydrolysis of the pheromone (an acetate) was silylated with TBDMSCl. The EI-mass spectrum of the silyl derivative (**84**) showed a diagnostic peak at m/z 199 ($C_5H_8OSiMe_2Bu$) presumably due to the splitting of the cyclobutane ring to yield a C_5 fragment (Bierl-Leonhardt et al. 1981).

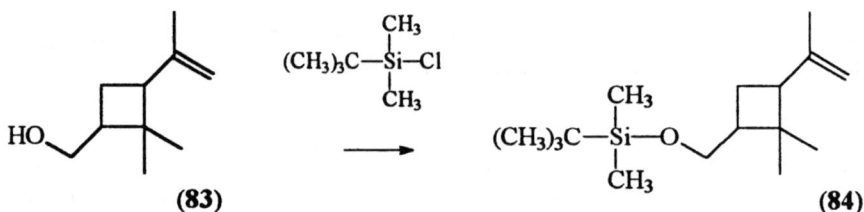

(83) → (84)

7.2.2.2. Acetylation of Hydroxy Groups

Acetylation is a useful microreaction for confirming the presence of alcohols, phenols, and amines. The reasons for preparing acyl derivative of a compound containing OH groups are very similar to those described for silylation. In fact, the reaction furnishes information to distinguish primary, secondary, or tertiary alcohols. If the compound under consideration has biological activity, a loss of activity on acetylation and its return on hydrolysis may indicate that the active substance is a phenol, or a primary or secondary alcohol.

Microscale acetylations are carried out even with a few nanograms of compound by procedures similar to those described by Huwyler (1972). For compounds trapped in glass capillaries by preparative GC, THF containing acetic anhydride and pyridine (1:1) can be added, and the reaction mixture is then examined by GC (Attygalle & Morgan 1984b). The reaction with primary alcohols is usually complete in 30 min; secondary alcohols require longer reaction times, while tertiary alcohols do not react (Huwyler 1972). The acetylation procedure has been used to establish the number of OH groups found in phenolic compounds present in the defensive secretion of an ant (Attygalle & Morgan 1988). The number of phenolic groups present can be estimated by the retention-time shift of a GC peak observed after the derivatization, or by GC-MS.

7.2.2.3. Trifluoroacetylation

In some cases, GC-MS analysis of trifluoroacetyl derivatives of alcohols and phenols provides more useful information than that from simple acetyl derivatives. Although trifluoroacetic anhydride is the frequently used reagent for this purpose, N-methyl-bis(trifluoroacetamide) (MBTFA) is a more convenient reagent since the reaction can be performed under milder and neutral conditions (Donike 1973). The same reagent can be used for trifluoroacetylation of primary and secondary amines (section 7.2.7.2).

7.2.2.4. Formation of Urethanes

Alcohols are converted to urethanes by treating with an alkylisocyanate. This derivatization is particularly useful to attach a chromophore to alcohols when HPLC methods are used for their isolation. The reaction is performed typically by heating the alcohol with an alkyl or arylisocyanate at 100°C for 1.5 h. At least for alkylisocyanates, the excess reagent can be removed by a stream of N_2 and the residue can be analyzed by GC or HPLC. The structure of (2E,6E)-3,7-dimethyldeca-2,6-dien-1,10-diol (**85**), isolated from the hairpencils of males of the queen butterfly (*Danaus gilippus berenice*), was confirmed by converting it to the *bis*-1-naphthyl urethane (**86**) followed by a melting-point comparison of the derivative with that obtained from an authentic sample (Meinwald et al. 1969).

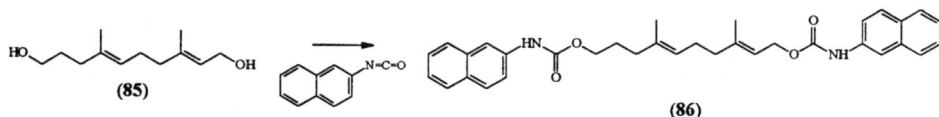

7.2.2.5. Bromination of hydroxy groups

Although this is a procedure a microchemist may rarely need, an interesting application of this technique has been described by Bjostad et al. (1996). The EI-mass spectrum of 6-methyl-3-nonanone, a pheromone of the caddis fly, was not sufficiently informative to suggest the position of the methyl group. To locate the methyl branching, the ketone was first reduced with $LiAlH_4$ to the secondary alcohol **87** (section 7.2.4.1). Using triphenylphosphine dibromide the alcohol was then transformed to the bromide (**88**), which was finally converted to the hydrocarbon 4-methylnonane. The mass spectrum of the hydrocarbon established the position of the methyl group.

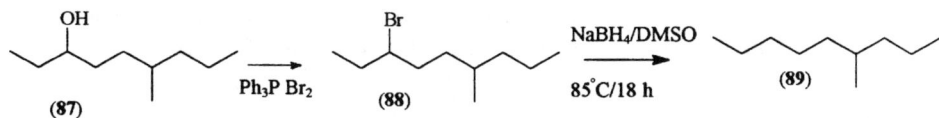

The procedure described by Hutchins et al. (1969) was used to convert the alkyl bromide to the hydrocarbon. The conversion is simple. The bromide is dissolved in dimethylsulfoxide (DMSO) and a small amount of sodium borohydride is added. The mixture is heated to 85°C for 18 h. After cooling, the mixture is extracted with hexane and the extracts are washed with water and examined by GC-MS. In general, this procedure can be used to determine the carbon skeleton of alcohols since they can be converted to bromides readily.

7.2.2.6. Preparation of Nicotinates

For unsaturated alcohols, the mass spectra of their nicotinate derivatives allow the localization of double-bond positions (Harvey 1984, 1988). This method is less useful than the DMDS procedure (section 7.2.1.6.3) for monounsaturated alcohols, but for polyunsaturated alcohols, it offers some promise. The mass spectral procedure is similar to that used for nitrogen-containing derivatives of fatty acids (described in detail in section 7.2.5.6). For the formation of nicotinates, the alcohols are treated with nicotinoyl chloride (**90**) hydrochloride in pyridine.

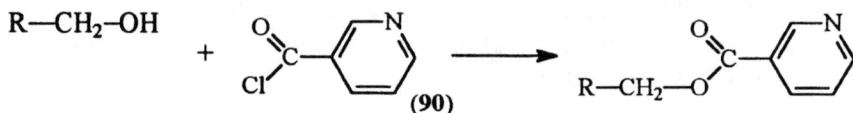

7.2.2.7. Oxidation of Alcohols

The oxidation of the hydroxy group to a carbonyl moiety is often performed in the structure elucidation of hydroxy compounds. Although many oxidizing agents are available for this purpose, mild reagents are preferred since chemical transformations of other functional groups that may be present must be avoided.

7.2.2.7.1. Oxidation of Primary and Secondary Alcohols by Pyridinium Chlorochromate

Pyridinium chlorochromate (PCC, **91**) oxidizes primary and secondary alcohols to carbonyl compounds (Corey & Suggs 1975). The alcohol is dissolved in dichloromethane, mixed with PCC in dichloromethane, and kept at room temperature for a few hours. After dilution with pentane and filtering through a short column of silica, the solution is ready for GC-MS analysis. For example, nanogram amounts of citronellol (**92**) can be oxidized to citronellal (**93**) in this way.

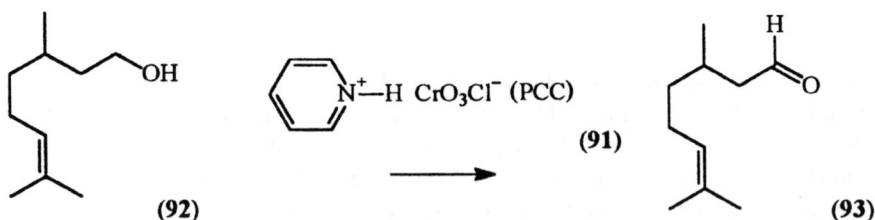

The reagent is particularly useful because it is not known to attack isolated C-C double bonds (Piancatelli et al. 1982). However, Doolittle et al. (1990) have reported difficulties with oxidation of conjugated triene alcohols with PCC. Although the desired triene aldehyde was obtained, the oxidizing agent also attacked the conjugated triene system. The amounts of by-products were so excessive that Doolittle et al. (1990) resorted to the periodinane procedure described below instead.

7.2.2.7.2. Oxidation of Primary and Secondary Alcohols by Dess-Martin Periodinane Reagent

The periodinane reagent (Dess & Martin 1983) is a mild and selective reagent for oxidizing primary and secondary alcohols to aldehydes and ketones. This reagent is preferred to PCC for the oxidation of highly conjugated alcohols (Doolittle et al. 1990), although the reagent is known to cause some isomerization and to explode violently if heated under confinement. Typically, the alcohol in dichloromethane is mixed with the periodinane reagent (**94**) in dichloromethane. After several hours at room temperature, the reaction mixture is washed with a solution of sodium thiosulfate/NaHCO$_3$ until the organic layer becomes clear, and the organic layer is removed for analysis. With this procedure conjugated alcohol (**95**) was oxidized to aldehyde (**96**).

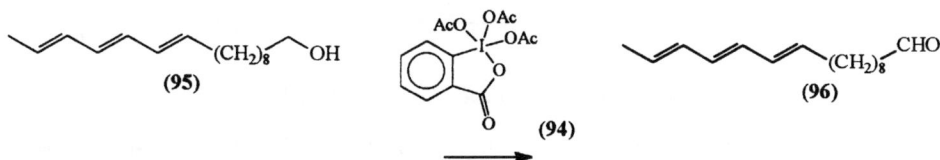

7.2.2.7.3. Oxidation of Primary and Secondary Alcohols by Chromium Trioxide CrO$_3$

In the identification of (10*E*)-10-tridecen-2-yl acetate as a pheromone component of the Hessian fly, the carbon-carbon double bond was first hydrogenated followed by hydrolysis of the ester group to obtain 2-tridecanol (**97**) (Foster et al. 1991). Subsequently, this secondary alcohol was oxidized to 2-tridecanone (**98**) by CrO$_3$/celite (Schwartz & Weirauch 1969).

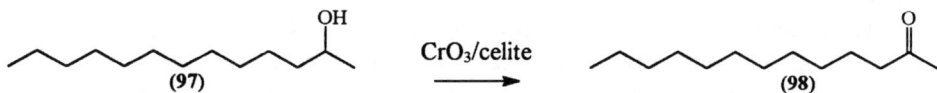

Similarly, in the identification of 3,7-dimethyl-2-pentadecanol, the alcohol of the sex pheromone of some pine sawflies, the compound was oxidized to the ketone with CrO$_3$-pyridine complex in dichloromethane (Jewett et al. 1976).

7.2.2.7.4. Oxidation of Primary Alcohols to Carboxylic Acids by Pyridinium Dichromate

The oxidation of primary alcohols to carboxylic acids is achieved under mild conditions with pyridinum dichromate (PDC, **99**) in dimethylformamide (DMF) (Corey & Schmidt 1979). Choice of solvent is crucial; if dichloromethane is used, the oxidation stops at the aldehyde stage. The alcohol is mixed with PDC/

DMF (10%, 20 μl) and after 12 h, the tarry brown mixture that results is mixed with water and extracted two to three times with hexane/ether (9:1) (the aqueous layer must be acidic, pH ~ 3). The combined hexane/ether phase is washed with water, dried over Na_2SO_4 and filtered through Celite™. Lanne et al. (1987) used the PDC/DMF reagent to oxidize alcohols present in the labial gland secretion of male bumble bees. For example, citronellol (92) is oxidized to citronellic acid (93) by this reagent. The reagent is also useful for the oxidation of secondary alcohols to ketones.

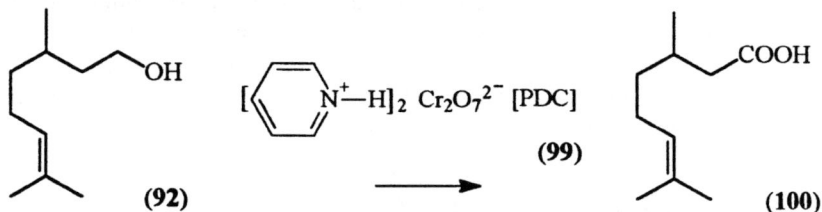

$$[\langle\!\!\!\!\begin{array}{c}\\ N\end{array}\!\!\!\!\overset{+}{-}H]_2 \; Cr_2O_7{}^{2-} \; [PDC]$$

(99)

(92) ⟶ (100)

7.2.2.8. Conversion of Alcohols to the Basic Carbon Skeleton

Alcohols can be converted to the basis carbon skeleton by catalytic hydrogenolysis (section 7.2.1.5). However, by the catalytic procedure, primary alcohols usually lose the CH_2OH group to give a hydrocarbon with one carbon atom less than the parent alcohol. Sometimes it is desirable to convert a primary alcohol to a hydrocarbon bearing the same number of carbon atoms. One way to achieve this is the procedure described in section 7.2.2.5, in which alcohols are converted to bromides, and then reduced by $NaBH_4$/DMSO to the basic hydrocarbon. Alternatively, this can be performed by converting the alcohol to a mesyl derivative that can then be reduced by lithium aluminum hydride ($LiAlH_4$). This is one of the recommended procedures for the localization of methyl groups in branched long-chain alcohols.

The alcohols are converted to mesyl derivatives by 1% methanesulfonyl chloride (mesyl chloride, 101) in pyridine. Excess mesyl chloride is destroyed by methanol, and the solvent is evaporated with a stream of nitrogen (Simon et al. 1990). The residue is dissolved in ether and treated with $LiAlH_4$ in dry ether for 3 h. After the addition of water, the hydrocarbon products are extracted into hexane and analyzed by GC-MS.

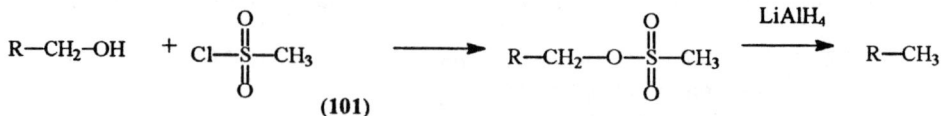

$$R{-}CH_2{-}OH \; + \; Cl{-}\overset{O}{\underset{O}{\overset{\|}{\underset{\|}{S}}}}{-}CH_3 \; \longrightarrow \; R{-}CH_2{-}O{-}\overset{O}{\underset{O}{\overset{\|}{\underset{\|}{S}}}}{-}CH_3 \; \xrightarrow{LiAlH_4} \; R{-}CH_3$$

(101)

7.2.2.9. Other Reactions of Alcohols

Glycols (1,2-diols) can be cleaved by CrO_3/H_2SO_4 (Jones reagent; Bowden et al. 1946; Harding et al. 1975) to give carboxylic acids. The compound is dissolved in acetone (100 µl) and treated with Jones reagent (10 µl). After 30 min at room temperature, aqueous sodium bisulfite is added and the mixture is extracted with ether. Evaporation of the ether layer gives a mixture of carboxylic acids which can be esterified and determined by GC-MS. For example, in the characterization of (8′Z,11Z)-3-(heptadeca-8′,11-dienyl)catechol dimethyl ether (**102**), the derivative **103** obtained from OsO$_4$ hydroxylation was subjected to oxidative cleavage by the Jones reagent and the acids **104** and **105** were identified after methylation (Jefferson & Wangchareontrakul 1986).

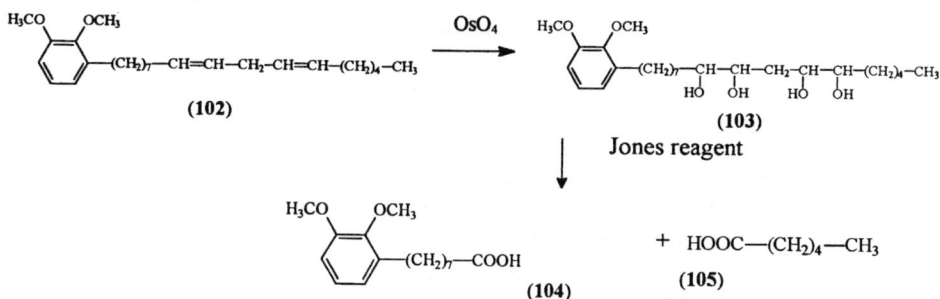

1,2-Diols can also be determined as cyclic acetals, ketals, or boronates. Blau and Dabre (1993a) provide an excellent review of this subject.

7.2.3. Reactions of Epoxides

Epoxides are often encountered in samples of biological significance such as pheromones and juvenile hormones. Furthermore, epoxide derivatives are sometimes used in structure elucidation of alkenes (section 7.2.1.6.5).

7.2.3.1. Deoxygenation with Potassium Selenocyanate

Deoxygenation of an epoxide ring to give the corresponding alkene can be accomplished with potassium selenocyanate with overall retention of configuration (Clive 1978). This reaction may be combined with a bioassay. Loss of activity after treatment with potassium selenocyanate and regain of activity by reepoxidation with *m*-CPBA indicates the presence of an epoxide group in the biologically active material. Also, the reaction may be used to convert an epoxide to a known alkene for retention time and spectral comparisons.

For example, Müller et al. (1988) used this reaction in the structure elucidation of caudoxirene (**106**). (*E*)-Viridiene (**107**), a known natural product, was produced when caudoxirene was treated with potassium selenocyanate.

caudoxirene (106)   (E)-viridiene (107)

Typically, the epoxide to be transformed is mixed with KSeCN in methanol (20 μl, 5 mg/ml). The mixture is sealed and heated to 65°C for 2 h. After cooling to room temperature, the mixture is analyzed by GC-MS.

7.2.3.2. Cleavage to Carbonyl Compounds

Techniques for locating an epoxide ring system in a carbon chain are of considerable value. The EI- or particularly the CI-mass spectrum of an epoxide is often sufficient for this purpose. However, for structural confirmation, epoxides may be converted to more easily recognizable entities. Periodic acid cleaves an epoxide to the corresponding carbonyl compounds (Mizuno et al. 1969). For example, cleavage of disparlure (108), the sex pheromone of the Gypsy moth, affords 6-methylheptanal (109) and undecanal (110), which can be identified by GC-MS.

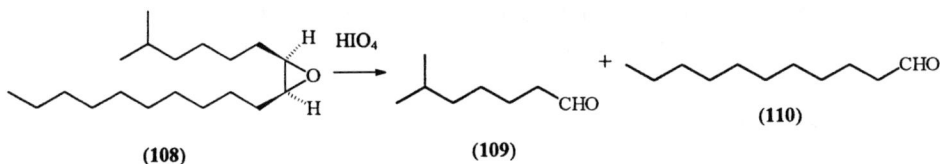

There are many applications of this reaction in natural product chemistry. The trisubstituted epoxide structural moiety present in a cembranoid diterpene (111) isolated from a South Pacific soft coral, was confirmed by periodic acid oxidation to the diketoaldehyde (112) (Ravi & Faulkner 1978).

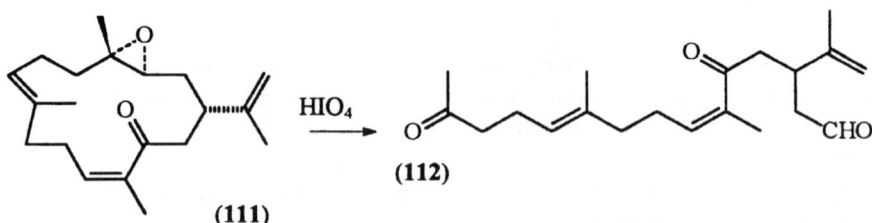

For the cleavage, the epoxide is dissolved in a chlorinated solvent or ether (50 μl) and mixed with a small amount of dry, powdered periodic acid (HIO_4) (Bierl et al. 1971; Bierl-Leonhardt et al. 1980b). The mixture is shaken for 5–10

min and an aliquot of the supernatant liquid can be analyzed by GC-MS. However, for nanogram amounts of sample, a reaction gas chromatographic procedure is preferred (Attygalle & Morgan 1984a). In this method, a small amount of HIO_4 is impregnated into a solid support such as methylsilicone coated silica and placed in the insert linear of the injection port of the gas chromatograph, or in a precolumn, and the epoxide to be cleaved is injected in the usual manner. The cleavage takes place almost instantaneously and gas chromatographic peaks are observed for the products. Note that this procedure also cleaves carbon-carbon double bonds to a certain extent.

7.2.3.3. Hydrolysis to Diols

Epoxides are easily hydrolyzed to diols, which can be further derivatized and analyzed by GC-MS. For example, in the characterization of (3Z,6Z)-3,6-*cis*-9,10-epoxyheneicosadiene (**113**), a component of the sex pheromone of the salt marsh caterpillar moth, the carbon-carbon double bonds of the compound were hydrogenated and the epoxide (**114**) obtained was hydrolyzed to a diol (**115**) (Hill & Roelofs 1981). The hydrolysis was accomplished by treating with 0.5% H_2SO_4 in 50% aqueous THF for 4 h at room temperature. The reaction mixture was diluted, and the product was extracted into petroleum ether. After silylation (section 7.2.2.1), the mass spectrum of the derivative (**116**) showed two prominent peaks at *m/z* 215 and 257, resulting from the cleavage of the bond between carbon atoms 9 and 10, and established the location of the original epoxide (Hill & Roelofs 1981).

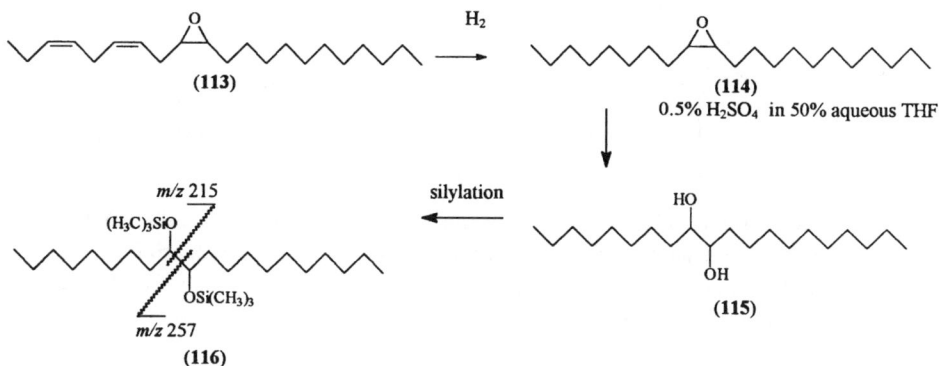

7.2.3.4. Other Reactions of Epoxides

Many ring opening reactions of analytical significance are known (Attygalle & Morgan 1988) but most of these methods are not widely applied.

7.2.4. Reactions of Aldehydes and Ketones

Long-chain aldehydes and ketones are often found in complex mixtures of semio-chemicals isolated from plants and animals. Frequently, these aldehydes and ketone are unsaturated. Unsaturated aldehydes are of particular interest since they are components of many insect pheromone mixtures. Usually it is unnecessary to derivatize aldehydes or ketones to improve gas chromatographic properties. The mass spectra of ketones often bear sufficient information to locate the carbonyl group. On the other hand, the mass spectra of unsaturated aldehydes are rather featureless, and the intensity of molecular ions are notably weak. Recognition of an unsaturated aldehyde entirely on the basis of its EI-mass spectrum is not an easy task even for an expert. For example, the upper mass region (above m/z 50) of a spectrum of a monounsaturated aldehyde appears very similar to that of a linear diunsaturated alcohol. In order to obtain conclusive data, it is often necessary to resort to derivatization.

7.2.4.1. Reduction of the Carbonyl Group to a Hydroxy Group

Reduction to the corresponding alcohols is helpful in the characterization of carbonyl compounds. The reagents generally used for reducing aldehydes and ketones are lithium aluminum hydride ($LiAlH_4$) or sodium borohydride ($NaBH_4$). However, $NaBH_4$ is preferred because $LiAlH_4$ also reduces other functional groups such as esters. For the identification of the sex pheromone of the caddisfly, the keto group of the pheromone, 6-methyl-3-nonanone (**117**), was reduced by $LiAlH_4$ to give the secondary alcohol (**87**) (Bjostad et al. 1996) (see section 7.2.2.5).

The reduction of chiral ketones produces diastereomeric alcohols that are usually separable by GC-MS. For example, an analysis of the reduction mixture of the ketone (**118**) by gas chromatography showed two peaks for the two diastereomers (**119**) (Stanley 1979).

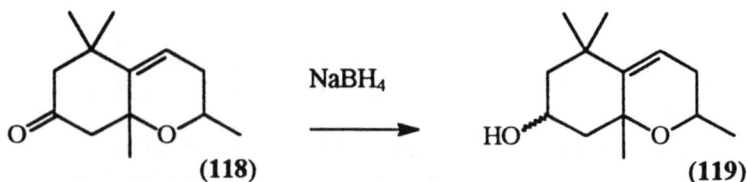

Reduction with NaBH$_4$ is often used in characterization of sugars. Aldoses or ketoses are reduced to polyalcohols (alditols). The advantage of this reduction is that each sugar produces a single derivative. For example, in the identification of lurlene (**78**), the sugar (**79**) obtained by the hydrolysis of the pheromone was reduced with NaBH$_4$ and the alditol (**120**) obtained was derivatized with BSTFA (**74**) for comparison with authentic standards (Jaenicke & Marner 1995).

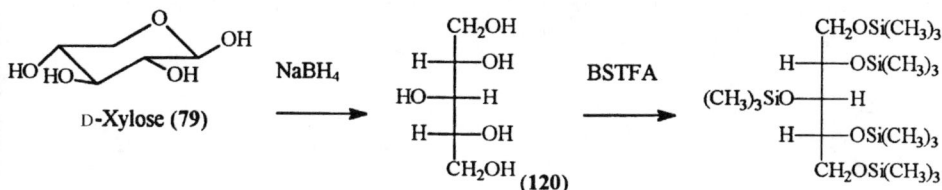

Typically, the sugar sample is mixed with aqueous NaBH$_4$ (50 µl, 100 mg/ ml) and kept for 2 h at room temperature (Fox et al. 1989). Excess NaBH$_4$ is destroyed with acetic acid/methanol (2 ml, 1:200 vol/vol) and the sample is evaporated to dryness. The evaporation step is repeated four times to remove traces of residual borate. The sample is allowed to dry for 3 h and derivatized with BSTFA (section 7.2.2.1).

7.2.4.2. Reduction of the Carbonyl Group to a Methylene Group

The carbonyl group of ketones and aldehydes can be converted to a methylene group by the Wolff-Kishner procedure. For example, in the determination of the stereochemistry of 1-(6-methylpiperidyl)propan-2-one (**121**), a defensive alkaloid

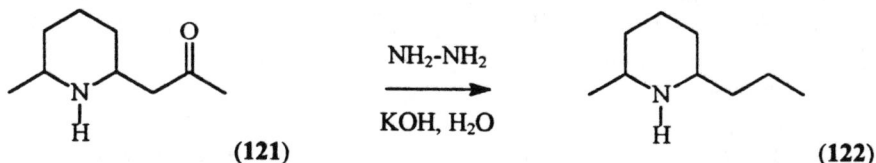

of a coccinellid beetle, the ketone was reduced by the Wolff-Kishner procedure. For the reduction, the carbonyl compounds are treated with 10% hydrazine hydrate in ethanol (20 µl) containing a trace of formic acid (Brown & Moore 1982). After 30 min, the solvent was evaporated, and the residue was mixed with 10% KOH in triethylene glycol (20 µl). The tube was sealed and heated at 200°C for 30 min (with DMSO solvent, the reaction can be conducted at lower temperatures). The product was diluted with water and extracted with chloroform. The *cis* stereochemistry of 1-(6-methylpiperidyl)propan-2-one from the beetles was established by a comparison of the GC retention time of the reduced product (**122**) with that of an authentic sample.

7.2.4.3. Formation of N,N-dimethylhydrazones

An easy way to recognize peaks from aldehydes and ketones in a gas chromatogram is to convert these carbonyl compounds into *N,N*-dimethylhydrazone derivatives. The derivatization reaction is straightforward and requires essentially only the addition of *N,N*-dimethylhydrazine (**123**) to an extract. This derivatization is particularly useful in identifying carbonyl compounds resulting from an ozonolysis reaction (Attygalle et al. 1989).

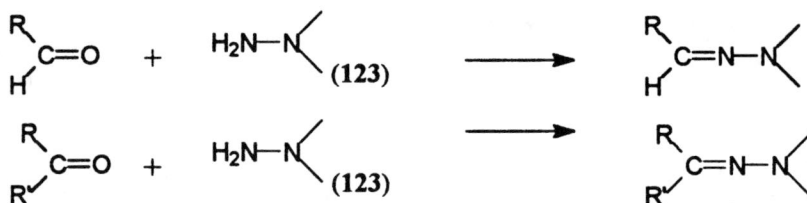

Aldehydes react rapidly with *N,N*-dimethylhydrazine at room temperature, and the reaction is near quantitative. In fact, aldehydes (in hexane) can be derivatized simply by mixing with neat reagent in the barrel of the GC syringe. Ketones react somewhat slowly, and only modest yields are reported. It has been reported that hindered ketones, such as 2,6-dimethyl-4-heptanone, are particularly difficult to derivatize by this procedure (McDaniel & Howard 1985).

The mass spectra of *N,N*-dimethylhydrazones not only show easily recognizable molecular ions to provide molecular weight information but also a number of other important peaks that allow meaningful structural deductions (Goldsmith & Djerassi 1966; Attygalle et al. 1989). For easy interpretation, the *N,N*-dimethylhydrazone spectra obtained from straight-chain aldehydes can be divided into several categories.

7.2.4.3.1. Derivatives of Saturated Aldehydes and Unsaturated Compounds with First Carbon-Carbon Double Bond Separated by at Least Four Methylene Groups from the Aldehyde Group

Under electron-impact conditions, these derivatives undergo a γ-hydrogen transfer by a McLafferty rearrangement to yield an ion of *m/z* 86 (**124**). A peak also is observed for the loss of 44 mass units from the molecular ion (Fig. 7.8A). In fact, in the spectra of saturated and unsaturated aldehydes with the first carbon-carbon double bond remote from the aldehyde group, the *m/z* 86 peak is often the most intense peak of the spectrum. However, when a carbon-carbon double bond is present at positions 2, 3, 4, 5, or 6, the spectra of the derivatives become diagnostic for the position of the double bond.

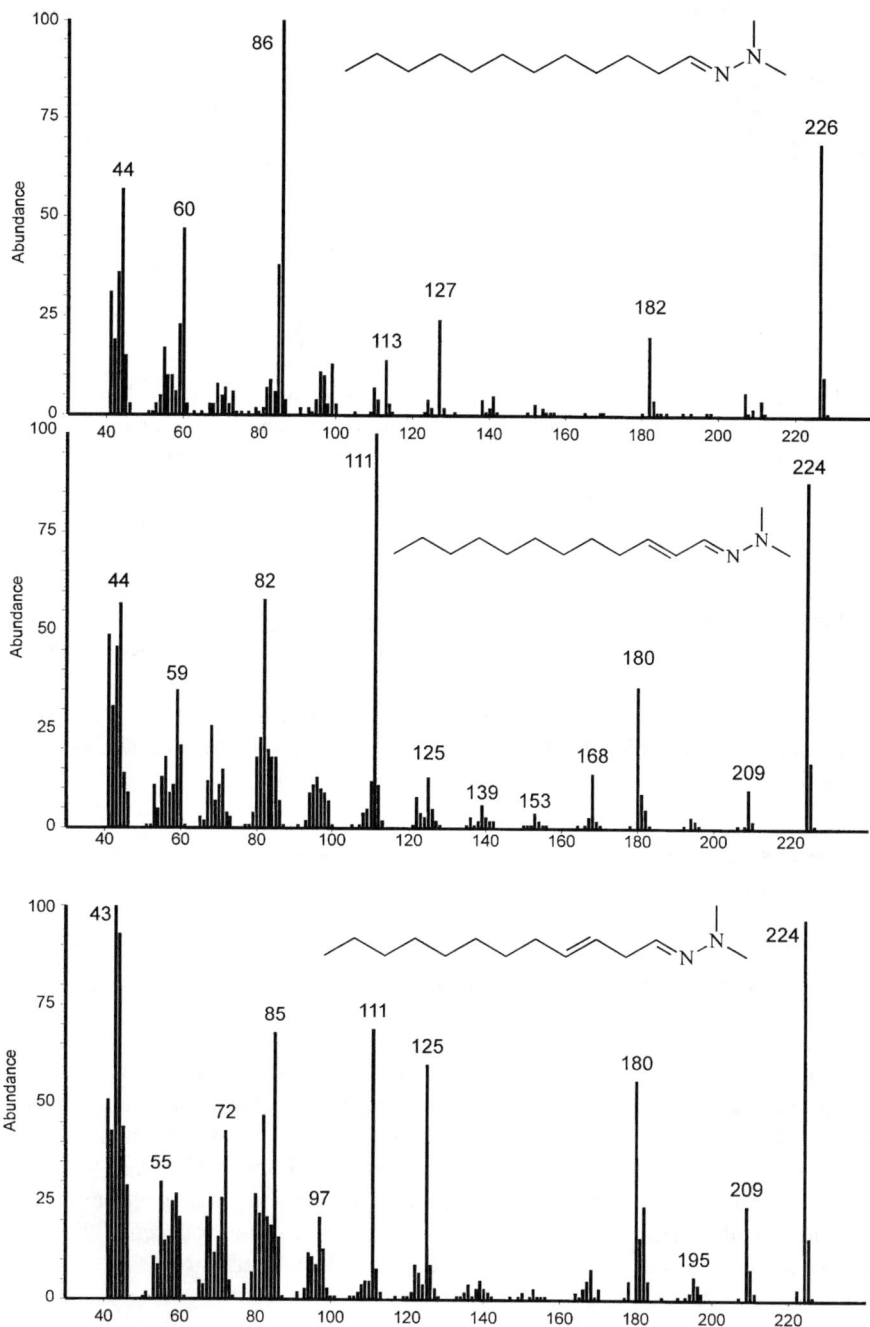

Figure 7.8.   Electron-impact mass spectra (70 eV) of *N,N*-dimethylhydrazones of dodeca-
nal (**A**), (*E*)-2-dodecenal (**B**), and (*E*)-3-dodecenal (**C**).

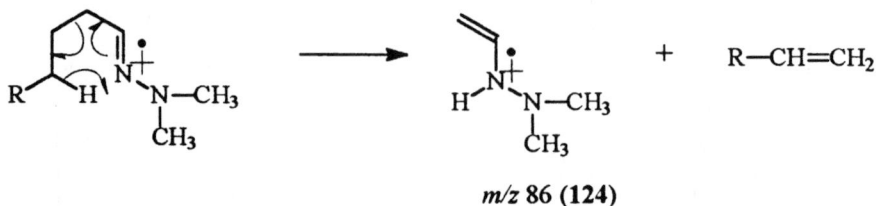

m/z 86 (124)

7.2.4.3.2. Derivatives of α,β-Unsaturated Aldehydes

The spectra obtained from derivatives of conjugated aldehydes do not show a prominent peak at m/z 86 because the McLafferty rearrangement is hindered, although γ-hydrogen atoms are present. The spectra are characterized by an intense ion at m/z 111 (125) from an allylic fission of the molecular ion. Since the resulting ion is resonance stabilized, it often represents the base peak of the spectrum (Fig. 7.8B).

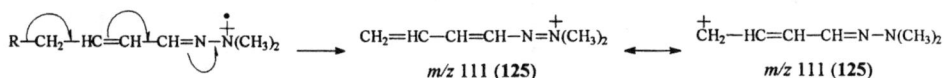

m/z 111 (125)   m/z 111 (125)

Moreover, the allylic fission is still observed when the conjugation is further extended. For example, the base peak (126) in the spectrum of (2E,4E)-2,4-decadienal is observed at m/z 137.

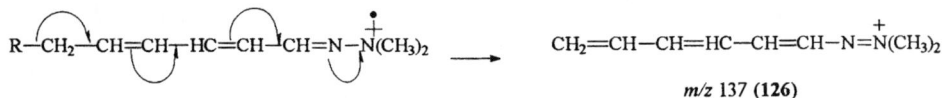

m/z 137 (126)

7.2.4.3.3. Derivatives of Unsaturated Aldehydes with a Double Bond at the C-3 Position

It is known that β,γ-unsaturated aldehydes exhibit poor gas chromatographic properties possibly because these unstable compounds exist in tautomeric forms (127, 128). Apparently, the N,N-dimethylhydrazone derivatives of 3-alkenals behave in a similar manner. The derivatives give broad GC peaks with two maxima and a valley in between. Figure 7.8C shows the mass spectrum corresponding to the major GC peak of the derivative of (E)-3-dodecenal. The two peaks at m/z 111 and 125 are diagnostic for 3-alkenal derivatives.

(127)   (128)

7.2.4.3.4. Derivatives of Unsaturated Aldehydes with a Double Bond at the C-4 Position or Small Saturated Aldehydes

The spectra of *N,N*-dimethylhydrazones of small aldehydes such as butyralde-hyde show a base peak at *m/z* 85, instead of *m/z* 86, since a γ-hydrogen transfer from a methyl group does not happen as readily as that from a secondary carbon atom. In the spectrum of the valeraldehyde derivative, however, both *m/z* 85 (100%) and 86 (90%) peaks are prominent (Goldsmith & Djerassi 1966). Appar-ently, when a γ-hydrogen transfer is not possible due to structural constraints, such as the presence of a double bond at the C-4 position, the base peak of the spectrum is observed at *m/z* 85 (**129**) (Fig. 7.9A) which can be attributed to the following allylic cleavage:

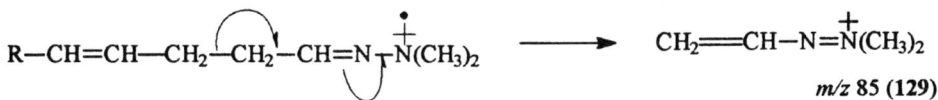

$$R-CH=CH-CH_2-CH_2-CH=N-\overset{+}{N}(CH_3)_2 \longrightarrow CH_2=CH-N=\overset{+}{N}(CH_3)_2$$

$$m/z\ 85\ (\mathbf{129})$$

For example, the spectrum of the derivative from (4*E*,6*E*,11*Z*)-hexadecatrienal, a sex pheromone component of the eri-silk moth, showed a peak at *m/z* 85 suggesting a carbon-carbon double bond at the 4 position (Bestmann et al. 1989).

7.2.4.3.5. Derivatives of Unsaturated Aldehydes with a Double Bond at the C-5 or C-6 Position

The mass spectra of derivatives obtained from 5- and 6-alkenals exhibit a fragment for M-68 and M-82, due to the loss of C_5H_8 or C_6H_{10} from the parent ion, respectively. The spectra of (5*E*)-dodecenal and (6*E*)-dodecenal derivatives depicted in Figure 7.9B and Figure 7.9C show clearly the expected fragments at *m/z* 156 and 142, respectively (Attygalle et al. 1998).

7.2.4.4. Reactions with O-Methylhydroxylamine

The formation of oximes with *O*-methylhydroxylamine (methoxyamine, **130**) is another useful procedure for aldehydes and ketones. For the derivatization, the carbonyl compounds are dissolved in pyridine and heated to 100°C for 15 min with *O*-methylhydroxylamine hydrochloride (Mason et al. 1990). Although the reaction mixture can be analyzed directly, preferably the mixture is washed with dilute $CuSO_4$ and the products extracted into ether before analysis. The reaction affords a mixture of *syn*- and *anti*-isomers which may coelute on GC analysis.

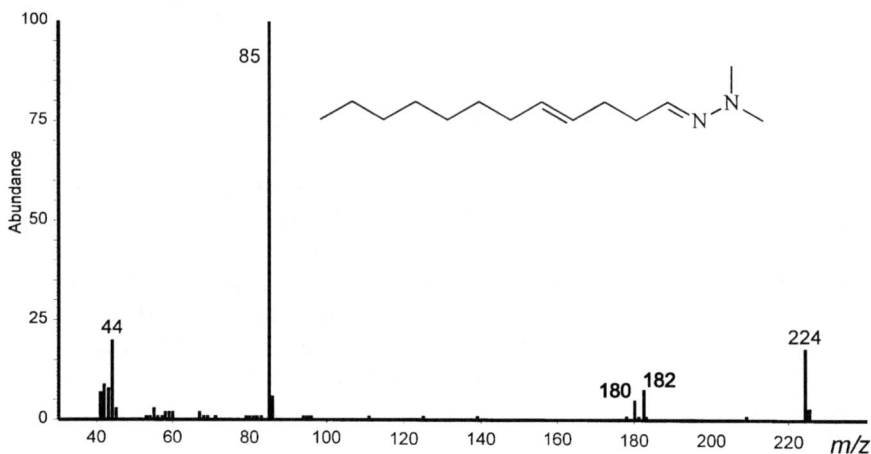

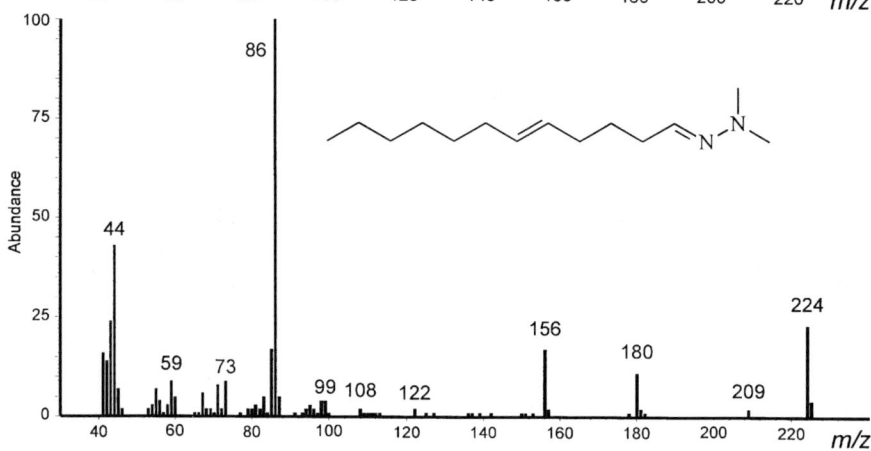

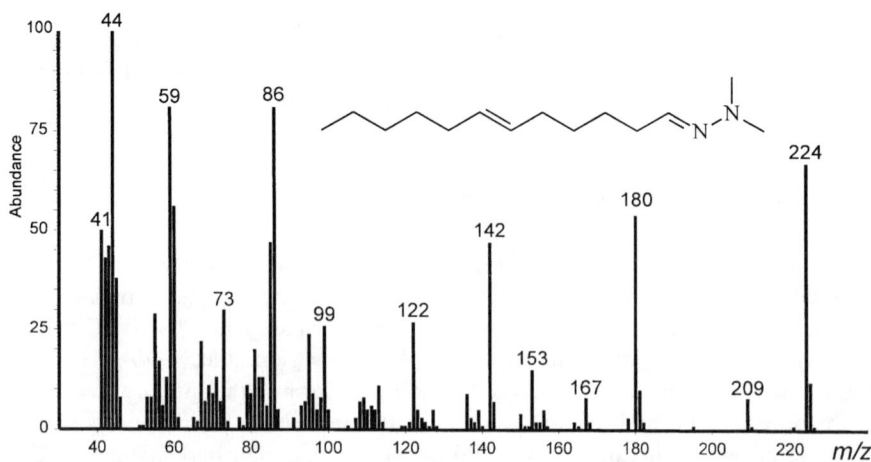

Figure 7.9. Electron-impact mass spectra (70 eV) of *N,N*-dimethylhydrazones of (*E*)-4-dodecenal, (*E*)-5-dodecenal, and (*E*)-6-dodecenal.

$$\underset{\text{(130)}}{R-\overset{\text{H}}{\underset{}{C}}=O} \quad + \quad H_2N-O-CH_3 \quad\longrightarrow\quad R-\overset{\text{H}}{\underset{}{C}}=N-O-CH_3$$

The method is particularly useful in conjunction with the DMDS procedure for the location of double bonds in unsaturated carbonyl compounds. For example, the mass spectrum of the *O*-methyl oxime of the DMDS derivative of (Z)-24-tritriaconten-2-one (**131**) showed, in addition to a molecular ion at *m/z* 599 (10%), diagnostic signals at *m/z* 173 and 426, establishing the double bond at C-24 of the underivatized ketone (Mason et al. 1990).

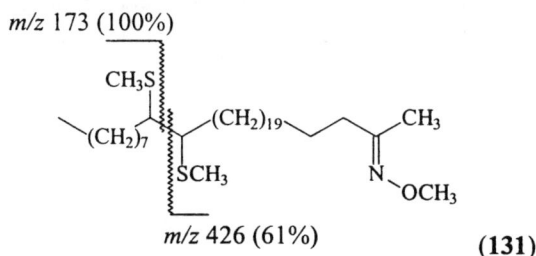

m/z 173 (100%)

m/z 426 (61%)   (**131**)

For low-molecular-weight aldehydes such as formaldehyde, derivatization with *O*-benzylhydroxylamine is recommended since benzyl oximes are less volatile than *O*-methyl oximes (Ollett et al. 1986).

7.2.4.5. Reactions of 1,2-Dicarbonyl Compounds

1,2-Dicarbonyl compounds are rather unstable and consequently direct gas chromatography is not recommended for their determination. Compounds such as glyoxal are formed when conjugated compounds are subjected to ozonolysis. Glyoxal (**132**) and other 1,2-dicarbonyl compounds can be analyzed after derivatization with *O*-phenylenediamine (**133**) to quinoxalines (**134**) (Moore & Brown 1971, Cronin & Gilbert 1973). The reagent (10% in ethyl acetate) is added to the ozonides and heated at 100°C for a few minutes followed by direct analysis by GC-MS.

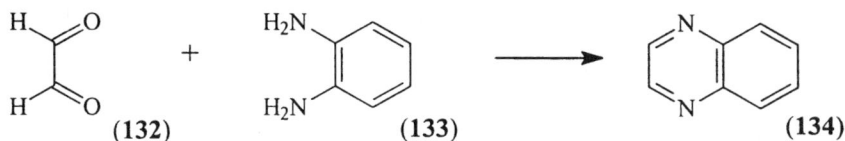

(**132**)   (**133**)   (**134**)

7.2.4.6. Reactions of 1,3-Dicarbonyl Compounds

An unusual group of compounds, 3-keto aldehydes, have been characterized from the defensive secretion of termites (Prestwich & Collins 1981). Since these unstable compounds exist in tautomeric forms (**135, 136**), they exhibit poor gas chromatographic properties. However, the conversion of these compounds to isoxazole derivatives (**137**) with hydroxylamine not only improves GC peak shapes but also provides more useful mass spectra.

(**135**)   (**136**)

(Prestwich & Collins 1981). The compound to be derivatized is dissolved in ethanol, treated with K_2CO_3 and hydroxylamine hydrochloride and heated at 80°C for 2 h. After the addition of dilute HCl and water, the mixture is extracted with hexane/ether (1/1) and the organic layer is analyzed by GC-MS.

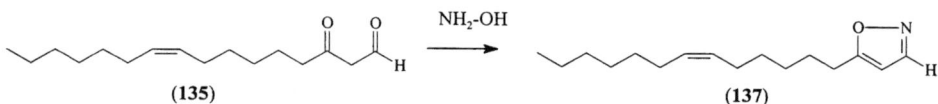

(**135**)   (**137**)

In general, 1,3-dicarbonyl compounds can be converted to pyrazole derivatives by the reaction with hydrazine (Moore & Brown 1971). Although mass spectra of pyrazoles show stronger molecular ions compared to those of isoxazoles, the latter are recommended because they give sharper GC peaks.

7.2.4.7. Reactions of 1,4-dicarbonyl Compounds

1,4-Dicarbonyl compounds (**138**) are converted to pyridazinines (**139**) with hydrazine (**140**) (Moore & Brown 1971). Pyridazinines are easily recognized by GC-MS since the spectra show a loss of two mass units from the molecular ion due to facile aromatization of the ring.

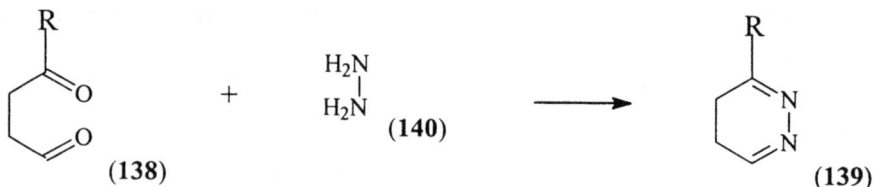

(**138**)   +   (**140**) $\longrightarrow$ (**139**)

7.2.5. Reactions of Carboxylic Acids

A number of carboxylic acids are known among semiochemicals. Although carboxylic acids can be analyzed by GC on well-deactivated columns, derivatization is highly recommended since it reduces adsorption losses, improves peak shapes, and acts as a diagnostic test for carboxylic acids.

7.2.5.1. *Reduction of the Carboxyl Group*

Similar to the reaction described for aldehydes and ketones, carboxylic acids can be reduced by lithium aluminum hydride in dry ether to primary alcohols. The reaction is somewhat slow, but near quantitative yields can be obtained with sufficient reaction times (24 h).

7.2.5.2. *Formation of Methyl esters*

Formation of methyl esters is a well-established derivatization procedure for carboxylic acids. The EI-mass spectra of the esters provide molecular weight information and also indicate whether substitution or unsaturation is present at carbon atoms in the vicinity of the carboxyl group. For example, methyl esters of 3-hydroxy acids show an intense diagnostic ion at *m/z* 103, which in fact is the base peak of the spectrum, indicating the position of the hydroxy group. 3-Hydroxydecanoic acid and its homologs are found in the metapleural gland secretions of attine ants (Nasciemento et al. 1996). The mass spectrum of methyl 3-hydroxydecanoate (**141**) shows the base peak at *m/z* 103. In general, methyl ester formation is a highly recommended method to locate a hydroxy group present in the carbon chain of the fatty acid. Although the intensity of the diagnostic ion diminishes when the hydroxyl group is remote from the carboxyl group, unambiguous conclusions can still be made from the mass spectrum (Nasciemento et al. 1996).

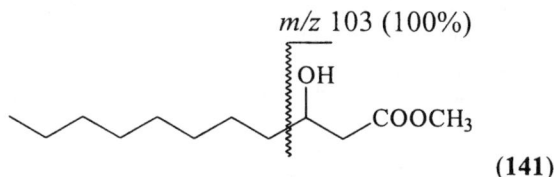

m/z 103 (100%)

(**141**)

Similarly, the formation of the methyl esters helps to locate the position of methyl groups in methyl-branched carboxylic acids. For example, the mass spectrum of methyl 4-methylhexanoate (**143**) obtained from the hydrogenated product of (3Z)-4-methyl-3,5-hexadienoic acid (**142**) established the position of the methyl group (Frenzel et al. 1990).

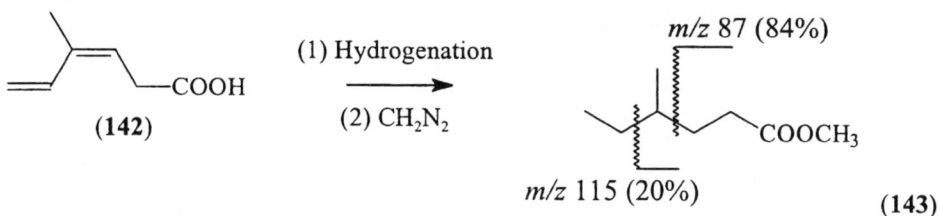

(**142**) (1) Hydrogenation (2) CH$_2$N$_2$

m/z 87 (84%)

m/z 115 (20%) (**143**)

Several methods are available for the formation of methyl esters from carboxylic acids. One of the quickest is the reaction with diazomethane (CH_2N_2). This reagent is a gas at room temperature. It is usually used in dilute ether or tetrahydrofuran solutions. One convenient method to make a solution of CH_2N_2 is to add a drop of dilute KOH to an ethereal solution of *N*-methyl-*N'*-nitro-*N*-nitrosoguanidine (MNNG). The ether layer, which becomes yellow immediately, is withdrawn with a syringe and added to the fatty acid sample until a slight yellow color persists. The mixture is then evaporated and the residue is dissolved in hexane for GC-MS analysis. *Diazomethane is highly toxic and explosive, and must be handled with due caution in a fume hood.* Excess diazomethane can be destroyed with a dilute solution of acetic acid. Reaction of the acid in benzene/methanol or ether/methanol with (trimethylsilyl)diazomethane, available from Aldrich as a hexane solution, is a safer alternative (Hashimoto et al. 1981).

Other methods for the formation of methyl esters include methanol-sulfuric acid, methanol-hydrochloric acid (Hornstein et al. 1960), and boron trifluoride methanol (Metcalfe & Schmitz 1966). These methods should be used with caution since they may result in undesired reactions, such as addition of methanol to double bonds. Blau and Dabre (1993b) provide an excellent summary of esterification reactions for carboxylic acids.

7.2.5.3. *Formation of Ethyl Esters*

Methods for the formation of ethyl esters are similar to those described for methyl esters. For example, diazoethane can be prepared from *N*-ethyl-*N'*-nitro-*N*-nitrosoguanidine. Another mild and convenient method is the treatment of acids with *N*,*N*-dimethylformamide diethyl acetal (**144**) (Vorbrüggen 1963) in a solvent such as dry ether or THF. The reaction mixture can be used directly or worked up by washing with water.

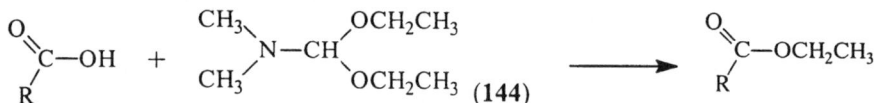

$$
\underset{R}{\overset{O}{\parallel}}C\!-\!OH \;+\; \underset{CH_3}{\overset{CH_3}{\diagdown}}N\!-\!CH\underset{OCH_2CH_3}{\overset{OCH_2CH_3}{\diagup}}\;(144) \longrightarrow \underset{R}{\overset{O}{\parallel}}C\!-\!OCH_2CH_3
$$

7.2.5.4. *Formation of Pentafluorobenzyl Esters*

Pentafluorobenzyl esters (**145**) of carboxylic acids offer a number of advantages. In addition to decreasing the polarity of acids, detectability is increased to picogram levels when a GC electron capture detector (ECD) is used. Also, the mass and gas phase infrared spectra of the derivatives offer additional information about the structures of the parent compounds. The derivatization procedure is simple; the acids in hexane are mixed with aqueous tetrabutyl ammonium hydroxide and sodium hydroxide, pentafluorobenzyl bromide is added and the contents are mixed well. After about 30 min, water is added and the products are extracted into hexane (Attygalle & Morgan 1986). Triethylamine or *N*-ethyl-piperidine can also be used as the base for the derivatization (Attygalle et al. 1991a).

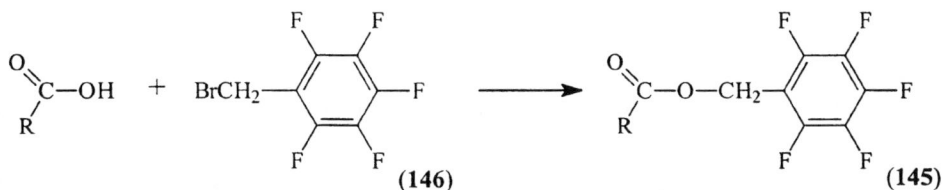

$$
\underset{R}{\overset{O}{\underset{\|}{C}}}\!-\!OH \;+\; BrCH_2\!-\!\!\!\!\!\bigcirc\!\!\!\!\!-F \quad\longrightarrow\quad \underset{R}{\overset{O}{\underset{\|}{C}}}\!-\!O\!-\!CH_2\!-\!\!\!\!\!\bigcirc\!\!\!\!\!-F
$$

(146) **(145)**

To distinguish the isomers of low-molecular-weight carboxylic acids such as tiglic, angelic, and senecioic acids from their mass spectra alone is not easy. However, the mass spectra of the corresponding pentafluorobenzyl esters are unique (Attygalle et al. 1991a). Likewise, the determination of formic acid directly by gas chromatography is not straightforward since this acid shows virtually no response with the flame-ionization detector (FID). Although pentafluorobenzyl formate gives a good response, it coelutes with the excess derivatization reagent, pentafluorobenzyl bromide, on nonpolar GC columns. This problem can be circumvented by using a polar column such as Carbowax®, on which the pentafluorobenzyl formate and pentafluorobenzyl bromide peaks are well separated. An alternative procedure is to use 1-(bromomethyl)-2,3,5,6-tetrafluoro-4-(trifluoromethyl)benzene instead of pentafluorobenzyl bromide.

Many applications of this procedure for the determination of acids are known (Attygalle et al. 1991b). For example, the method was adapted in the assay of minute quantities of indole-3-acetic acid from plant material (Epstein & Cohen 1981). In a comparative study, Markham et al. (1980) found that the PFB procedure is better than methylation for the detection of phenylacetic acid in plant tissues.

7.2.5.5. *Formation of p-Bromophenacyl Esters*

The method is usually used for the determination and isolation of fatty acids by HPLC methods (Durst et al. 1975), but derivatives of low-molecular-weight acids are amenable to GC analysis as well (Umeh 1971). The method is particularly useful for the determination of formic and other short chain acids in biological tissues. Weatherston et al. (1979) used this procedure for determination of formic acid in the larval defensive secretion of *Schizura concinna*. The derivatives were prepared by heating the acids with p-bromophenacyl bromide **(147)**. The derivatives **(148)** were extracted with an organic solvent after the addition of water.

$$R-\overset{O}{\underset{\|}{C}}-OH \quad + \quad Br-CH_2-\overset{O}{\underset{\|}{C}}\text{———}\langle\text{———}\rangle\text{———}Br \quad\longrightarrow\quad R-\overset{O}{\underset{\|}{C}}-O-CH_2-\overset{O}{\underset{\|}{C}}\text{———}\langle\text{———}\rangle\text{———}Br$$

(147)   (148)

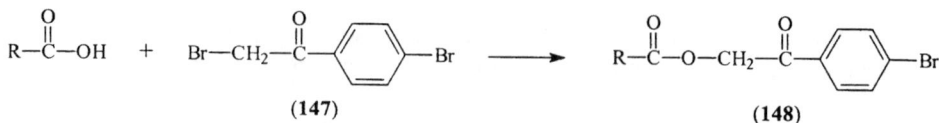

7.2.5.6. Formation of Nitrogen-Containing Derivatives

Fatty acids can be converted to a number of nitrogen-containing derivatives. For unsaturated fatty acids, the EI-mass spectra of such derivatives are more informative than those of methyl esters. Compounds that do not bear a charge-stabilizing functional group at a site remote to the double bond are known to undergo extensive double-bond scrambling under electron-impact conditions in the mass spectrometer. Tertiary amides such as pyrrolidides (**149**) (Vetter et al. 1971), and compounds such as picolinyl (Harvey 1982) (**151**), piperidyl (**152**), and morpholinyl (**153**) esters, have the ability to localize the positive charge on the nitrogen-containing functional group and thereby retard the double-bond isomerizations that take place before fragmentation in the mass spectrometer.

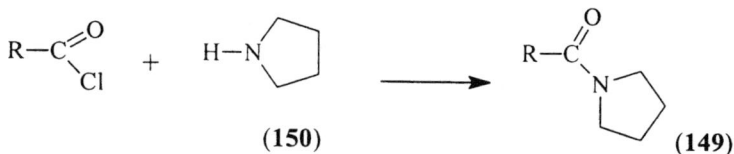

$$R-C\overset{\displaystyle O}{\underset{\displaystyle Cl}{}} \quad + \quad H-N\langle\rangle \quad\longrightarrow\quad R-C\overset{\displaystyle O}{\underset{\displaystyle N}{}}\langle\rangle$$

(150)   (149)

The EI-mass spectra of pyrrolidide derivatives obtained from monounsaturated fatty acids are simple and readily allow the location of the double bond (Andersson & Holman 1974). The spectra usually show a series of peak clusters separated by 14 mass units. However, this periodic pattern is interrupted when a carbon-carbon double bond is present. Thus, an increment of 26 mass units, instead of 28, between the first and third consecutive ion clusters suggest the position of a double bond. However, when the double bond is at carbon 2, 3, or 4, or the terminal or penultimate positions, the spectra fail to indicate the position unequivocally, although the spectra are unique to each isomer and can be used to compare with authentic standards. The method has been applied to polyunsaturated and methyl-branched fatty acids as well (Andersson et al. 1975).

In a comparative study, Jutila & Jalonen (1988) found that the diagnostic ions in the spectra of 4-hydroxy-1-methylpiperidyl esters (**154**) prepared from 4-hydroxy-1-methylpiperidine (**155**) to be more intense than those observed in esters of 1-piperidineethanol (**156**) and 4-(2-hydroxyethyl)morpholine (**157**).

$$R-C\overset{O}{\underset{Cl}{}} \quad + \quad HO-CH_2-CH_2-N\bigcirc \quad \longrightarrow \quad R-C\overset{O}{\underset{O-CH_2-CH_2-N\bigcirc}{}}$$

1-piperidineethanol (**156**) (**152**)

$$R-C\overset{O}{\underset{Cl}{}} \quad + \quad HO-CH_2-CH_2-N\bigcirc O \quad \longrightarrow \quad R-C\overset{O}{\underset{O-CH_2-CH_2-N\bigcirc O}{}}$$

4-(2-hydroxyethyl)morpholine
(**157**) (**153**)

$$R-C\overset{O}{\underset{Cl}{}} \quad + \quad HO-\bigcirc N-CH_3 \quad \longrightarrow \quad R-C\overset{O}{\underset{O-\bigcirc N-CH_3}{}}$$

4-hydroxy-1-methylpiperidine
(**155**) (**154**)

$$R-C\overset{O}{\underset{Cl}{}} \quad + \quad HO-CH_2-\bigcirc N \quad \longrightarrow \quad R-C\overset{O}{\underset{O-CH_2-\bigcirc N}{}}$$

3-(hydroxymethyl)pyridine (**158**) (**151**)

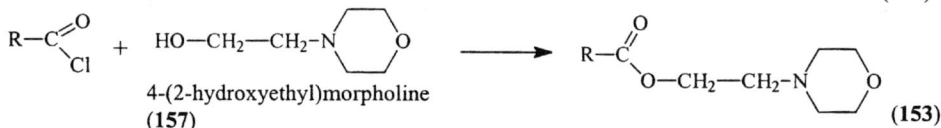

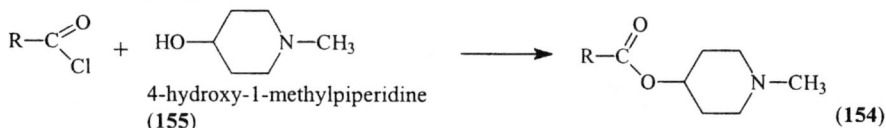

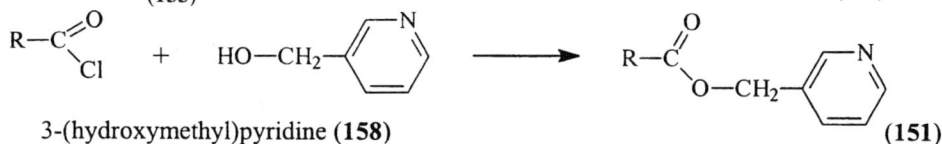

Similar results have been obtained with 3-picolinyl derivatives (**151**) (Harvey 1982; Christie et al. 1987) prepared from 3-(hydroxymethyl)pyridine (3-pyridyl-carbinol, 3-pyridinemethanol, **158**). Apparently, 3-picolinyl derivatives (**151**) offer better promise for locating methyl branchings, and double- and triple-bond positions (Christie et al. 1988) since the diagnostic ions are more abundant than those of pyrrolidides.

Pyrrolidides are prepared by treatment of acid chlorides with pyrrolidine (**150**). Alternatively, excellent yields are obtained by the aminolysis of the corresponding methyl esters. The esters are heated with pyrrolidine, with acetic acid catalysis, and the reaction mixture can be analyzed directly (Andersson & Holman 1974).

7.2.6. Reactions of Esters

Many known semiochemicals are esters, and the EI-mass spectra often exhibit diagnostic ions to suggest the acid moiety of the ester (Table 7.1). However, characterization of the alcohol portion is not as straightforward. Hence, esters are often subjected to microchemical reactions as an aid to structural elucidation.

7.2.6.1. Hydrolysis

Mild base hydrolysis is a diagnostic test for esters; if active peaks disappear, or biological activity changes on base hydrolysis, the active compound is virtually certain to be an ester. Amides are hydrolyzed by a base only under much more drastic conditions (strong base and elevated temperatures). Activity lost by hydrolysis can be restored by reesterification with the appropriate acid or acid chloride (Hendersen et al. 1972, 1973). Similarly, lactones can be hydrolyzed and then relactonized. In the characterization of 13-(1-methylpropyl)tridecanolide (159) from the lipids of the Argentine ant, Cavill et al. (1980) hydrolyzed the lactone to 13-hydroxy-14-methylhexadecanoic acid which was esterified to 160 and investigated by GC-MS. The spectrum of the methyl ester showed a significant peak at m/z 243 for the fragment arising for the α-cleavage adjacent to the carbon atom attached to the hydroxyl group.

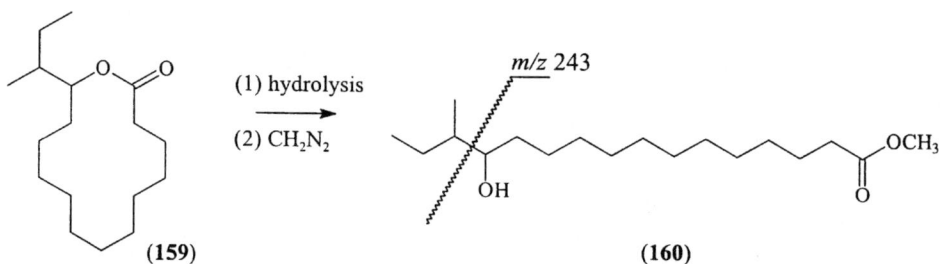

Typically, esters are mixed with 10% aqueous NaOH (5 M) in ethanol. Heating to 100°C for 1 h in a sealed tube may be required. After cooling, the mixture is acidified with dilute HCl and extracted with ether/hexane.

7.2.6.2. Reduction

Esters are reduced by lithium aluminum hydride to alcohols. In the identification of a macrocyclic lactone from the metasternal gland secretion of a beetle (*Phoracantha synonyma*), Moore and Brown (1976) reduced the compound (161) with LiAlH$_4$ to tetradeca-1,13-diol (162).

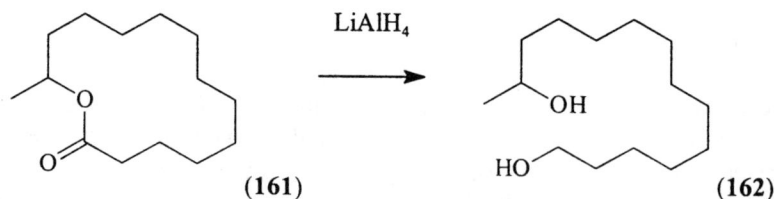

7.2.6.3. Transesterification

Esters are readily transesterified, replacing the alcohol moiety of the original ester with the alcohol used in excess as the solvent. This reaction is typically employed in the characterization of fatty acyl moieties of glycerides and other lipids. For example, triglycerides (163) are transesterified to the corresponding fatty acid methyl esters (FAMEs, 164). Several methods can be used. In one procedure, the fat sample is dissolved in methanol/hexane (4/1, 100 μl) and heated in a sealed tube with acetyl chloride (10 μl) at 100°C for 1 h (Fay & Richli 1991; Stránský & Jursik 1996). After cooling to room temp, aqueous K$_2$CO$_3$ is carefully added (foams!) until the mixture is basic, and the FAMEs are extracted with hexane. Similarly, acid methanolysis can be performed by heating a sample to 80°C with dry HCl/methanol (Lanne et al. 1987). After 1 h, the mixture is extracted with hexane, and the hexane extract is washed with dilute NaHCO$_3$.

Base methanolysis can also be used. The sample is treated with a solution of KOH in dry methanol (0.5 M, 25 μl) and the mixture is held at room temperature for 30 min. The mixture is neutralized with dilute HCl, then extracted with hexane.

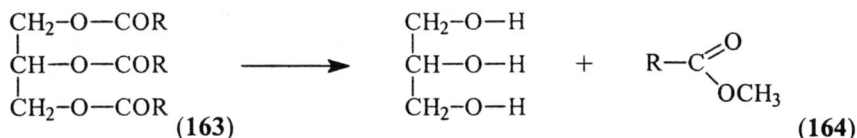

CH$_2$-O—COR
|
CH—O—COR ⟶ CH$_2$-O—H
| |
CH$_2$-O—COR   CH—O—H   +   R—C(=O)OCH$_3$
(163) |
            CH$_2$-O—H

(164)

Analogous to methods described for free fatty acids (section 7.2.5.6), FAMEs can be converted to a number of nitrogen-containing derivatives. The pyrrolidide derivatives (**149**) are prepared by heating FAMEs with pyrrolidine (**150**) and acetic acid at 100°C for 30 min. After the addition of dilute HCl, the amides are extracted with dichloromethane.

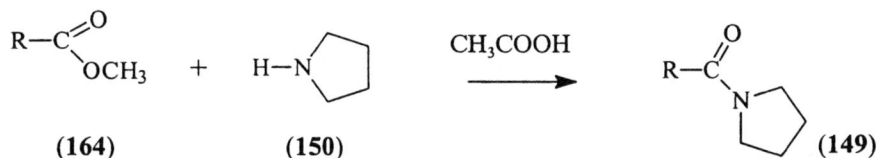

R—C(=O)OCH$_3$ +   H—N⟨pyrrolidine⟩ $\xrightarrow{\text{CH}_3\text{COOH}}$ R—C(=O)N⟨pyrrolidine⟩

(164)   (150)   (149)

Fabriás et al. (1989) used the formation of pyrrolidides for confirmation of the identities of monounsaturated fatty acids in the sex pheromone gland of the processionary moth, *Thaumetopoea pityocampa*.

3-Picolinyl derivatives (**151**) can also be produced by transesterification (Harvey 1982; Christie et al. 1987). The method appears to be useful for location of carbon-carbon triple bonds and methyl branches. However, when mixtures of isomeric methyl-branched compounds are derivatized, it is difficult to obtain reliable mass spectra because these derivatives usually coelute on GC and the composite mass spectrum does not allow unambiguous determination of the branch points (Simon et al. 1990). When such difficulties are encountered, it is better to convert the acids to alcohols and eventually to hydrocarbons to determine the branch positions (section 7.2.2.8).

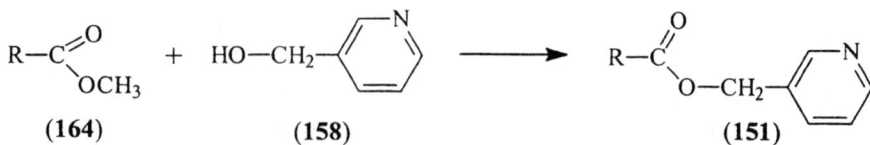

(164)   (158)   (151)

Mass spectrometric analysis of 4,4-dimethyloxazoline (DMOX) derivatives (165) formed with 2-amino-2-methylpropanol (166) (Yu et al. 1989) is another useful technique for unsaturated acids. The derivatives are formed by heating FAMEs with 2-amino-2-methylpropanol at 180°C for 12 h (Fay & Richli 1991). These derivatives show much better gas chromatographic properties than pyrrolidide or picolinyl derivatives. The mass spectrometric fragmentation patterns are similar to those of pyrrolidide derivatives, however, relative intensities of diagnostic ions are stronger in DMOX spectra. The method has been applied to conjugated fatty acids (Spitzer et al. 1994) and to compounds bearing up to six carbon-carbon double bonds, although the localization of double bonds proximal to the oxazoline ring is difficult. The spectra are useful also for determining the location of methyl branchings (Yu et al. 1988) and acetylenic bonds (Zhang et al. 1989) in fatty acids.

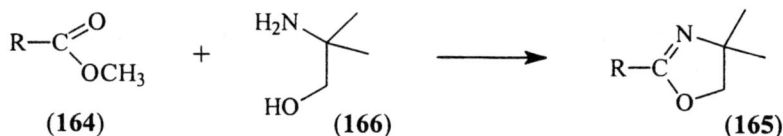

(164)   (166)   (165)

7.2.7. Reactions of Amines

Amines are polar compounds that can exhibit poor chromatographic properties. Their reactions are analogous to those of alcohols, and amides are formed from 1° and 2° amines under conditions similar to those used in ester formation, and so the derivatization reactions will not be discussed in detail.

7.2.7.1. Acetylation

Peddler et al. (1976) used acetylation to characterize the secondary amine nature of pyrrolidine alkaloids found in ant venoms. Acetylation was also the key step employed to show the presence of primary hydroxy and amino groups in 1-(2-hydroxyethyl)-2-(12-aminotridecyl) pyrrolidine (29) (Attygalle et al. 1993b). The gas-phase infrared spectrum of the diacetyl derivative (167) showed two strong absorptions in the carbonyl region, one at 1710 cm^{-1} confirming the secondary carboxamide, and another at 1760 cm^{-1} from the acetate ester group.

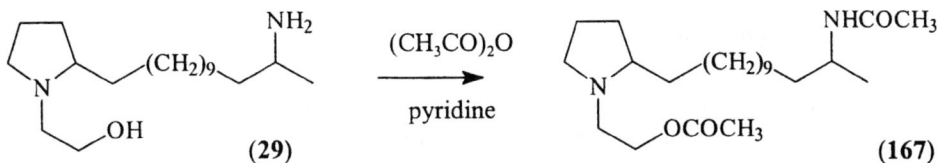

In polycyclic bridge-head amine oxides, the formation of *N*-acetates from *N*-oxides of the bridge-head nitrogen has been used to establish configurations at the ring junctions. The treatment of such amines with acetic anhydride leads to the formation of enamines due to the loss of elements of water. LaLonde et al. (1971) demonstrated that this conversion, known as the Polonovski transformation (Russel & Mikol 1968), involves first the formation of an *N*-acyloxyammonium acetate, which immediately loses predominantly any *trans* ring-junction hydrogen atoms to give an immonium ion. The immonium ion rapidly eliminates a proton to give an enamine. In general, the enamine is formed by an elimination of ring-junction hydrogen atoms *trans* to the *N*-oxide oxygen atom.

Tursch et al. (1975) used this procedure to establish the stereochemistry at the ring junctions of coccinelline (**168**), one of the defensive alkaloids of ladybird beetles. Coccinelline (*cis-trans-cis* ring fusion) in dichloromethane, when treated with acetic anhydride or ethyl chloroformate gave the unstable enamine (**169**) (*cis-cis*). Catalytic hydrogenation of this enamine gave a separable mixture of myrrhine (**170**) (90%) and precoccinelline (**171**) (10%), two compounds known to bear *cis-cis-cis* and *cis-trans-cis* configurations, respectively. In this way, it was shown that in coccinelline only one hydrogen atom is present *trans* to the *N*-oxide oxygen atom. A subsequent loss of a hydrogen from either side leads to the same enamine.

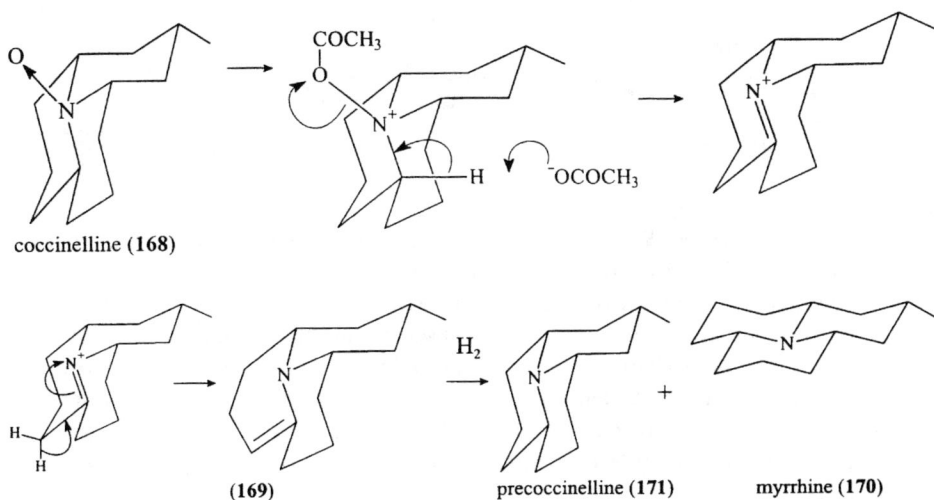

coccinelline (**168**)

(**169**)   precoccinelline (**171**)   myrrhine (**170**)

7.2.7.2. Trifluoroacetylation

Sometimes trifluoroacetylation of primary and secondary amines, followed by GC-MS analysis of the derivatives, provides more information than simple acetylation. Trifluoroacetic anhydride or N-methyl-*bis*(trifluoroacetamide) (MBTFA, **172**) can be used (Donike 1973). The very volatile by-products formed, trifluoroacetic acid or methyltrifluoroacetamide (**173**), respectively, can be removed by a slow stream of nitrogen. A few microliters of MBTFA or the anhydride are added to the amines and the mixture is heated at 60–100°C for about 30 min. The reaction mixture can be analyzed directly by GC-MS.

R-NH$_2$ + CH$_3$—N$\left(\begin{array}{l} \text{C—CF}_3 \\ \text{C—CF}_3 \end{array}\right)$ $\longrightarrow$ R-NH-COCF$_3$ + CH$_3$-NH-COCF$_3$

MBTFA (**172**)   (**173**)

For example, in the characterization of 2-(12-aminotridecyl)-pyrrolidine (**174**), the *bis*-trifluoroacetyl derivative (**175**) showed a base peak at m/z 166 from the expected α-cleavage. In contrast, in the spectrum of the diacetyl derivative of **174,** although an intense signal is observed at m/z 112, the base peak is still observed at m/z 70 similar to that of the underivatized compound (Attygalle et al. 1993b).

In another application, Weitz et al. (1986) used trifluoroacetylation to determine morphine and codeine from mammalian brain.

Heptafluorobutyramides of amines can be formed in a similar manner. The amine is mixed with heptafluorobutyric anhydride and the mixture is warmed to about 50°C for about 10 min. After the addition of aqueous NaHCO$_3$, the mixture is extracted with hexane and analyzed by GC-MS. For example, in the identification of 2-(5-hexenyl)-5-(8-nonenyl)pyrrolidine (**64**), and related compounds from ants heptafluorobutyramide derivatives were formed (Jones et al. 1982) (section 7.2.1.6.4).

7.2.7.3. Pentafluorobenzyl Derivatives

Pentafluorobenzyl bromide (PFBB) reacts with primary and secondary amines. This reaction leaves the hydroxy groups of alcohols and phenols intact, in contrast to acylation reactions which derivatize both OH and NH groups. In fact, this derivatization is quite helpful in locating an amino group in a carbon chain because the mass spectrum of the derivative clearly indicates the position of the amino group. For example, the presence of a -CH(NH$_2$)-CH$_3$ moiety in 1-(2-hydroxyethyl)-2-(12-aminotridecyl) pyrrolidine (**29**) was shown by the *m/z* 224 peak due to an α-cleavage observed in the mass spectrum of the derivative (**176**) (Attygalle et al. 1993b).

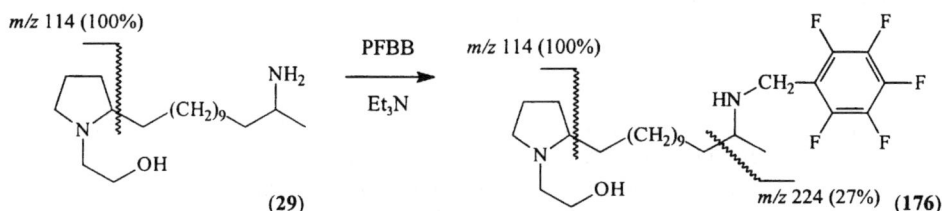

7.2.7.4. N-Methylation of Amines

Small amounts of *N*-methyl amines are sometimes required for comparison with natural substances. They are readily made from primary amines by reaction with 28% formaldehyde (20 μl) and formic acid (20 μl), warmed in a closed vial for 10 h. The solution is basified with 10% KOH and the products extracted with dichloromethane (Jones et al. 1982). For example, the *N*-methyl derivative (**178**) of 2-(5-hexyl)-5-(nonyl)pyrrolidine (**177**) was prepared in this way. Direct reaction of the amine with methyl iodide would have given the nonvolatile tetrasubstituted amine salt.

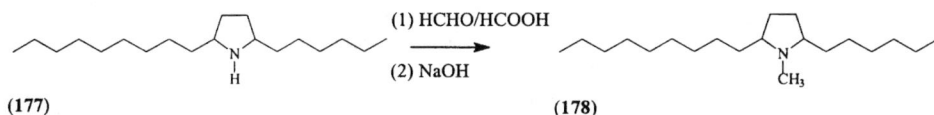

7.2.7.5. Oxidation of Piperidines

t-Butylhypochlorite converts 2-methyl-6-alkylpiperidines to the corresponding piperideines (Brand et al. 1972). This is a useful reaction to synthesize small amounts of material for retention time and spectral comparisons. For example, treatment of *cis*-2-methyl-6-undecylpiperidine (**179**) results in two products, 2-methyl-6-undecyl-Δ^{1,6}-piperideine (**180**) and its Δ^{1,2}-isomer (**181**). The reaction is conducted by mixing aqueous *t*-butylhypochlorite (2 μl) with the piperidine (about 1 μg). After 1 h at room temperature, the reaction mixture is diluted with dichloromethane and basified with aqueous NaOH. The layers are mixed well and the organic layer is removed for analysis.

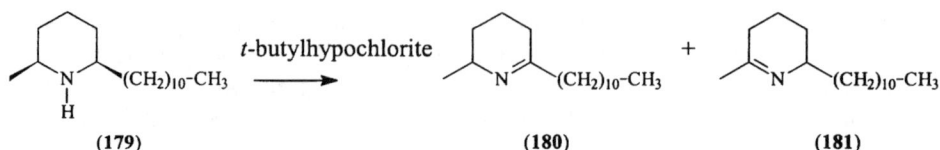

(179)   (180)   (181)

The mass spectra of the two isomers are significantly different. The base peak of the first-eluting 1,6-isomer is at m/z 111, while that of the later-eluting 1,2-isomer is at m/z 110.

7.2.7.6. Reduction of Imines

Cyclic imines such as piperideines and pyrrolines are often found in defensive secretions of insects. Reduction by sodium borohydride provides proof for the presence of the C=N linkage in these compounds (Pedder et al. 1976). For example, 2,5-disubstituted pyrrolines (e.g., **183** → **184** + **185**) give equal amounts of *cis* and *trans* isomers with the *cis* isomer usually eluting first on GC. On the other hand, reduction of disubstituted piperideines provides overwhelmingly the *cis* form (e.g., **180** → **179** + **182**). The use of NaBD$_4$ instead of NaBH$_4$ provides a way to distinguish the reduction products from endogenous saturated components that may be present in a natural mixture before the reduction.

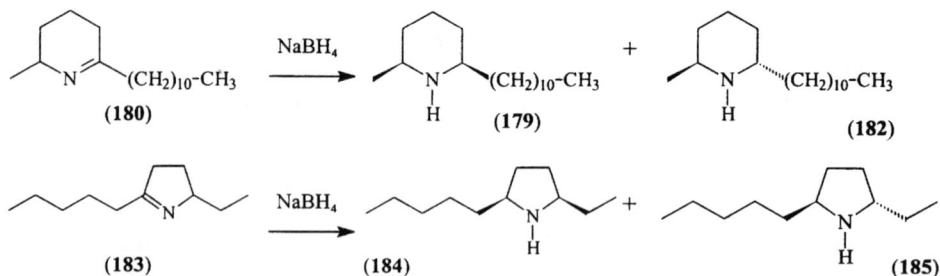

(180)   (179)   (182)

(183)   (184)   (185)

7.3. Conclusion

Many chemical ecologists interested in the characterization of semiochemicals may encounter problems that require the use of reactions discussed in this chapter. While the intention of this chapter is not to provide a comprehensive review of all available methods, the selected examples discussed here may stimulate chemists to explore other microchemical methods. These procedures, in conjunction with GC-MS, behavioral, or other analyses can often provide sufficient data for complete structure elucidation of many unknown semiochemicals and thereby avoid the necessity of the large-scale isolations required for NMR analysis.

7.4. Acknowledgment

Many helpful suggestions of Professors Jerrold Meinwald and Jocelyn Millar, and the assistance of Mr. Kithsiri Herath with many figures, are gratefully acknowledged.

7.5. References

Abley, P., D. McQuillin, D.E. Minnikin, K. Kusamran, K. Maskens & N. Polgar. 1970. Location of olefinic links in long-chain esters by methoxymercuration-demercuration followed by gas chromatography-mass spectrometry. Chem. Commun. pp. 348–349.

Adhikary, B.A. & R.A. Harkness. 1969. Determination of carbon skeletons of microgram amounts of steroids and sterols by gas chromatography after their high temperature catalytic reduction. Anal. Chem. **41**:470–476.

Andersson, B.A. & R.T. Holman. 1974. Pyrrolidides for mass spectrometric determination of the position of the double bond in monounsaturated fatty acids. Lipids **9**:185–190.

Andersson, B.A., W.W. Christie & R.T. Holman. 1975. Mass spectrometric determination of positions of double bonds in polyunsaturated fatty acid pyrrolidides. Lipids **10**:215–219.

Aplin, R.T. & L. Coles. 1967. A simple procedure for location of ethylenic bonds by mass spectrometry. Chem. Commun. pp. 858–859.

Asakawa, Y., R. Matsuda, & M. Tori. 1986. Hydroxylation of menthols and cineoles with *m*-chloroperbenzoic acid. Experientia **42**:201–203.

Attygalle, A.B. 1994. Gas phase infrared spectroscopy in characterization of unsaturated natural products. Pure Appl. Chemistry **66**:2323–2326.

Attygalle, A.B., K.B. Herath & J. Meinwald. 1998. Cycloalkene budding: A unique rearrangement observed in the mass spectra of N,N-dimethylhydrazones of unsaturated aldehydes. J. Org. Chem. **63**:(issue 2, Jan 24, in press).

Attygalle, A.B. & E.D. Morgan. 1982. Structures of homofarnesene and bishomofarnesene isomers from *Myrmica* ants. J. Chem. Soc. Perkin Trans. I. pp. 949–951.

Attygalle, A.B. & E.D. Morgan. 1983. Reaction gas chromatography without solvent, for identification of nanogram quantities of natural products. Anal. Chem. **55**:1379–1384.

Attygalle, A.B. & E.D. Morgan. 1984a. Determination of oxirane ring position in epoxides at the nanogram level by reaction gas chromatography. Anal. Chem. **56**:1530–1533.

Attygalle, A.B. & E.D. Morgan. 1984b. Chemical reactions with nanogram quantities of compounds collected from GC effluent. J. Chromatogr. **290**:321–330.

Attygalle, A.B. & E.D. Morgan. 1986. Versatile microreactor and extractor. Anal. Chem. **58**:3054–3058.

Attygalle, A.B. & E.D. Morgan. 1988. Pheromones in nanogram quantities: Structure determination by combined microchemical and gas chromatographic methods. Angew. Chem. Int. Ed. Engl. **27**:460–478.

Attygalle, A.B., A. Zlatkis & B.S. Middleditch. 1989. Derivatization with 1,1-dimethylhydrazine for identification of carbonyl compounds resulting from ozonolysis. J. Chromatogr. **472**:284–289.

Attygalle, A.B., J. Meinwald, J.K. Liebherr & T. Eisner. 1991a. Sexual dimorphism in the defensive secretion of a carabid beetle. Experientia **47**:296–299.

Attygalle, A.B., T. Eisner & J. Meinwald. 1991b. Biosynthesis of methacrylic and isobutyric acids in a carabid beetle, *Scarites subterraneus.* Tetrahedron Lett. **32**:4849–4852.

Attygalle, A.B., K.D. McCormick, C.L. Blankespoor, T. Eisner & J. Meinwald. 1993a. Azamacrolides: a new family of alkaloids from a coccinelid beetle. Proc. Natl. Acad. Sci. USA. **90**:5204–5208.

Attygalle, A.B., S.-C. Xu, K.D. McCormick, C.L. Blankespoor, T. Eisner & J. Meinwald. 1993b. Alkaloids of the Mexican bean beetle, *Epilachna varivestis* (Coccinelidae). Tetrahedron **49**:9333–9342.

Attygalle, A.B., G.N. Jham & J. Meinwald. 1993c. Determination of double bond position in some unsaturated terpenes and other branched compounds by alkylthiolation. Anal. Chem. **65**:2528–2533.

Attygalle, A.B., A. Svatoš, C. Wilcox & S. Voerman. 1994. Gas-phase infrared spectroscopy for determination of double-bond configuration of monounsaturated compounds. Anal. Chem. **66**:1696–1703.

Attygalle, A.B., G.N. Jham, A. Svatoš, R.T.S. Frighetto, J. Meinwald, E. Vilela, F.A. Ferrara & M.A. Uchoa-Fernandes. 1995a. Microscale, random reduction: Application to the characterization of (3*E*,8*Z*,11*Z*)-3,8,11-tetradecatrienyl acetate, a new lepidopteran sex pheromone. Tetrahedron Lett. **36**:5471–5474.

Attygalle, A.B., A. Svatoš, C. Wilcox & S. Voerman. 1995b. Gas-phase infrared spectroscopy for the determination of double-bond configuration of some polyunsaturated pheromones and related compounds. Anal. Chem. **67**:1558–1567.

Attygalle, A.B., G.N. Jham, A. Svatoš, R.T.S. Frighetto, F.A. Ferrara, M.A. Uchôa-Fernandes, E. Vilela & J. Meinwald. 1996. (3*E*,8*Z*,11*Z*)-3,8,11-Tetradecatrienyl acetate, a sex attractant for the tomato pest *Scrobipalpuloides absoluta.* Biorg. Med. Chem. **4**:305–314.

Ayasse, M., W. Engels, A. Hefetz, G. Lübke & W. Francke. 1990. Ontogenic patterns in amounts and proportions of Dufour's gland volatile secretions in virgin and nesting queens of *Lasioglossum malachurum* (Hymenoptera: Halictidae). Z. Naturforsch. **45C**:709–714.

Baker, R., J.W.S. Bradshaw & W. Speed. 1992. Methoxymercuration-demercuration and mass spectrometry in the identification of the sex pheromone of *Panolis flammea,* the pine beauty moth. Experientia **38**:233–234.

Beroza, M. & B.A. Bierl. 1966. Apparatus for ozonolysis of microgram to milligram amounts of compound. Anal. Chem. **38**:1976–1977.

Beroza, M. & B.A. Bierl. 1967. Rapid determination of olefin position in organic compounds in microgram range by ozonolysis and gas chromatography. Anal. Chem. **39**:1131–1135.

Beroza, M. & B.A. Bierl. 1969. Ozone generator for microanalysis. Mikrochim. Acta pp. 720–723.

Beroza, M. & R. Sarmiento. 1963. Determination of the carbon-skeleton and other structural features of organic compounds by gas chromatography. Anal. Chem. **35**:1353–1357.

Beroza, M. & R. Sarmiento. 1966. Apparatus for reaction gas chromatography. Instantaneous hydrogenation of unsaturated esters, alcohols, ethers, ketones, and other compound types, and determination of their separation factors. Anal. Chem. **38**:1042–1047.

Bestmann, H.J., T. Brosche, K.H. Koschatzky, K. Michaelis, H. Platz, O. Vostrowsky & W. Knauf. 1980. Pheromone XXX. Identifizierung eines neuartiges Pheromonekomplexes aus der Graseule *Scotia exclamationis*. Tetrahedron Lett. **21**:747–750.

Bestmann, H.J., A.B. Attygalle, J. Glasbrenner, R. Riemer & O. Vostrowsky. 1987. The absolute configuration of the ant alarm pheromone Manicone. Angew. Chemie Int. Ed. Engl. **26**:784–785.

Bestmann, H.J., A.B. Attygalle, J. Schwarz, W. Garbe, O. Vostrowsky & I. Tomida. 1989. Identification and synthesis of female sex pheromone of eri-silk worm, *Samia cynthia ricini* (Lepidoptera: Saturniidae). Tetrahedron Lett. **30**:2911–2914.

Bierl, B.A., M. Beroza & M.H. Aldridge. 1971. Determination of epoxide position and configuration at the microgram level and recognition of epoxides by reaction thin-layer chromatography. Anal. Chem. **43**:636–641.

Bierl-Leonhardt, B.A. & E.D. DeVilbiss. 1980. Structure identification of terpene-type alcohols at microgram levels. Anal. Chem. **53**:936–938.

Bierl-Leonhardt, B.A., D.S. Moreno, M. Schwartz, H.S. Foster, J.R. Plimmer & E.D. Devilbiss. 1980a. Identification of the pheromone of the comstock mealybug. Life Sci. **27**:399–402.

Bierl-Leonhardt, B.A., E.D. Devilbiss & J.R. Plimmer. 1980b. Location of double-bond position in long-chain aldehydes and acetates by mass spectral analysis of epoxide derivatives. J. Chromatogr. Sci. **18**:364–367.

Bierl-Leonhardt, B.A., D.S. Moreno, M. Schwartz, J. Fargerlund & J.R. Plimmer. 1981. Isolation, identification and synthesis of the sex pheromone of the citrus mealybug, *Planococcus citri* (Risso). Tetrahedron Lett. **22**:389–392.

Billen, J.P.J., R.P. Evershed, A.B. Attygalle, E.D. Morgan & D.G. Ollett. 1986. Contents of Dufour glands of workers of three species of *Tetramorium* (Hymenoptera: Formicidae). J. Chem. Ecol. **12**:669–685.

Bjostad, L.B., D.K. Jewett & D.L. Brigham. 1996. Sex pheromone of caddisfly *Herperophylex occidentalis* (Banks) (Trichoptera: Limnephilidae). J. Chem. Ecol. **22**:103–121.

Blau, K. 1993. Practical considerations. *In:* Handbook for Derivatives for Chromatography, 2nd ed., eds. K. Blau & J.M. Halket, pp. 349–356, Wiley, New York.

Blau, K. & A. Dabre. 1993a. Formation of cyclic derivatives. *In:* Handbook for Derivatives for Chromatography, 2nd ed., eds. K. Blau & J.M. Halket, pp. 141–155, Wiley, New York.

Blau, K. & A. Dabre. 1993b. Esterification. *In:* Handbook for Derivatives for Chromatography, 2nd ed., eds. K. Blau & J.M. Halket, pp. 11–30, Wiley, New York.

Blau, K. & J.M. Halket. 1993. Handbook for Derivatives for Chromatography, 2nd ed, Wiley, New York.

Blau, K. & G.S. King. 1977. Handbook for Derivatives for Chromatography. Heyden, London.

Blomquist, G.J., R.W. Howard, C.A. McDaniel, S. Remaley, L.A. Dwyer & D.R. Nelson. 1980. Application of methoxymercuration demercuration followed by mass spectrometry as a convenient microanalytical technique for double-bond location in insect-derived alkenes. J. Chem. Ecol. **6**:257–269.

Bouchoux, G., Y. Hoppilliard, P. Jaudon & J.-M. Pechine. 1987. Determination of epoxide position in aliphatic chain by negative ion chemical ionization mass spectrometry. Spectros. Int. J. **5**:247–252.

Bowden, K., I.M. Heibron, E.R.H. Jones & B.C.L. Weedon. 1946. Researches on acetylenic compounds. Part I. The preparation of acetylenic ketones by oxidation of acetylenic carbinols and glycols. J. Chem. Soc. pp. 39–45.

Braconnier, M.F., J.C. Braekman, D. Daloze & J.M. Pasteels. 1985. (Z)-1,17-Diaminooctadec-9-ene, a novel aliphatic diamine from Coccinellidae. Experientia. **41**:519–520.

Brand, J.M., M.S. Blum, H.M. Fales & J.G. MacConnell. 1972. Fire ant venoms: Comparative analysis of alkaloidal components. Toxicon **10**:259–271.

Brown, W.V. & B.P. Moore. 1982. The defensive alkaloids of *Cryptolaemus montrouzieri* (Coleoptera: Coccinellidae). Aust. J. Chem. **35**:1255–1261.

Brownlee, R.G. & R.M. Silverstein. 1968. A micro-preparative gas chromatograph and a modified carbon skeleton determinator. Anal. Chem. **40**:2077–2079.

Budzikiewicz, H. 1988. Chapter 1, *In:* Studies in Natural Products Chemistry, Vol. 2, ed. Atta-ur-Rahman, pp. 3–18, Elsevier, Amsterdam.

Burk, M.J., R.H. Crabtree & D.V. McGrath. 1986. Identification and determination of millimolar C_6–C_8 alkenes in the corresponding alkanes. Anal. Chem. **58**:977–978.

Buser, H.-R., H. Arn, P. Guerin & S. Rauscher. 1983. Determination of double bond position in monounsaturated acetates by mass spectrometry of dimethyl disulfide adducts. Anal. Chem. **55**:818–822.

Carballeira, N.M., F. Shalabi & C. Cruz. 1994. Thietane, tetrahydrothiophene and tetrahydrothiopyran formation in reaction of methylene interrupted dienoates with dimethyl disulfide. Tetrahedron Lett. **35**:5575–5578.

Carlsen, P.H.J., T. Katsuki, V.S. Martin & K.B. Sharpless. 1981. A greatly improved procedure for ruthenium tetroxide catalyzed oxidations of organic compounds. J. Org. Chem. **46**:3936–3938.

Carlson, D.A., C.-S. Roan, R.A. Yost & J. Hector. 1989. Dimethyl disulfide derivatives of long chain alkenes, alkadienes, and alkatrienes for gas chromatography/mass spectrometry. Anal. Chem. **61**:1564–1571.

Caserio, M.C., C.L. Fisher & J.K. Kim. 1985. Boron trifluoride catalyzed addition of disulfide to alkenes. J. Org. Chem. **50**:4390–4393.

Cavill, G.W.K., E. Houghton, F.J. McDonald & P.J. Williams. 1976. Isolation and characterization of dolichodial and related compounds from the Argentine ant, *Iridomyrmex humulis.* Insect Biochem. **6**:483–490.

Cavill, G.W.K., N.W. Davies & F.J. McDonald. 1980. Characterization of aggregation factors and associated compounds from the Argentine ant, *Iridomyrmex humulis.* J. Chem. Ecol. **6**:371–384.

Christie, W.W., E.Y. Brechany & R.T. Holman. 1987. Mass spectra of the picolinyl esters of isomeric mono and dienoic fatty acids. Lipids **22**:224–228.

Christie, W.W., E.Y. Brechany & M.S.F. Lie Ken Jie. 1988. Mass spectra of the picolinyl ester derivatives of some isomeric dimethylene-interrupted octadecadiynoic acids. Chem. Phys. Lipids. **46**:225–229.

Christie, W.W., E.Y. Brechany & V.K.S. Shukla. 1989. Analysis of seed oils containing cyclopentenyl fatty acids by combined chromatographic procedures. Lipids **24**:116–120.

Cimino, G., A. Passeggio, G. Sodano, A. Spinella & G. Villiani. 1991. Alarm pheromone from the Mediterranean opisthobranch *Haminoea navicula.* Experientia **47**:61–63.

Clive, D.L.J. 1978. Modern organoselenium chemistry. Tetrahedron **34**:1049–1132.

Cookson, R.C., I.D.R. Stevens & C.T. Watts. 1966. A new reagent for oxidation of alcohols to ketones in neutral solution at room temperature. Chem. Commun. pp. 744.

Corey, E.J. & G. Schmidt. 1979. Useful procedures for the oxidation of alcohols involving pyridinium dichromate in aprotic media. Tetrahedron Lett. pp. 399–402.

Corey, E.J. & J.W. Suggs. 1975. Pyridinium chlorochromate. An efficient reagent for oxidation of primary and secondary alcohols to carbonyl compounds. Tetrahedron Lett. pp. 2647–2650.

Corey, E.J., W.L. Mock & D.J. Pasto. 1961. Chemistry of diimide. Some new systems for the hydrogenation of multiple bonds. Tetrahedron Lett. pp. 347–352.

Cronin, D.A. & J. Gilbert. 1973. Hydrogenation and ozonolysis of submicrogram quantities of unsaturated organic compounds eluted from gas chromatographic columns. J. Chromatogr. **87**:387–400.

Davison, V.L. & H.J. Dutton. 1966. Microreactor chromatography. Quantitative determinations of double bond positions by ozonization-pyrolysis. Anal. Chem. **38**:1302–1305.

Dess, D.B. & J.C. Martin. 1983. Readily accessible 12-I-5 oxidant for the conversion of primary and secondary alcohols to aldehydes and ketones. J. Org. Chem. **48**:4155–4156.

Donike, M. 1973. Acylierung mit bis(Acylamiden). *N*-Methyl-bis(trifluoracetamid) und bis(trifluoracetamid), zwei neue Reagenzien zur trifluoracetylierung. J. Chromatogr. **78**:273–279.

Doolittle, R.E., J.H. Tumlinson & A. Proveaux. 1985. Determination of double bond position in conjugated dienes by chemical ionization mass spectrometry with isobutane. Anal. Chem. **57**:1625–1630.

Doolittle, R.E., A. Brabham & J.H. Tumlinson. 1990. Sex pheromone of *Manduca sexta* (L). A stereoselective synthesis of (10*E*,12*E*,14*Z*)-10,12,14-hexdecatrienal and isomers. J. Chem. Ecol. **16**:1131–1153.

Dunkelblum, E., S.H. Tan & P.J. Silk. 1985. Double-bond location in monounsaturated fatty acids by dimethyl disulfide derivatization and mass spectrometry: application to analysis of fatty acids in pheromone glands of four Lepidoptera. J. Chem. Ecol. **11**:265–277.

Durst, H.D., M. Milano, E.J. Kikta Jr., S.A. Connelly & E. Grushka. 1975. Phenacyl esters of fatty acids via crown ether catalysis for enhanced ultraviolet detection in liquid chromatography. Anal. Chem. **47**:1797–1801.

Edwards, J.P. & J. Chambers. 1984. Identification and source of a queen-specific chemical in the Pharaoh's ant *Monomorium pharaonis*. J. Chem. Ecol. **10**:1731–1747.

Einhorn, J., H. Virelizier, A.L. Gemal & J.C. Tabet. 1985. Direct determination of double bond position in long chain conjugated dienes by t.$C_4H_9^+$-chemical ionization mass spectrometry. Tetrahedron Lett. **26**:1445–1448.

Einhorn, J., H. Virelizier & J.C. Tabet. 1987. New reagents to locate conjugated double bonds in the gas phase. Spectros. Int. J. **5**:171–182.

Einhorn, J., P. Menassieu, C. Malosse & P.-H. Ducrot. 1990. Identification of the sex pheromone of the maritime pine scale *Matsucoccus feytaudi*. Tetrahedron Lett. **31**:6633–6636.

Epstein, E. & J.D. Cohen. 1981. Electron-capture gas chromatographic detection of indole-3-acetic acid from plants. J. Chromatogr. **209**:413–420.

Evershed, R.P. 1993. Advances in silylation. *In*: Handbook for Derivatives for Chromatography, 2nd ed., eds. K. Blau & J.M. Halket, pp. 51–108, Wiley, New York.

Fabriás, G., G. Arsequell & F. Camps. 1989. Sex pheromone precursors in the processionary moth *Thaumetopoea pityocampa* (Lepidoptera: Thaumetopoeae). Insect Biochem. **19**:177–181.

Fay, L. & U. Richli. 1991. Location of double-bonds in polyunsaturated fatty acids by gas chromatography-mass spectrometry after 4,4-dimethyloxazoline derivatization. J. Chromatogr. **541**:89–98.

Foster, S.P., M.O. Harris & J.G. Millar. 1991. Identification of the sex pheromone of the Hessian fly, *Mayetiola destructor* (Say). Naturwissenchaften **78**:130–131.

Fox, A., S.L. Morgan & J. Gilbart. 1989. Preparation of alditol acetates and their analysis by gas chromatography (GC) and mass spectrometry (MS). *In:* Analysis of Carbohydrates by GLC and MS, eds. C.J. Biemann & G.D. McGinnis, pp. 87–117, CRC Press, Boca Raton, FL.

Francis, G.W. 1981. Alkylthiolation for the determination of double-bond position in unsaturated fatty acid esters. Chem. Phys. Lipids. **29**:369–374.

Francis, G.W. & K. Veland. 1981. Alkylthiolation for the determination of double-bond positions in liner alkenes. J. Chromatogr. **219**:379–384.

Frenzel, M., Dettner, W. Boland & P. Erbes. 1990. Identification and biological significance of 4-methyl-3Z,5-hexadienoic acid produced by males of the gall-forming tephritids *Urophora cadui* (L.) and *Urophora stylata* (Fab.) (Diptera: Tephritidae). Experientia **46**:542–547.

Goldsmith, D. & C. Djerrasi. 1966. Mass spectrometry in structural and stereochemical problems. CXIII. Specific hydrogen rearrangements in the fragmentation of hydrazones. J. Org. Chem. **31**:3661–3666.

Griepink, F.C., T.A. van Beek, M.A. Posthumus, A. de Groot, J.A. Visser & S. Voerman. 1996. Identification of the sex pheromone of *Scrobipalpa absoluta;* determination of double bond positions in triply unsaturated straight chain molecules by means of dimethyl disulfide derivatization. Tetrahedron Lett. **37**:411–414.

Guerrero, A., F. Camps, J. Coll, M. Riba, J. Einhorn & C. Descoins. 1981. Identification of a potential sex pheromone of the processionary moth *Thaumetopoea pityocampa.* Tetrahedron Lett. **22**:2013–2016.

Guss, P.L., J.H. Tumlinson, P.E. Sonnet & J.R. McLaughlin. 1983. Identification of a female-produced sex pheromone from the southern corn rootworm, *Diabrotica undecimpunctata howardi* Barber. J. Chem. Ecol. **9**:1363–1375.

Hall, D.R., P.S. Beevor, R. Lester, R.G. Poppi & B.F. Nesbitt. 1975. Synthesis of the major sex pheromone of the Egyptian cotton leafworm *Spodoptera littoralis* (Biosd). Chem. Ind. pp. 216–217.

Hall, D.R., P.S. Beevor, R. Lester & B.F. Nesbitt. 1980. (E,E)-10,12-Hexadecadienal: A component of the sex pheromone of the spiny bollworm, *Earias insulana* (Biosd.) (Lepidoptera: Noctuidae). Experientia **36**:152–154.

Harding, K.E., L.M. May & K.F. Dick. 1975. Selective oxidation of allylic alcohols with chromic acid. J. Org. Chem. **40**:1664–1665.

Harvey, D.J. 1982. Picolinyl esters as derivatives for the structural determination of long chain branched and unsaturated fatty acids. Biomed. Environ. Mass Spectrum. **9**:33–38.

Harvey, D.J. 1984. Picolinyl derivatives for the characterization of cyclopropane fatty acids by mass spectrometry. Biomed. Mass Spectrom. **11**:187–192.

Harvey, D.J. 1988. Identification of long-chain fatty acids and alcohols from human cerumen by the use of picolinyl and nicotinate esters. Biomed. Environ. Mass Spectrom. **18**:719–723.

Hashimoto, N., T. Aoyama & T. Shioiri. 1981. New methods and reagents in organic synthesis. 14. A simple efficient preparation of methyl esters with trimethylsilyldiazomethane (TMSCHN$_2$) and its application to gas chromatographic analysis of fatty acids. Chem. Pharm. Bull. **29**:1475–1478.

Heath, R.R., G.E. Burnsed, J.H. Tumlinson & R.E. Doolittle. 1980. Separation of a series of positional and geometrical isomers of olefinic aliphatic primary alcohols and acetates by capillary gas chromatography. J. Chromatog. **189**:199–208.

Henderson, H.E., F.L. Warren, O.P.H. Augustyn, B.V. Burger, D.F. Schneider, P.R. Boshoff, H.S.C. Spies & H. Geertsema. 1972. Sex pheromones. Cis-dec-5-en-1-yl 3-methylbutanoate as the pheromone from the pine emperor moth (*Nudaurelia cytherea cytherea* Fabr.) Chem. Commun. pp. 686–687.

Henderson, H.E., F.L. Warren, O.P.H. Augustyn, B.V. Burger, D.F. Schneider, P.R. Boshoff, H.S.C. Spies & H. Geertsema. 1973. Isolation and structure of the sex-pheromone of the moth, *Nudaurelia cytherea cytherea.* J. Insect Physiol. **19**:1257–1264.

Henrick, C.A., W.E. Willy, J.W. Baum, T.A. Baer, B.A. Garcia, T.A. Mastre & S.M. Chang. 1975. Stereoselective synthesis of alkyl (2*E*,4*E*)- and (2*Z*,2*E*)-trimethyl-2,4-dodecadienoates. Insect growth regulators with juvenile hormone activity. J. Org. Chem. **40**:1–7.

Hill, A.S. & W.L. Roelofs. 1981. Sex pheromone of the saltmarsh caterpillar moth, *Estigmene acrea.* J. Chem. Ecol. **7**:655–668.

Ho, H.Y., Y.T. Tao, R.S. Tsai, Y.L. Wu, H.K. Tseng & Y.S. Chow. 1996. Isolation, identification, and synthesis of sex pheromone components of female tea cluster caterpillar, *Andraca bipunctata* Walker (Lepidoptera: Bombycidae) in Taiwan. J. Chem. Ecol. **22**:271–285.

Hoff, J.E. & E.D. Feit. 1964. New technique for functional group analysis in gas chromatography. Syringe reactions. Anal. Chem. **36**:1002–1008.

Hogge, L.R. & J.G. Millar. 1987. Characterization of unsaturated aliphatic compounds by GC/mass spectrometry. *In:* Advances in Chromatography, Vol. 27, eds. J.C. Giddings, E. Grushka & P.R. Brown, pp. 299–351, Marcel Dekker, New York.

Hogge, L.R., E.W. Underhill & J.W. Wong. 1985. The determination of position and geometry of unsaturation in multi-unsaturated aliphatic compounds by GC-MS. J. Chromatogr. Sci. **23**:171–175.

Horiike, M. & C. Hirano. 1984. Characterization of double bond positions in dodecenols by electron impact mass spectrometry. Biomed. Mass Spectrom. **11**:145–148.

Horiike, M. & C. Hirano. 1987. Location of double bonds in tetradecenyl acetates by electron impact mass spectrometry. Biomed. Environ. Mass Spectrom. **14**:183–185.

Horiike, M., N. Miyata & C. Hirano. 1981. An empirical approach to the location of double bonds in tetradecen-1-yl acetates by mass spectrometry. Biomed. Mass Spectrom. **8**:41–42.

Horiike, M., S. Oomae & C. Hirano. 1986. Determination of the double bond positions in three dodecenol (Z) isomers by electron impact mass spectrometry. Biomed. Environ. Mass Spectrom. **13**:117–120.

Horiike, M., S.Y. Gu & C. Hirano. 1990. Location of double-bond position in tetradecenols without chemical modification by mass spectrometry. Org. Mass Spectrom. **25**:329–332.

Hornstein, I., J.A. Alford, L.E. Elliott & P.E. Crowe. 1960. Determination of free fatty acids in fat. Anal. Chem. **32**:540–541.

Howard, R.W., C.A. MacDaniel & G.J. Blomquist. 1978. Cuticular hydrocarbons of the eastern subterranean termite, *Reticulitermes flavipes* (Kollar) (Isoptera: Rhinotermitidae). J. Chem. Ecol. **4**:233–245.

Hurter, J., E.F. Boller, E. Städler, B. Blattmann, H.-R. Buser, N.U. Bosshard, L. Damm, M.W. Kozlowski, R. Schöni, F. Raschdorf, R. Dahinden, E. Schlumpf, H. Fritz, W.J. Richter & J. Schreiber. 1987. Oviposition-deterring pheromone in *Rhagoletis cerasi* L.: purification and determination of the chemical constitution. Experientia **43**:157–163.

Hutchins, R.O., D. Hoke, J. Keogh & D. Koharski. 1969. Sodium borohydride in dimethyl sulfoxide or sulfolane: convenient systems for selective reduction of primary, secondary, and certain tertiary halides and tosylates. Tetrahedron Lett. **40**:3405–3498.

Huwyler, S. 1972. Ultramikromethoden. Acetylierung von Alkoholen. Experientia **28**:718–719.

Jaenicke, L. & F.-J. Marner. 1995. Lurlene, the sexual pheromone of the green flagellate *Chlamydomonas allensworthii.* Liebigs Ann. Chem. pp. 1343–1345.

Jefferson, A. & S. Wangchareontrakul. 1986. Urushiol, laccol, thistol and phenylalkyl catechol compounds in Burmese lac from *Melanorrhoes usitata.* J. Chromatogr. **367**:145–154.

Jewett, D.M., F. Matsumura & H.C. Coppel. 1976. Sex pheromone specificity in the pine sawflies: interchange of acid moieties in an ester. Science **192**:51–53.

Jones, T.H., M.S. Blum, R.W. Howard, C.A. McDaniel, H.M. Fales, M.B. Dubois & J. Torres. 1982. Venom chemistry of ants in the genus *Tetramorium.* J. Chem. Ecol. **8**:285–300.

Jones, T.H., M.S. Blum & H.G. Robertson. 1990. Novel dialkylpiperidines in the venom of the ant *Monomorium delagoense.* J. Natural Prod. **53**:429–435.

Jutila, M. & J. Jalonen. 1988. Piperidyl and morpholinyl esters as derivatives for the structural determiantion on long-chain fatty acids by mass spectrometry. Biomed. Environ. Mass Spectrom. **17**:433–436.

Knapp, D.R. 1979. Handbook of Analytical Derivatization Reactions, Wiley, New York.

Kobayashi, M., T. Koyama, K. Ogura, S. Seto, F.J. Ritter & I.E.M. Brüggemann-Rotgans. 1980. Bioorganic synthesis and absolute configuration of faranal. J. Am. Chem. Soc. **102**:6602–6604.

LaLonde, R.T., E. Auer, C.F. Wong & V.P. Muralidharan. 1971. The Polonovski transformation of (+)-nupharidine. A study of the stereochemistry and utility in synthesis. J. Am. Chem. Soc. **93**:2501–2506.

Lanne, B.S., M. Applegren, G. Bergström & C. Löfstedt. 1985. Determination of the double bond position in monounsaturated acetates from their mass spectra. Anal. Chem. **57**:1621–1625.

Lanne, B.S., G. Bergström, A.-B. Wassgren & B. Törnbäck. 1987. Biogenetic pattern of straight chain marking compounds in male bumblebees. Comp. Biochem. Physiol. **88B**:631–636.

Lanne, B.S., G. Bergström & L. Löfqvist. 1988. Dufour gland alkenes from the four ant species: *F. polyctena, F. lugubris, F. truncorum,* and *F. uralensis.* Comp. Biochem. Physiol. **91B**:729–734.

Lemieux, R.U. & E. von Rudloff. 1955. Periodate-permanganate oxidations. 1. Oxidation of olefins. Can. J. Chem. **33**:1701–1709.

Leonhardt, B.A. & E.D. DeVilbiss. 1985. Separation and double-bond determination of nanogram quantities of aliphatic monounsaturated alcohols, aldehydes and carboxylic acid methyl esters. J. Chromatogr. **322**:484–490.

Leonhardt, B.A., E.D. DeVilbiss & J.A. Klun. 1985. Gas chromatographic-mass spectrometric indication of double bond position in monounsaturated primary acetates and alcohols without derivatization. Org. Mass. Spectrom. **18**:9–11.

Ma, T.S. & A.S. Ladas. 1976. Organic Functional Group Analysis by Gas Chromatography. Academic Press, London.

Malosse, C. & J. Einhorn. 1990. Nitric oxide chemical ionization mass spectrometry of long-chain unsaturated alcohols, acetates, and aldehydes. Anal. Chem. **62**:287–293.

Markham, G., D.G. Lichty & F. Wightman. 1980. Comparative study of derivatization procedure for the quantitative determination of the auxin, phenylacetic acid, by gas chromatography. J. Chromatogr. **192**:429–433.

Mason, R.T., T.H. Jones, H.M. Fales, L.K. Pannell & D. Crews. 1990. Characterization, synthesis, and behavioral responses to sex attractiveness pheromones of red-sided garter snakes (*Thanophis sirtalis parietalis*). J. Chem. Ecol. **16**:2353–2369.

McDaniel, C.A. & R.W. Howard. 1985. Mass spectral determination of aldehydes, ketones, and carboxylic acids using 1,1-dimethylhydrazine. J. Chem. Ecol. **11**:303–310.

McDonough, L.M. & D.A. George. 1970. Gas chromatographic determination of the cis-trans isomer content of olefins. J. Chromatogr. Sci. **8**:158–161.

McDonough, L.M., H.G. Davis, P.S. Chapman & C.L. Smithhisler. 1993. Response of male codling moths (*Cydia pomonella*) to components of conspecific female sex pheromone glands in flight tunnel tests. J. Chem. Ecol. **19**:1737–1748.

Meinwald, J., Y.C. Meinwald & P.H. Mazzocchi. 1969. Sex pheromone of the queen butterfly: Chemistry. Science **164**:1174–1175.

Metcalfe, L.D. & A.A. Schmitz. 1966. Rapid preparation of fatty esters from lipids for gas chromatographic analysis. Anal. Chem. **38**:514–51.

Middleditch, B.S. 1989. Analytical Artifacts. Elsevier, Amsterdam.

Mizuno, G.R., E.C. Ellison & J.R. Chiapault. 1969. Rapid micromethod for locating the oxirane group in 1,2-epoxides. Microchem. J. **14**:227–234.

Moore, B.P. & W.V. Brown. 1971. Gas-liquid chromatographic identification of ozonolysis fragments as a basis for micro-scale structure determinations. J. Chromatogr. **60**:157–166.

Moore, B.P. & B.V. Brown. 1976. The chemistry of the metasternal gland secretion of the eucalypt longicorn *Phoracantha synonyma* (Coleoptera: Cerambycidae). Aust. J. Chem. **29**:1365–1374.

Müller, D.G., W. Boland, U. Becker & T. Wahl. 1988. Caudoxirene, the spermatozoid-releasing and attracting factor in the marine brown alga *Perithalia caudata* (Phaeophyceae, Sporochnales). Biol. Chem. Hoppe-Seyler **369**:655–659.

Murata, Y., H.J.C. Yeh, L.K. Pannell, T.H. Jones, H.M. Fales & R.T. Mason. 1991. New ketodienes from the integumental lipids of the Guam brown tree snake, *Boiga irregularis*. J. Natural Prod. **54**:233–240.

Nasciemento, R.R.D., E. Shoeters, E.D. Morgan, J. Billen & D.J. Stradling. 1996. Chemistry of metapleural gland secretions of three attine ants, *Atta sexdens rubropilosa, Atta cephalotes,* and *Acromyrmex octospinosus.* J. Chem. Ecol. **22**:987–1000.

Nelson, D.R., C.L. Fatland & J.E. Baker. 1984. Mass spectral analysis of epicuticular *n*-alkadienes in three *Sitophilus* weevils. Insect Biochem. **14**:435–444.

Nesbitt, B.F., P.S. Beevor, R.A. Cole, R. Lester & R.G. Poppi. 1973. Synthesis of both geometric isomers of the major sex pheromone of the red bollworm moth. Tetrahedron Lett. pp. 4669–4670.

Nesbitt, B.F., P.S. Beevor, R.A. Cole, R. Lester & R.G. Poppi 1975. The isolation and identification of the female sex pheromone of the red bollworm moth, *Diparopsis castania*. J. Insect Physiol. **21**:1091–1096.

Nickell, E.C. & O.S. Privett. 1966. A simple, rapid micromethod for the determination of the structure of unsaturated fatty acids via ozonolysis. Lipids 1:166–170.

Ollett, D.G., A.B. Attygalle & E.D. Morgan. 1986. Micro-chemical method for determining formaldehyde, lower carbonyl compounds and alkylidene end groups in the nanogram range using Keele micro-reactor. J. Chromatogr. 367:207–212.

Pedder, D.J., H.M. Fales, T. Jaouni, M.S. Blum, J. MacConnell & R.M. Crewe. 1976. Constituents of the venom of a South African fire ant (*Solenopsis punctaticeps*). Tetrahedron 32:2275–2279.

Piancatelli, G., A. Scettri, & M. D'auria. 1982. Pyridinium chlorochromate: A versatile oxidant in organic synthesis. Synthesis pp. 245–258.

Pierce, A.E. 1968. Silylation of Organic Compounds. Pierce Chemical Company, Rockford, IL.

Prestwich, G.D. & M.S. Collins. 1981. 3-Oxo-(Z)-9-hexadecenal: an unusual enolic β-keto aldehyde from a termite soldier defense secretion. J. Org. Chem. 46:2383–2385.

Ravi, B.N. & D.J. Faulkner. 1978. Cembranoid diterpenes from a south Pacific soft coral. J. Org. Chem. 43:2127–2131.

Rossi, R., A. Carpita, M.G. Quirici & C.A. Veracini. 1982. Insect pheromone components. Use of ^{13}C NMR spectroscopy for assigning the configuration of C=C double bonds of monoenic or dienic pheromone components and for quantitative determination of Z/E mixtures. Tetrahedron 38:639–644.

Russel, G.A. & G.J. Mikol. 1968. Acid-catalyzed rearrangements of sulfoxides and amine oxides. The Plummerer and Polonovski reactions. In: Mechanisms of Molecular Migrations, Vol. 1, ed. B.S. Thygarajan, pp. 157–207, Interscience, New York.

Schildknecht, H., B. Weber & K. Dettner. 1983. Über Arthropodabwehrsoffe, LXV. Die chemische Ökologie des Grundscwimmers *Laccophilus minustus*. Z. Naturforsch. 38b:1678–1685.

Schmitz, B. & R.A. Klein. 1986. Mass spectrometric localization of carbon-carbon double bonds: a critical review of recent methods. Chem. Phys. Lipids 39:285–311.

Schulz, S., W. Francke & M. Boppré. 1988. Carboxylic acids from hairpencils of male *Amauris* butterflies (Lep.: Danainae). Biol. Chem. Hoppe-Seyler 369:633–638.

Schwartz, D.P. & J.T. Weirauch. 1969. A chromic acid column procedure for the quantitative oxidation of secondary alcohols at the micromole level. Microchem. J. 14:597–602.

Schwartz, M., G.F. Graminski & R.M. Waters. 1986. Insect pheromone synthesis of (E)-13,13-dimethyl-11-tetradecen-1-ol acetate via a thiophenol-mediated olefin inversion. J. Org. Chem. 51:260–263.

Schwartz, N.N. & J.H. Blumbergs. 1964. Epoxidations with *m*-chloroperbenzoic acid. J. Org. Chem. 29:1976–1979.

Scribe, P., J. Guezennec, J. Dagaut, C. Pepe & A. Saliot. 1988. Identification of the position and the stereochemistry of the double bond in monounsaturated fatty acid methyl esters by gas chromatography/mass spectrometry of dimethyl disulfide derivatives. Anal. Chem. 60:928–931.

Scribe, P., C. Pepe, A. Barouxis, C. Fuche, J. Dagaut & A. Saliot. 1990. Détermination de la position de l'insaturation des mono-ènes par chromatographie en phase gazeuse capillaire-spectrométrie de masse (CGS/SM) des dérivés diméthyl-disulfures: application à l'analyse d'un mélange complexe d'alcènes. Analusis **18**:284–288.

Simon, E., W. Kern & G. Spiteller. 1990. Localization of the branch in monomethyl branched fatty acids. Biomed. Environ. Mass Spectrom. **19**:129–136.

Sonnet, P.E. 1974. A practical synthesis of the sex pheromone of the pink bollworm. J. Org. Chem. **39**:3793–3794.

Sonnet, P.E. 1980. Olefin inversions. Tetrahedron **36**:557–604.

Spitzer, V., F. Marx & K. Pfeilsticker. 1994. Electron impact mass spectra of the oxazoline derivatives of some conjugated diene and triene C_{18} fatty acids. J. Am. Oil Chem. Soc. **71**:873–876.

Stanley, G. 1979. Dehydration, reduction and oxidation reactions, and the synthesis of reference compounds. J. Chromatogr. **178**:487–493.

Starr, R.C., F.J. Marner & L. Jaenicke. 1995. Chemoattraction of male gametes by a pheromone produced by female gametes of *Chlamydomonas*. Proc. Natl. Acad. Sci. USA **92**:641–645.

Stránský, K. & T. Jursik. 1996. Simple quantitative transesterification of lipids. Fett/Lipids **98**:65–71.

Svatoš A. & A.B. Attygalle. 1997. Characterization of vinyl groups conjugated to carbon-carbon double bonds in volatile compounds by GC-FT/IR. Anal. Chem. **69**:1827–1836.

Takano, I., Y. Kuwahara, R.W. Howard & T. Suzuki. 1989. Location of double bond positions in conjugated alkadienes by electron impact mass spectrometry of 3,4-dimethylthio-substituted thiolane derivatives. Agric. Biol. Chem. **53**:1413–1415.

Tanaka, Y., H. Sato, A. Kageyu & T. Tomita. 1985. Separation of geranylgeraniol isomers by high-performance liquid chromatography and identification by ^{13}C nuclear magnetic resonance spectroscopy. J. Chromatogr. **347**:275–283.

Tonini, C., G. Cassani, P. Massardo, G. Guglielmetti & P.L. Castellari. 1986. Study of sex pheromone of leopard moth, *Zeuzera pyrina* L. Isolation and identification of three components. J. Chem. Ecol. **12**:1545–1558.

Tori, M. 1988. Application of 2D NMR techniques to structure determination of natural products. *In:* Studies in Natural Products Chemistry, Vol. 2, ed. Atta-ur-Rahman, pp. 81–114, Elsevier, Amsterdam.

Tumlinson, J.H. & R.R. Heath. 1976. Structure elucidation of insect pheromones by microchemical methods. J. Chem. Ecol. **2**:87–99.

Tumlinson, J.H., R.R. Heath, & R.E. Doolittle. 1974. Application of chemical ionization mass spectrometry of epoxides to the determination of olefin position in aliphatic chains. Anal. Chem. **46**:1309–1312.

Tursch, B., D. Daloze, J.C. Breakman, C. Hootele & J.M. Pasteels. 1975. Chemical ecology of Arthropods-X. The structure of myrrhine and the biosynthesis of coccinelline. Tetrahedron **31**:1541–1543.

Umeh, E.O. 1971. The separation and determination of low-molecular-weight (C_2–C_{10})

straight chain carboxylic acids in dilute aqueous solution by the gas chromatography of their *p*-bromophenacyl and *p*-phenylphenacyl esters. J. Chromatogr. **56**:29–36.

van Look, G., G. Simchen & J. Herberle. 1995. Silylating Agents. Fluka Chemical Corp., Ronkonkoma, NY.

Vékey, K., G. Baàn & K.R. Jennings. 1988. Identification of *Z* and *E* isomers by chemical ionization. Biomed. Environ. Mass Spectrom. **16**:267–268.

Vetter, W., W. Walther & M. Vecchi. 1971. Pyrrolidide als Derivate für die Strukturaufklärung aliphatischer und alicyclischer Carbonsäuren mittels Massenspektrometrie. Helv. Chim. Acta **54**:1500–1605.

Vincenti, M., G. Guglielmetti, G. Cassani & C. Tonini. 1987. Determination of double bond position in diunsaturated compounds by mass spectrometry of dimethyl disulfide derivatives. Anal. Chem. **59**:694–699.

Vorbrüggen, H. 1963. The reaction of carboxylic acids and phenols with amide-acetals. Angew. Chem. Int. Ed. Engl. **2**:211–212.

Vostrowsky, O. & K. Michaelis. 1981. Methoxymerkurierung-Demerkurierung von Pheromonen zur Bestimmung der Doppelbindungsposition. Z. Naturforsch. **36B**:402–1005.

Vostrowsky, O., K. Michaelis & H.J. Bestmann. 1981. Methoxymerkurierung-Demerkurierung zur Bestimmung der Doppelbindungspositionen doppelt ungesättigter Lepidopteren-Pheromone. Liebigs Ann. Chem. pp. 1721–1724.

Weatherston, I., J.E. Percy, L.M. MacDonald & J.A. MacDonald. 1979. Morphology of the prothoracic defensive gland of *Schizura concinna* (J.E. Smith) (Lepidoptera: Notodontidae) and the nature of its secretion. J. Chem. Ecol. **5**:165–177.

Weitz, C.J., L.I. Lowney, K.F. Faull, G. Feistner & A. Goldstein. 1986. Morphine and codeine from mammalian brain. Proc. Natl. Acad. Sci. USA **83**:9784–9788.

Yamamoto, K., A. Shibahara, T. Nakayama & G. Kajimoto. 1991a. Double-bond localization in heneicosapentaenoic acid by a gas chromatography/mass spectrometry (GC/MS) method. Lipids **26**:948–950.

Yamamoto, K., A. Shibahara, T. Nakayama & G. Kajimoto. 1991b. Determination of double-bond position in methylene interrupted dienoic fatty acids by GC-MS as their dimethyl disulfide adducts. Chem. Phys. Lipids **60**:39–50.

Yamaoka, R., M. Tokoro & K. Hayashiya. 1987. Determination of geometric configuration in minute amounts of highly unsaturated termite trail pheromone by capillary gas chromatography in combination with mass spectrometry and Fourier-transform infrared spectroscopy. J. Chromatogr. **399**:259–267.

Young, D.C., P. Vouros, B. Decosta & M.F. Holich. 1987. Location of conjugated diene position in an aliphatic chain by mass spectrometry of the 4-phenyl-1,2,4-triazoline-3,5-dione adduct. Anal. Chem. **59**:1954–1957.

Young, D.C., P. Vouros & M.F. Holich. 1990. Gas chromatography-mass spectrometry of conjugated dienes by derivatization with 4-methyl-1,2,4-triazoline-3,5-dione. J. Chromatogr. **522**:295–302.

Yu, Q.T., B.N. Liu & Z.H. Huang. 1988. Location of methyl branchings in fatty acids: fatty acids in uropygial secretion of Shanghai duck by GC-MS of 4,4-dimethyloxazoline derivatives. Lipids **23**:804–810.

Yu, Q.T., B.N. Liu, J.Y. Zhang & Z.H. Huang. 1989. Location of double bonds in fatty acids of fish oil and rat testis lipids. Gas chromatography-mass spectrometry of the oxazoline derivatives. Lipids **24**:79–83.

Zhang, J.Y., X.J. Yu, H.Y. Wang, B.N. Liu, Q.T. Yu & Z.H. Huang. 1989. Location of triple bonds in fatty acids from the kernel oil of *Pyrularia edulis* by GC-MS of their 4,4-dimethyloxazoline derivatives. J. Am. Oil Chem. Soc. **66**:256–259.

8

Separation of Enantiomers and Determination of Absolute Configuration

Kenji Mori

8.1. Introduction

The absolute configuration of a chiral molecule (i.e., a molecule that is not superimposable on its mirror image) refers to the actual three-dimensional struc-

Me(CH₂)₉$\diagup$(CH₂)₅Me

(S)-1
EAG

Me(CH₂)₇⋮⌢⌢⌢(CH₂)₃Me

(5S,9S)-2
F

Me(CH₂)₅⋮⌢⌢⌢⌢(CH₂)₃Me

(5R,11S)-3
F

CH₂=CH(CH₂)₁₁$\diagup$(CH₂)₃Me

(S)-4
F

(45 : 55)
(R)-5   (S)-5
CGC[a)]

(S)-6
B

(3R,4S)-7
B

(R)-8
D-GC[b)]

(R)-9
B

(3S,4S)-10
CGC[c)]

(S)-11   +   (S)-12
(10 : 3)
F, CGC[d)]

(4S,5S)-13
CGC[c)], EAD

(4S,5S)-14
CGC[c)], EAD, B

(R)-15
CGC[e)]

(S)-16
D-CGC[f)]

(R)-17
D-GC[b, g)], CGC[c)]

(1R,2S)-18
EAD,D-CGC[h)]

(S)-19
D-CGC[i)]

(R)-20
D-CGC[f)]

(R)-21
D-CGC[f)]

(2R,6R,10R)-22
B, EAG

(R)-23
CGC[c)]

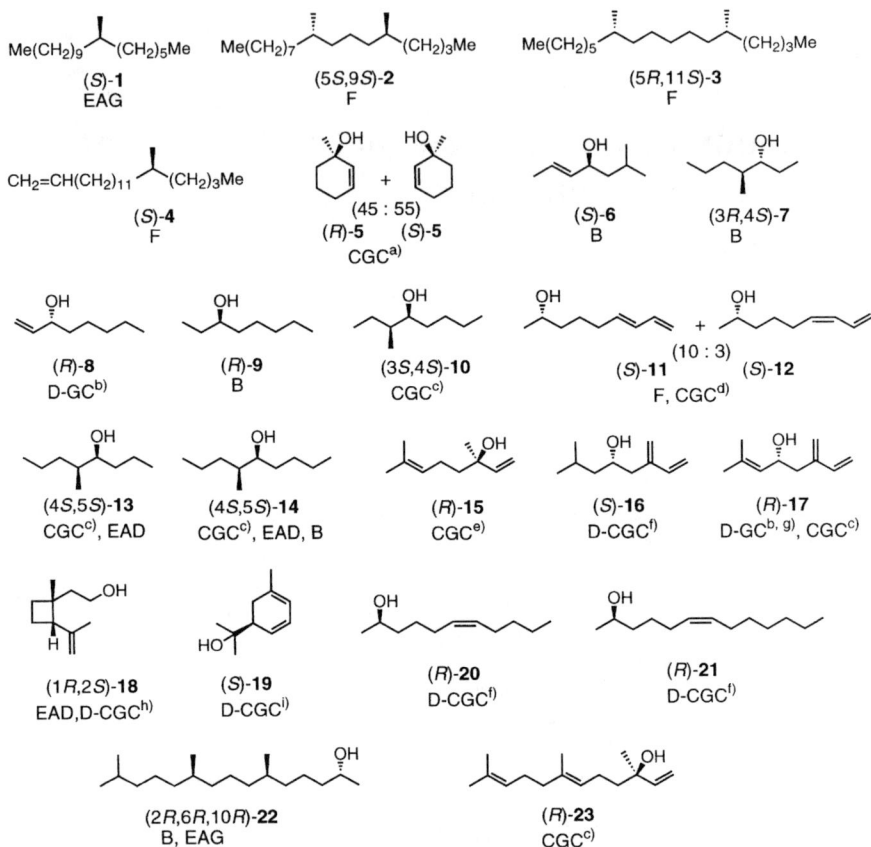

ture of the molecule in space. Because the biological activity of a chiral compound is usually critically dependent on its absolute configuration, the determination of configuration is a crucial part of structure identification, and of a subsequent understanding of a molecule's function. The absolute configurations of many pheromones have been determined by the methods described in this chapter. Figures 8.1 to 8.5 show the structures of chiral pheromones (organized by chemical classes) with established absolute configurations, along with the methods used to determine the absolute configurations, and the species from which the compounds have been isolated.

Fundamentals of organic stereochemistry are described in any college level textbook, and so are not described in detail here. For our purposes, we briefly review the fundamentals of the assignment of absolute configurations. Those wishing an in-depth treatment of the subject are referred to Eliel's authoritative monograph (Eliel et al. 1994).

Absolute configurations are specified by the Cahn-Ingold-Prelog R,S conven-

◄ *Figure 8.1.* Determination of the absolute configuration of pheromone hydrocarbons and alcohols. Methods of investigation are shown under the structural formula by abbreviations common to Figures 8.1 to 8.5 as follows: B, bioassay; CD, circular dichroism; CGC, gas chromatography employing a chiral stationary phase; D, derivatization; EAD, coupled GC-electroantennographic detection; EAG, electroantennogram; F, field test; GC, gas chromatography employing an achiral stationary phase; IR, infrared spectroscopy; mmp, mixed melting point determination; NMR, nuclear magnetic resonance spectroscopy; ORD, optical rotatory dispersion. In addition, a), Chirasil Dex®; b), (S)-MeCH(OAc)COCl, GC; c), Cyclodex®-B; d), Octakis(6-O-methyl-2,3-di-O-pentyl)-γ-cyclodextrin; e), CP-cyclodextrin β-236-M-19; f), *i*-PrNCO, Chirasil® L-Val; g), (S)-Camphanoyl chloride, GC; h), (CF₃CO)₂O, Lipodex® C; i), 2,6-di-O-methyl-3-O-pentyl-β-cyclodextrin.

Names of the insects that release the pheromones **1** to **23** are listed here together with the references for the determination of the absolute configurations: **1**, *Lambdina fiscellaria* (Li et al. 1993b); **2**, *Leucoptera scitella* (Tóth et al. 1989b); **3**, *Lambdina fiscellaria* and *L. fiscellaria lugubrosa* (Li et al. 1993a); **4**, *Lyonetia clerkella* (Sato et al. 1986); **5**, *Dendroctonus pseudotsugae* (Lindgren et al. 1992); **6** (rhynchophorol), *Rhynchophorus palmarum* (Oehlschlager et al. 1992); **7**, *Leptogenys diminuta* (Steghaus-Kovâc et al. 1992); **8**, *Ahasverus advena* (Pierce et al. 1991); **9**, *Myrmica scabrinoides* and *M. rubra* (Cammaerts & Mori 1987); **10**, (phoenicol), *Rhynchophorus phoenicis* (Mori et al. 1993b; Perez et al. 1994; Rochat et al. 1995); **11** and **12**, *Stigmella malella* (Tóth et al. 1995); **13** (cruentatol), *Rhynchophorus cruentatus* (Perez et al. 1994); **14**, *Rhynchophorus bilineatus* (Oehlschlager et al. 1995), *R. vulneratus,* and *Metamasius hemipterus* (Mori et al. 1993c); **15**, (linalool), *Holotrichia parallela* (Leal et al. 1993b); **16** (ipsenol), *Ips paraconfusus* (Oertle et al. 1990); **17** (ipsdienol), *Ips pini* (Lanier et al; 1980; Miller et al. 1989; Seybold et al. 1995a, 1995b); **18** (grandisol), *Pissodes strobi* and *P. nemorensis* (Hibbard & Webster 1993), *Pityogenes quadridens* and *Pityophthorus pityographus* (Francke et al. 1995); **19**, *Ips sexdentatus* (Francke et al. 1995); **20** (nostrenol), *Euroleon nostras* and *Grocus bore* (Baeckström et al. 1989); **21**, *Synclysis baetica* and *Acanthaclicis occitanica* (Bergström et al. 1992); **22**, *Corcyra cephalonica* (Mori et al. 1991); **23** (nerolidol), *Amblypelta lutescens* (Aldrich et al. 1993).

tions, and are represented on paper by projection formulas such as Fischer projections. To describe the absolute configuration, one must first apply the Cahn-Ingold-Prelog sequence rules, as follows.

1. Rank the atoms directly attached to the chiral center and assign priorities in order of decreasing atomic number.

2. If a decision about priority cannot be reached by applying rule 1, compare atomic numbers of the second (and if necessary, subsequent) atoms in each substituent, continuing until the first point of difference.

3. Multiply bonded atoms are considered as an equivalent number of singly bonded atoms, for example, CH=O = −C(−O)H-O-C.

Following these rules, assign priorities to the four substituent groups attached to a chiral carbon, and place the molecule, as shown in the examples in Figure 8.6,

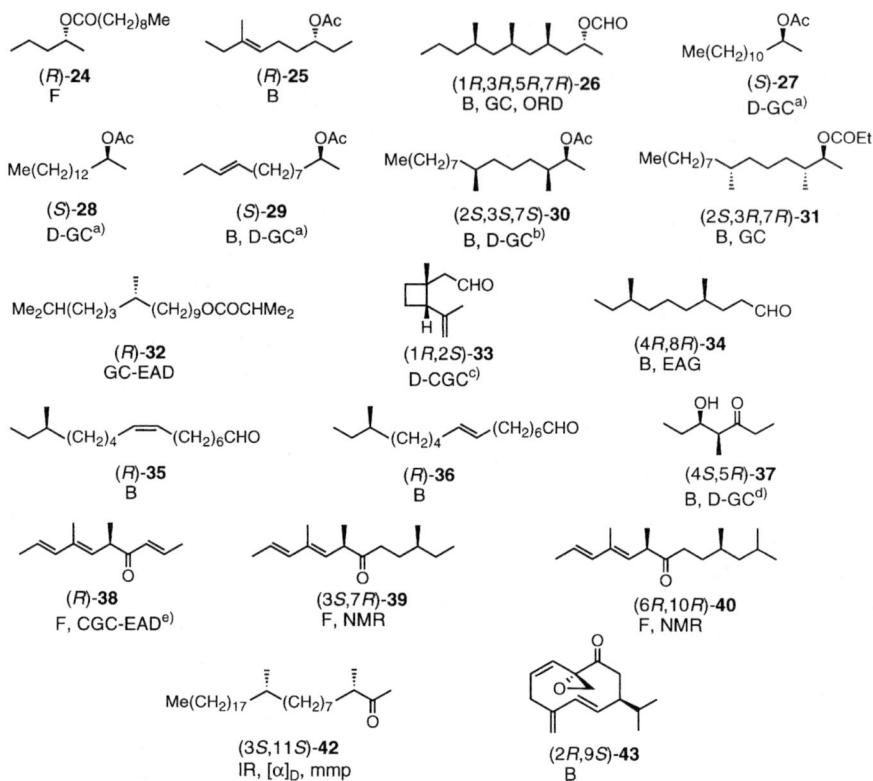

Figure 8.2. Determination of the absolute configuration of pheromone acetates, propionate, aldehydes, and ketones. Here, a), KOH/MeOH or LiAlH$_4$, (S)-MeCH(OAc)COCl, GC; b), the corresponding alcohol (diprionol = precursor of **30**) was analyzed, C$_6$F$_5$COCl, GC; c), LiAlH$_4$, (CH$_3$CO)$_2$O, Lipodex®-C; d), (S)-MeCH(OAc)COCl, GC; e), Cyclodex-B.

Names of the insects that release the pheromones **24** to **43** are listed here together with the references for the determination of the absolute configurations: **24**, *Oiketicus kirbyi* (Rhainds et al. 1994); **25** (quadrilure), *Cathartus quadricollis* (Pierce et al. 1988); **26** (lardolure) *Lardoglyphus konoi* and *Carpoglyphus lactis* (Mori & Kuwahara 1986; Kuwahara et al. 1991); **27**, *Drosophila mulleri* (Bartelt et al. 1989); **28**, *Drosophila busckii* (Schaner et al. 1989); **29**, *Mayetiola destructor* (Millar et al. 1991); **30**, *Neodiprion sertifer* (Olaifa et al. 1988, Wassgren & Bergström 1995); **31**, *Diprion similis* (Olaifa et al. 1988); **32**, *Euproctis pseudoconspersa* (Ichikawa et al. 1995); **33** (grandisal), *Pissodes strobi* and *P. nemorensis* (Hibbard & Webster 1993); **34** (tribolure), *Tribolium castaneum* (Levinson & Mori 1983); **35** and **36** [(Z)- and (E)-trogodermal] *Trogoderma glabrum, T. inclusum,* and *T. variabile* (Silverstein et al. 1980); **37** (sitophilure), *Sitophilus oryzae* and *S. zeamais* (Walgenbach et al. 1987; Levinson et al. 1990); **38**, *Matsucoccus josephi* (Dunkelblum et al. 1995); **39**, *Matsucoccus feytaudi* (Jactel et al. 1994); **40** (matsuone), *Matsucoccus thumbergianae* (Mori & Harashima 1993b; Park et al. 1994); **42**, *Blattella germanica* (Mori & Takikawa 1990); **43** (periplanone-A), *Periplaneta americana* (Kuwahara & Mori 1990).

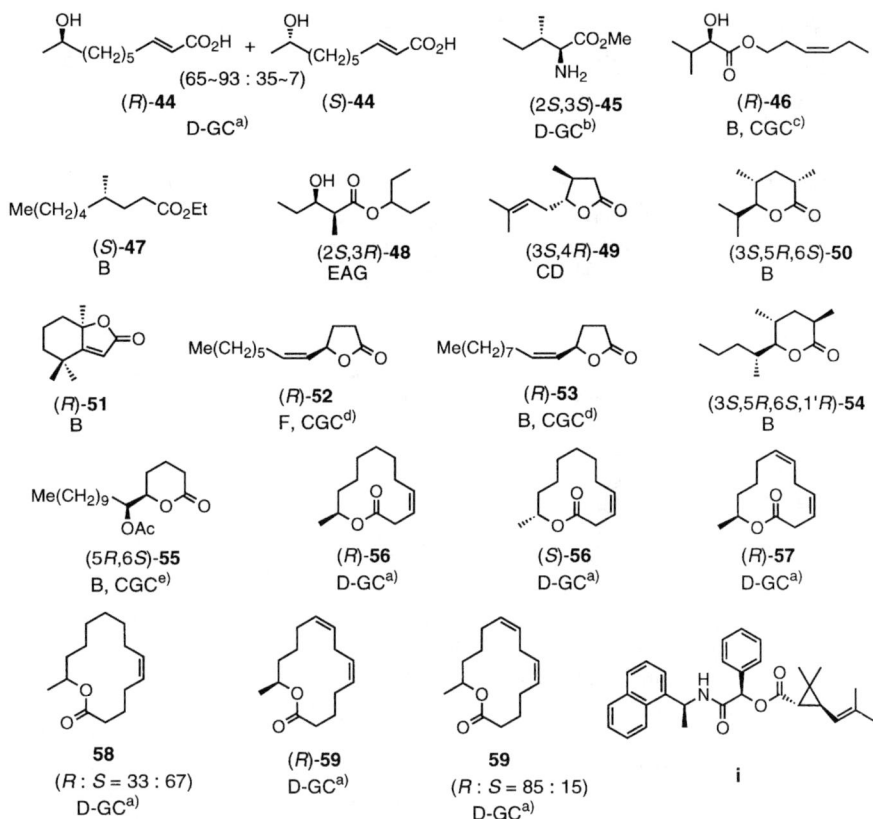

$(65\sim93 : 35\sim7)$

(R)-44   (S)-44
D-GC[a)]

(2S,3S)-45
D-GC[b)]

(R)-46
B, CGC[c)]

Me(CH₂)₄ ... CO₂Et

(S)-47
B

(2S,3R)-48
EAG

(3S,4R)-49
CD

(3S,5R,6S)-50
B

(R)-51
B

Me(CH₂)₅ ...

(R)-52
F, CGC[d)]

Me(CH₂)₇ ...

(R)-53
B, CGC[d)]

(3S,5R,6S,1'R)-54
B

Me(CH₂)₉ ...
OAc

(5R,6S)-55
B, CGC[e)]

(R)-56
D-GC[a)]

(S)-56
D-GC[a)]

(R)-57
D-GC[a)]

58
(R : S = 33 : 67)
D-GC[a)]

(R)-59
D-GC[a)]

59
(R : S = 85 : 15)
D-GC[a)]

i

Figure 8.3. Determination of the absolute configuration of pheromone acid, esters and lactones. Here, a), MeOH, BF₃·OEt₂, (S)-MeCH(OAc) COCl, GC; b), (S)-MeCH(OAc)-COCl, GC; c), Cyclodex®-B; d), Chiraldex® GTA; e), chiral stationary phase = **i**.

Names of the insects that release the pheromones **44** to **59** are listed here together with the references for the determination of the absolute configurations: **44**, *Apis mellifera* (Slessor et al. 1990); **45** (L-isoleucine methyl ester), *Holotrichia parallela* (Leal et al. 1992); **46**, *Pristhesancus plagipennis* (James et al. 1994); **47**, *Oryctes rhinoceros* (Hallett et al. 1995); **48** (sitophilate), *Sitophilus granarius* (Levinson et al. 1990); **49** (eldanolide), *Eldana saccharina* (Vigneron et al. 1982); **50**, *Macrocentrus grandii* (Swedenborg et al. 1994); **51** (dihydroactinidiolide), *Solenopsis invicta* (J.H. Tumlinson, unpublished); **52** (buibuilactone), *Anomala cuprea* (Leal et al. 1993a), *A. octiescostata* (Leal et al. 1994a), *A. albopilosa sakishimana* (Leal et al. 1994b); **53** (japonilure), *Popillia japonica* (Tumlinson et al. 1977), *Anomala cuprea* (Leal et al. 1993a), *A. albopilosa sakishimana* (Leal et al. 1994b); **54** (invictolide), *Solenopsis invicta* (J.H. Tumlinson, unpublished); **55**, *Culex pipiens fatigans* (Laurence et al. 1985); (R)-**56**, *Oryzaephilus mercator* (Oehlschlager et al. 1987); (S)-**56**, *Cryptolestes ferrugineus* (Oehlschlager et al. 1987); **57**, *Oryzaephilus mercator* (Oehlschlager et al. 1987); **58**, *Cryptolestes pusillus* (Oehlschlager et al. 1987); (R)-**59**, *Oryzaephilus surinamensis* (Oehlschlager et al. 1987); **59**, *Cryptolestes turcicus* (Oehlschlager et al. 1987).

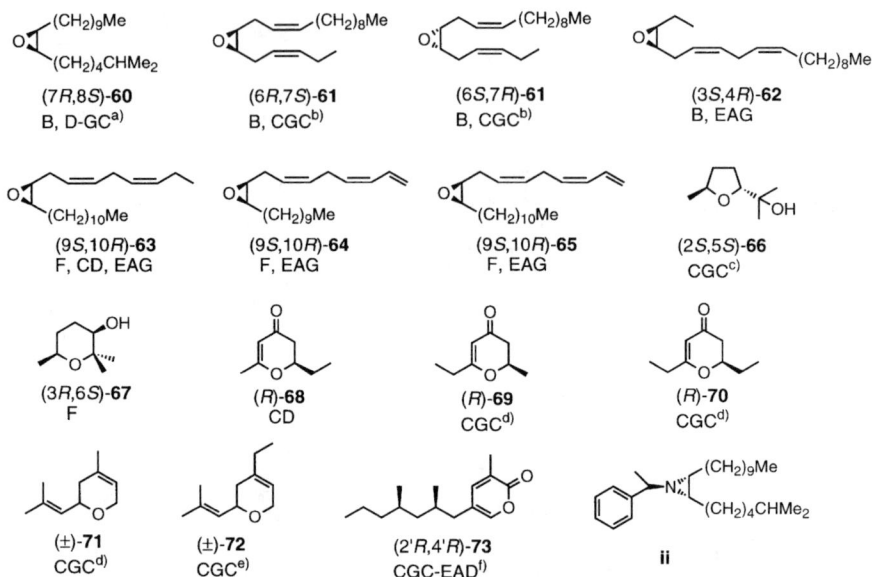

Structures with labels:

(7R,8S)-60
B, D-GC[a)]

(6R,7S)-61
B, CGC[b)]

(6S,7R)-61
B, CGC[b)]

(3S,4R)-62
B, EAG

(9S,10R)-63
F, CD, EAG

(9S,10R)-64
F, EAG

(9S,10R)-65
F, EAG

(2S,5S)-66
CGC[c)]

(3R,6S)-67
F

(R)-68
CD

(R)-69
CGC[d)]

(R)-70
CGC[d)]

(±)-71
CGC[d)]

(±)-72
CGC[e)]

(2'R,4'R)-73
CGC-EAD[f)]

ii

Figure 8.4. Determination of the absolute configurations of pheromones with an oxygen heterocycle. Here, a), derivatization of **60** (100 mg) to the aziridine **ii** and GC; b), Heptakis(2,6-di-*O*-methyl-3-*O*-pentyl)-β-cyclodextrin; c), Chirasil® L-Val or Chirasil Dex® or octakis(3-*O*-acetyl-2,6-di-*O*-pentyl)-γ-cyclodextrin; d), XE-60-L-Val® (*S*)-α-phenethylamide; e), Mn(II) bis[3-heptafluorobutanoyl-(1*R*)-camphorate]; f), Chiraldex® GTA (= trifluoroacetylated γ-cyclodextrin).

Names of the insects that release the pheromones **60** to **73** are listed here together with the references for the determination of the absolute configurations: **60** (disparlure), *Lymantria dispar* (Oliver & Waters 1995); (6*R*,7*S*)-**61**, *Colotois pennaria* (Szöcs et al. 1993); (6*S*,7*R*)-**61**, *Erannis defoliaria* (Szöcs et al. 1993); **62**, *Boarmia selenaria* (Cossé et al. 1992); **63**, *Estigmene acrea* (Mori & Ebata 1986); **64**, *Hyphantria cunea* (Tóth et al. 1989a); **65**, *Hyphantria cunea* (Tóth et al. 1989a); **66** (pityol), *Pityophthorus pityographus* (Francke et al. 1987); *Conophthorus resinosae* (Pierce et al. 1995), *Conophthorus coniperda* (Birgersson et al. 1995); **67** (vittatol), *Pteleobius vittatus* (Klimetzek et al. 1989); **68** (hepialone), *Hepialus californicus* (Kubo et al. 1985); **69** and **70**, *Hepialus hecta* (Schulz et al. 1990); **71** and **72**, *Euroleon nostras* and *Grocus bore* (Baeckström et al. 1989), *Synclysis baetica* and *Acanthaclisis occitanica* (Bergström et al. 1992); **73** (supellapyrone), *Supella longipalpa* (Leal et al. 1995).

so that the group of the lowest priority (usually hydrogen) is pointing away from you. If the curved arrow drawn from the highest to the second-highest to third-highest priority substituent is clockwise (see the configuration of pheromone **53**), the chiral center has the *R* configuration. If the arrow is counterclockwise (see the configuration of pheromone **89**), the center has the *S* configuration.

In the determination of absolute configuration, there are actually three related

(3R,4S)-**74**
CGC[a)]

(1S,5R)-**75**
CGC[b)]

(1R,5S,7R)-**76**
B

(1S,2R,5R)-**77**
CGC[c)]

(1R,4S,5R,7R)-**78**
B, CGC[d)]

(1R,3S,5R)-**79**
CGC[e)]

(1R,3S,5R)-**80**
CGC[f)]

(5S,7S)-**81**
CGC[g)]

(5R,7S)-**82**
CGC[b)]

(R)-**83**   +   (S)-**83**
B, CGC[g, h)]

(2S,6R,8S)-**84**
CGC[g)]

(2S,6R,8S)-**85**
CGC[g)]

(2S,4R,6R,8S)-**86**
CGC[i)]

(8RS,15R)-**87**
NMR

88
NMR, X-ray

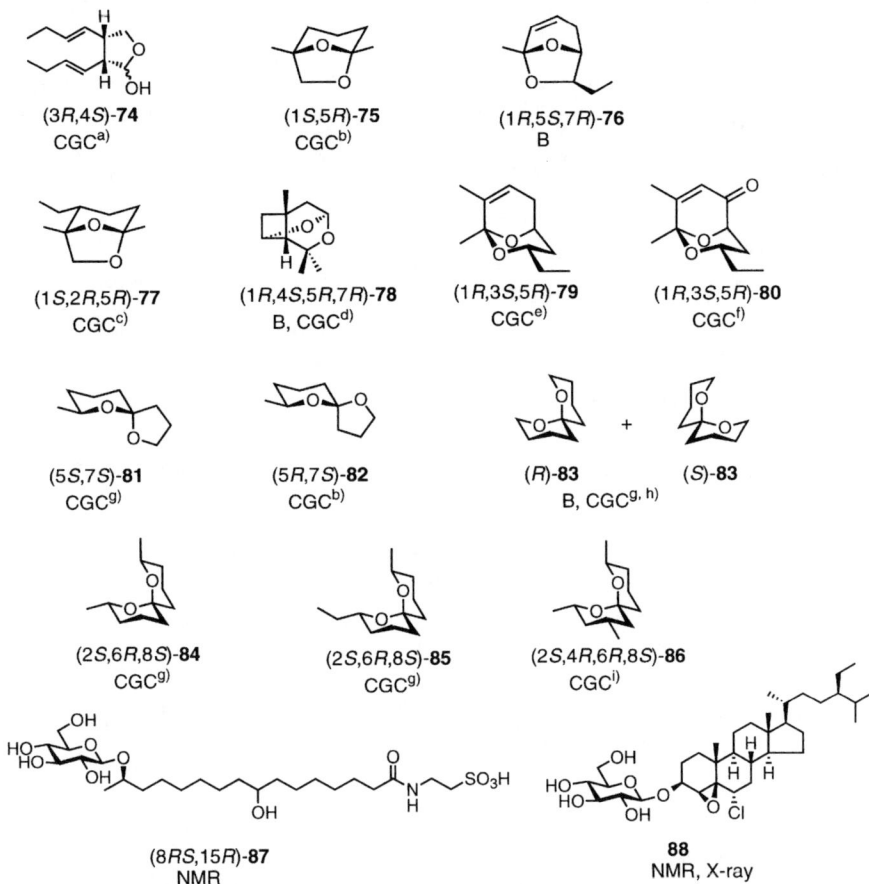

Figure 8.5. Determination of the absolute configurations of pheromone hemiacetals, acetals, spiroacetals, and glucosides. Here, a), Cyclodex®-B (permethylated β-cyclodextrin); b), 2,3,6-tri-*O*-hexyl-α-cyclodextrin; c), 2,6,-di-*O*-methyl-3-*O*-pentyl-β-cyclodextrin; d), Cu(II) bis[3-heptafluorobutanoyl-(1*R*)-camphorate]; e), XE-60-L-Val®-(*S*)-α-phenethylamide; f), Ni(II) bis[heptafluorobutanoyl-(1*R*,5*S*)-pinan-4-oate]; g), Lipodex® A (2,3,6-tri-*O*-methyl-β-cyclodextrin) or Chirasil-Dex®; h), Ni(II) bis[2-heptafluorobutanoyl-(1*S*,5*S*)-4-methylthujonate]; i), β-Cyclodextrin-derived stationary phase.

Names of the insects which release the pheromones **74** to **88** are listed here together with the references for the determination of the absolute configurations: **74**, *Biprorulus bibax* (Mori et al 1992; James & Mori 1995); **75** (frontalin), *Dendroctonus simplex* (Francke et al. 1995); **76**, *Mus musculus* (Novotny et al. 1995); **77** (bicolorin), *Taphrorychus bicolor* (Francke et al. 1995); **78** (lineatin), *Trypodendron lineatum* (Schurig et al. 1982); **79** and **80**, *Hepialus hecta* (Schulz et al. 1990); **81**, *Bactrocera xanthodes* (Fletcher et al. 1992), *Conophthorus resinosae* (Pierce et al. 1995), *Conophthorus coniperda* (Birgersson et al. 1995); **82** as a minor component, *Conophthorus coniperda* (Birgersson et al. 1995); **83** (olean), *Bactrocera oleae* (Haniotakis et al. 1986), *Bactrocera cocuminatus* (Krohn et al. 1991); **84**, *Andrena wilkella* (Tengö et al. 1990), *Bactrocera cucumins* (Kitching et al. 1989); **85**, *Bactrocera nigrotibialis* (Perkins & Kitching 1990); **86**, *Cantao parentum* (Moore et al. 1994); **87**, *Rhagoletis cerasi* (Ernst & Wagner 1989); **88** (blattellastanoside A), *Blattella germanica* (Mori et al. 1993a).

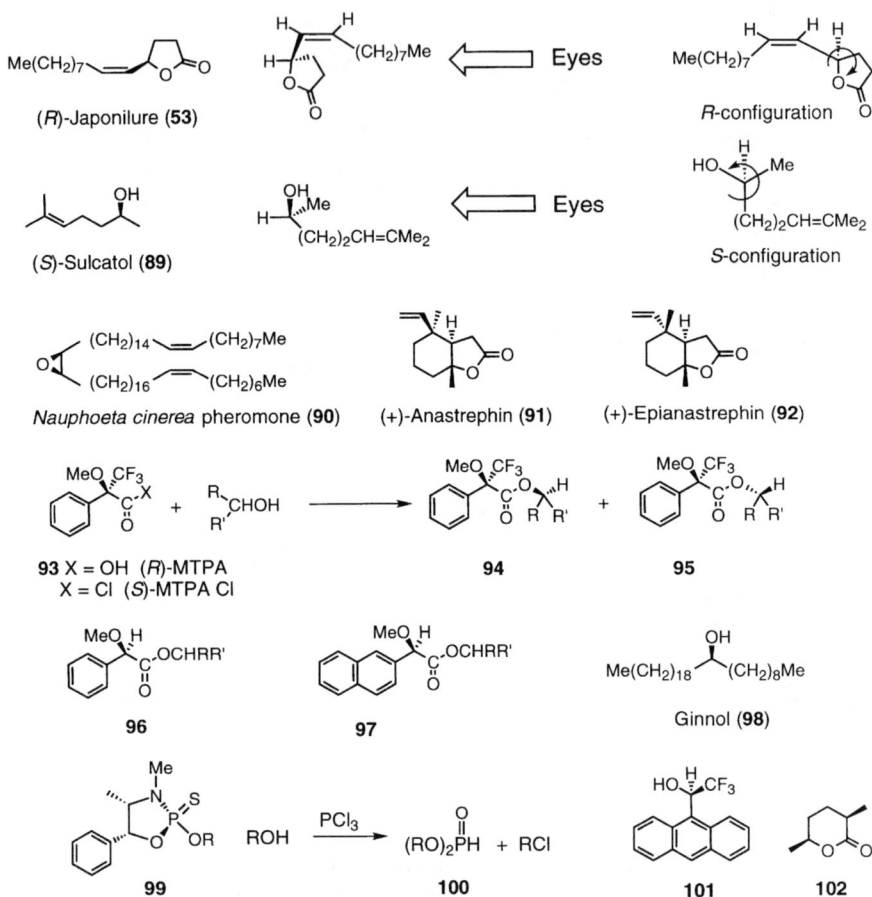

Figure 8.6. Structural formulas—1.

problems. First, the chiral compound must be isolated in chemically pure form; second, the enantiomeric purity must be determined; and third, the absolute configuration of at least one of the two enantiomers must be ascertained. The first problem is solved by application of the standard separation methods described elsewhere in this volume, while the second and third problems are the subject of this chapter. All three of these problems are often compounded by the trace (submicrogram) quantities of the natural material that may be available, which limits the methods that can be applied to determine the configuration.

Furthermore, there are two possible scenarios in the determination of the configuration of a chiral compound. The first and simplest case involves the determination of the configuration of a known compound for which standards of one and hopefully both enantiomers are available. In this case, a comparison of properties of the isolated compound with those of the standards, such as retention times on chiral stationary phases, usually suffices to determine the configuration and/or enantiomeric purity of the unknown compound. In the second, more difficult case, the isolated compound is previously unknown, and it is seldom possible to determine its absolute configuration without synthesizing an enantiomerically pure, or at least, enantiomerically enriched sample of known configuration by means of enantioselective synthesis (Mori 1992), or less frequently, by selectively and carefully degrading the molecule to chiral fragments for which standards are available.

The synthesis by unequivocal routes of one or both of the enantiomers of a newly identified compound and the determination of its absolute configuration represents the final step in the compound's identification. Enantioselective synthesis is achieved by starting from enantiomerically pure building blocks of known absolute configuration so that the stereochemistry of the final product can be unambiguously correlated with that of the starting material. Enantiomerically pure building blocks may be available naturally (for example, monosaccharides and their derivatives), or they may be generated from racemates, for example, by kinetic resolution with enzymes, or by formation of separable diastereomers, followed by regeneration of the enantiomers from the diastereomers. Alternatively, nonracemic chiral intermediates may be generated by induction of asymmetry by chiral catalysts (e.g., the Sharpless asymmetric epoxidation and dihydroxylation reactions). It has sometimes proven necessary to use X-ray crystallographic analysis to confirm the stereostructures of synthetic intermediates. For example, the stereochemistries of synthetic lineatin (**78**, Fig. 8.5; Mori et al. 1983), olean (**83**, Fig. 8.5; Mori et al. 1985), and blattellastanoside A (**88,** Fig. 8.5; Mori et al. 1993a) were assigned on the basis of the X-ray analyses of their synthetic intermediates. The importance of synthesis for the final confirmation of both structure and absolute stereochemistry cannot be overemphasized; the literature is replete with examples of compounds whose structures, proposed on the basis of spectral interpretation, required revision once syntheses were carried out.

Several general reviews are available on the analysis and separation of chiral organic molecules (Morrison 1983; Schurig 1985, 1986; Schreier et al. 1995). Stereochemical analysis of pheromones was thoroughly reviewed in Mori (1984), and therefore this chapter mainly summarizes advances since then.

8.2. Determination of Absolute Configuration by Chiroptical Methods

8.2.1. Specific Rotation

The concept of specific rotation (the rotation of the plane of polarized light by chiral compounds) was introduced in 1835 by the French physicist Biot. Specific rotation (defined below) is reported at a specific temperature T, and the units of $[\alpha]_D^T$ are actually not degrees (°), but 10^{-1} deg $\cdot$ cm^2 $\cdot$ g^{-1}.

$$[\alpha]_D^T = \frac{\text{observed rotation } (\alpha)}{\text{path length } l \text{ (dm)} \times \text{concentration of sample } c \text{ (g/ml)}}$$

In practice, the rotation is usually measured in cells of 1 dm length, with the compound dissolved in an achiral solvent. The sodium D line (589 nm) is usually but not always used, and the temperature, solvent, and concentration are reported. The magnitude of $[\alpha]_D^T$ varies greatly with structure. The largest value observed amongst insect pheromones, for example, was $[\alpha]_D = -574°$ (hexane) for periplanone A (2*R*,9*S*)–**43** (Fig. 8.2) produced by the American cockroach (Kuwahara & Mori 1990), while some compounds, such as the cockroach pheromone (**90**) (Fig. 8.6), produced by *Nauphoeta cinerea* (Mori & Argade 1994) have $[\alpha]_D \approx$ 0. If the sign and magnitude of the specific rotation of an enantiomerically pure pheromone is known through unambiguous synthesis of standards, and if the pheromone is available in an amount sufficient to measure its specific rotation, comparison of the sign of the specific rotation should enable the assignment of the absolute configuration of the product (Mori 1973).

To express the enantiomeric composition of a mixture of two enantiomers, the term enantiomeric purity is used, which is usually equal to optical purity.

$$\% \text{ enantiomeric purity} = \frac{M^+ - M^-}{M^+ + M^-} \times 100 = \% \text{ optical purity}$$

$$= \frac{[\alpha] \text{ of the mixture}}{[\alpha] \text{ of the pure enantiomer}} \times 100.$$

In this equation, M^+ is the mole fraction of the dextrorotatory enantiomer, here the predominant one, and M^- is the mole fraction of the levorotatory enantiomer.

However, there are several potential pitfalls with the measurement of specific rotations. First, chemical purity of the sample is critical, particularly for compounds with small optical rotations, because trace impurities with high rotations can lead to ambiguous or even wrong conclusions. Second, when trying to measure the specific rotation or enantiomeric composition accurately, even traces of achiral substances such as solvents will cause artificially low values due to dilution. Third, some chiral compounds appear to have specific rotations that are too low to be measured, as mentioned above. Because of these problems, measurement of optical rotation is neither the best nor the most reliable method for determination of either absolute configuration or enantiomeric composition.

8.2.2. Optical Rotatory Dispersion

Optical rotatory dispersion (ORD) refers to the change in optical rotation with wavelength. Because the magnitude of the specific rotation is usually larger at shorter wavelengths, ORD may be useful for study of the stereochemistry of materials available in limited amounts. Mori and Ikunaka (1987) studied the

ORD curves of some chiral spiroacetals, determining that positive plain ORD curves were exhibited by various spiroacetals with the *S* configuration at the spiro center, while the *R* enantiomers gave negative ORD curves (see Eliel et al. 1994 for definitions of terms). The measurement was possible with 20–30 μg of sample, but to our knowledge, ORD has not been used in the determination of the absolute configuration of the natural compounds.

8.2.3. Circular Dichroism

If a chiral pheromone possesses a carbonyl chromophore, its circular dichroism (CD) measurement can be informative, and CD spectra have been used in several cases (for details, see Eliel et al. 1994). For example, the absolute configuration of natural (3*S*,4*R*)-eldanolide (**49**, Fig. 8.3; Vigneron et al. 1982), (*R*)-hepialone (**68**, Fig. 8.4; Kubo et al. 1985), and the enantiomers of anestrephin and epianestrephin (**92** Mori & Nakazono 1988) were assigned by CD methods.

8.3. Formation of Diastereomeric Derivatives

Enantiomeric compounds may be derivatized with enantiomerically pure derivatizing agents to form diastereomeric derivatives, which will (at least in theory) have different physical, chemical, and spectroscopic properties. In practice, these differences vary from unmistakable to unobservable. Diastereomeric derivatives can be analyzed by chromatographic methods with standard achiral stationary or mobile phases (sections 8.5 and 8.6) or by methods such as nuclear magnetic resonance (NMR) (section 8.4), in which the sizes of peaks due to particular nuclei can be integrated and compared. Formation of diastereomeric derivatives for determination of enantiomeric purities is obviously contingent on a derivatizable group, such as an alcohol, amine, acid, or carbonyl, being present. Several further points also need to be considered. First, the enantiomeric purity of the chiral derivatizing agent is crucial to the accurate determination of enantiomeric purities of analytes. For example, if the derivatizing agent (e.g., with *R* configuration) is contaminated with a small amount of the *S*-enantiomer, then on reaction with the isolated compound (e.g., pure *R* configuration), two diastereomers will be generated (*R,R* and *R,S*). However, the NMR characteristics of the minor (*R,S*) diastereomer are indistinguishable from those of its enantiomer (*S,R*), which could lead the unwary to believe that the isolated compound was not enantiomerically pure, when in fact, the problem lies with an enantiomerically impure derivatizing reagent.

Second, the derivatizing conditions must not cause racemization in either the substrate or the derivatizing reagent. Third, the reaction rates for each enantiomer may be different, and so the reaction must be allowed to go to completion to ensure no discrimination between enantiomers. Fourth, the choice of derivatizing reagent will be determined to some extent by the method of analysis. Derivatives for gas chromatography (GC) must be volatile, limiting the molecular weight and/or polarity of the derivatizing reagent. Volatility is not a factor for derivatives for high-pressure liquid chromatography (HPLC), but detectability may be important and so a derivatizing reagent with a strong ultraviolet (UV) chromophore or other easily detected group may be desirable.

There are a number of commercially available (e.g., Aldrich, Fluka) derivatizing agents for alcohols, amines, and other classes of compounds, and more extensive lists appear in Allenmark (1988). Useful compilations of commercially available chiral reagents, classified according to the functional group being derivatized, can be found in the specialty "chiral compounds" catalogs from Fluka and Aldrich. Some of the more commonly used chiral derivatizing agents are listed in Table 8.1. Microscale reactions, procedures, and apparatuses are described in detail in chapter 7. Reactions are first carried out with the racemic and enantiomerically enriched synthetic standards to optimize reaction, workup, and analysis conditions. Derivatizing reactions performed with nanogram amounts of substrate use a large excess (100- to 1000-fold or more) of derivatizing reagent because much of the reagent may be degraded by traces of moisture. Microgram to milligram quantities of derivatizing reagents are most easily handled and transferred as solutions (0.1–1 M) in the dry reaction solvent. When the reaction is

Table 8.1. Chiral Derivatizing Reagents for Formation of Diastereomeric Derivatives for Chromatographic or NMR Analysis[a]

Substrate	Reagent	Source/Reference[b]
1,2° alcohol, 1,2° amines	α-methoxy-α-(trifluoromethyl)phenylacetic acid chloride (Mosher's reagent)	Fluka, Aldrich
	(S)-2-acetoxypropanoyl chloride (2-acetyl lactic acid chloride)	Fluka; Slessor et al. (1985)
	α-methoxy- or acetoxyphenylacetic acid (chloride)[c]	Fluka, Aldrich
	menthyl chloroformate	Fluka, Aldrich
	5-oxo-tetrahydrofuran-2-carboxylic acid (chloride)[c]	Fluka; Doolittle and Heath (1984)
	1-(1-naphthyl)ethyl- and 1-phenylethyl isocyanate	Fluka, Aldrich
Carboxylic acids	1-(1-naphthyl)ethylamine	Fluka, Aldrich
	1-(1-phenyl)ethylamine (α-methyl benzylamine)	Many
	chiral 2° alcohols (e.g., 2-octanol, 1-phenylethanol)	Fluka, Aldrich
Ketones	diisopropyl O,O'-bis(trimethylsilyl)-tartrate	Fluka, Aldrich
	1,4-dimethoxy-2,3-butanediol	Fluka

[a]Unless otherwise stated, reagents are available in both enantiomeric forms.

[b]Reagents may also be available from other sources.

[c]The commercially available acids must be converted to the acid chlorides (see text).

complete, some form of aqueous workup is usually used to remove excess reagent and reaction byproducts.

8.3.1. Representative Derivatization Procedures

8.3.1.1. Alcohols and Amines

Alcohols and amines are usually converted to esters or amides, respectively, by treatment with an excess of a enantiomerically pure acid chloride (Table 8.1) and pyridine or triethylamine in dry CH_2Cl_2, $CHCl_3$, or ether at room temp for ~1–12 h. A catalytic amount (1–10%) of 4-(N,N-dimethylamino)pyridine (DMAP) can be used to accelerate the reaction. The reaction is worked up, for example, by addition of water and hexane (or the reaction solvent), washing the organic layer several times with dilute $NaHCO_3$, once with water, and drying over Na_2SO_4. A wash with dilute HCl can also be used to remove traces of pyridine, Et_3N, and DMAP. If the excess derivatizing reagent proves resistant to hydrolysis (e.g., excess Mosher's acid chloride, Table 8.1), the excess can be destroyed by addition of 3-dimethylaminopropylamine. The resulting aminoamide is then removed by washing with dilute HCl (Dale & Mosher 1973). Carbonate and carbamate derivatives of alcohols and amines respectively are formed by reaction with menthyl chloroformate enantiomers (Table 8.1), using the same solvents and conditions.

If the derivatizing agents are only available as acids (Table 8.1), the acids are readily converted to the acid chlorides by treatment of the acid with 2–3 equivalents of thionyl chloride (room temperature, or warming in some cases) or oxalyl chloride with a trace of dimethyl formamide (DMF) catalyst at room temperature. Reactions are run overnight in a fumehood, venting the gases generated through a drying tube. The excess thionyl or oxalyl chloride are removed under partial vacuum or blown off with dry N_2, and the crude acid chlorides can be purified by distillation. However, there are reports of partial racemization during distillation of, or reactions with acid chlorides with active hydrogens on the α carbon (e.g., N-trifluoroacetyl-L-proline, Nichols et al. 1973; N-trifluoroacetyl-L-alanine; Kruse et al. 1979).

Alternatively, alcohols and amines can be derivatized directly with free acids, using coupling reagents such as dicyclohexylcarbodiimide (DCC). The acid is treated with an excess of DCC in dry $CHCl_3$ or CH_2Cl_2 (Kruse et al. 1979) or toluene (Ôi et al. 1981) and the amine or alcohol is added and allowed to react for 12 h at room temperature. With CH_2Cl_2 as the solvent, the by-product dicyclohexylurea precipitates, and the derivative solution can be pipetted off (Kruse et al. 1979). Reactions can be catalyzed with DMAP.

Isocyanates (Table 8.1) react with 1° or 2° amines and alcohols to form ureas or carbamates, respectively. Amines react readily (1 h at room temperature in CH_2Cl_2 or toluene), while alcohols react more slowly, and either heating (80°C;

Pirkle & Hoekstra 1974) or catalysis with DMAP may be required. Tertiary alcohols may require gentle heating (60°C) for several days with the neat reagent in a sealed tube (Rudmann & Aldrich 1987; Hashimoto & Corey 1981). The former authors reported racemization at higher temperatures.

8.3.1.2. Carboxylic Acids

Chiral acids are derivatized as esters or amides by preliminary conversion to the acid chloride by treatment with thionyl chloride or oxalyl chloride, followed by treatment with a chiral alcohol or amine (Table 8.1) and pyridine or Et$_3$N in halogenated solvents (see above). Alternatively, the acid and the alcohol or amine are mixed with coupling agents such as DCC; 1-hydroxybenzotriazole has been used to catalyze amide formation (Avgerinos & Hutt 1987).

8.3.1.3. Carbonyl Compounds

Chiral ketones and aldehydes are converted to diastereomeric derivatives by treatment with reagents derived from tartaric acid (Table 8.1). For example, acetal derivatives are readily prepared by treatment of carbonyls with 1,4-dimethoxy-2,3-butanediol in CCl$_4$, with *p*-toluenesulphonic acid catalysis (Langer & Seebach 1979). More recently, treatment of carbonyl compounds with diisopropyl *O,O'-bis*(trimethylsilyl)-tartrate at −78°C in CH$_2$Cl$_2$ with trimethylsilyl triflate catalyst, then warming to room temperature for 2–4 h has been employed (Knierzinger et al. 1990). The completed reaction is quenched with Et$_3$N, concentrated *in vacuo,* and the product purified by washing through a plug of silica with ether.

8.4. Determination of Absolute Configuration and Enantiomeric Purity by NMR Methods

The application of NMR techniques to the determination of enantiomeric purity of semiochemicals was discussed in detail by Mori (1984). Overall, a relatively large amount of sample is required (>100 µg to a few milligrams) because of the inherent insensitivity of NMR. The accuracy of the determination of the enantiomeric purity may also limited by the dynamic range and sensitivity of the NMR experiment to about ±1–2%.

Two different strategies can be used to determine optical purity by NMR methods. First, the sample can be placed in a chiral environment during the NMR experiment by using a chiral solvent, or chiral additives to the solvent such as chiral solvating or chiral lanthanide shift reagents. Second, the sample can be derivatized with a chiral derivatizing agent, generating diastereomers whose chromatographic and NMR spectral properties are different. The latter technique is most commonly used today, particularly as increases in magnetic fields of NMR spectrometers have resulted in increases in both NMR resolution and sensitivity.

8.4.1. Use of Chiral NMR Shift Reagents

NMR shift reagents are usually β-dicarbonyl complexes of lanthanide metals such as europium, praseodymium, and ytterbium. They are mild Lewis acids, which when they complex with the lone pair electrons of heteroatoms in a solute, substantially alter the magnetic field in their immediate vicinity. When the metal atom is complexed to chiral ligands, it can bind more strongly to one enantiomer of a solute than to the other, in essence creating transitory diastereomeric complexes. Thus, the chemical shifts of protons (or other nuclei) may be different for each diastereomeric complex, and the resolved NMR peaks can be integrated to determine the enantiomeric ratio. The formation and collapse of the diastereomeric complexes is fast on the NMR timescale, so less than one equivalent of the shift reagent is required. In practice, small sequential aliquots of shift reagent are added until satisfactory results are obtained.

However, NMR signals are broadened by the presence of the paramagnetic metals, and unfortunately the broadening effect increases as the square of the magnetic field, whereas the resolution of peaks increases only linearly with the field strength. Thus, at the high fields typical of modern NMR instruments, signals may show up only as broad humps in the baseline. However, with the increased resolution of high-field instruments, very small amounts of shift reagent may still provide some resolution without peaks becoming so broad as to be unusable. A number of chiral shift reagents are available from standard sources (e.g., Aldrich or Fluka), and several detailed examples of their application are covered in Mori (1984). However, the broadening effect at higher fields, coupled with the tremendous progress in chromatography with chiral stationary phases, has decreased the use of chiral shift reagents.

8.4.2. NMR of Diastereomeric Derivatives

NMR of samples in achiral solvents cannot be used to directly determine the absolute configuration of molecules. Chiral environments for the solute in the NMR experiment can be created by using chiral solvents or solvating agents, and these have in fact found limited use. For example, Pirkle's chiral solvating reagent (R)-$(-)$-2,2,2-trifluoro-1-(9-anthryl)ethanol (**101**, Pirkle et al. 1977) was used for the determination of the enantiomeric purity of a synthetic sample of the sex pheromone of the carpenter bee (**102**, Mori & Senda 1985). However, it is much more common and generally useful to determine configurations and enantiomeric purities indirectly by derivatization with a chirally pure derivatizing agent (say R), creating a diastereomer (e.g., R,R), which will in theory have different physical and chemical properties, including NMR chemical shifts, from the (S,R) diastereomer. Furthermore, any NMR-sensitive nucleus in the molecules can be observed. For example, if the amount of sample is limited, only ^1H or possibly ^{19}F NMR (for derivatizing agents such as Mosher's reagent [Table 8.1], which contain fluorines) may be possible. With larger amounts of sample, ^{13}C NMR can be used, with the advantage that ^{13}C signals are spread over 200 ppm or more (in contrast to ~10 ppm for ^{1}H) so that the chances of resolving signals from diastereomeric carbons is enhanced. A racemic synthetic standard and at least one of the two enantiomers of known configuration are normally required to unequivocally assign the configuration of the isolated compound.

Although the amount of sample required may be prohibitive (0.75–1 mmol), ^{31}P NMR can be used for the determination of the enantiomeric purities of alcohols. For example, either chiral (**99**, Johnson et al. 1984) or achiral (**100**, Feringa et al. 1985) phosphorus derivatives were prepared and analyzed by ^{31}P NMR. This method has been extended to primary alcohols in which the chiral center is remote from the functional group (e.g., citronellol, Alexakis et al. 1992), in which it may be difficult or virtually impossible to determine the enantiomeric purity by other methods.

In some cases, it has been possible to determine absolute stereochemistries from derivatized racemic compounds (i.e., mixtures of the two diastereomers) by careful consideration of the preferred conformations of the derivatives, and the resulting effects on chemical shifts (Dale & Mosher 1973). For example, the absolute stereochemistries of some secondary alcohols were determined after derivatization with chiral α-methoxy-α-trifluoromethylphenylacetic acid chloride **93** (Mosher's reagent, Table 8.1) to give the esters **94** or **95**, without having standards of known configuration. Thus, the absolute configurations of some terpenoidal alcohols of marine origin were successfully deduced from the NMR spectra of 2–7 mg of the derivatized alcohols (Ohtani et al. 1991). The absolute configurations of secondary alcohols have also been determined from the NMR spectra of their *O*-methylmandelate (**96**, Trost et al. 1986) or 2-napthylmethoxya-cetate derivatives (**97**, Kusumi et al. 1994). For example, the absolute configuration of ginnol (**98**, a constituent of *Ginkgo biloba* leaves) was determined by esterification with the (*R*)-and (*S*)-2-napthylmethoxyacetic acids, resulting in the marked upfield shift of the protons located on the same side of the molecule as the naphthalene ring, enabling the configuration to be deduced as *S*.

This general idea has been further extended to other secondary alcohols in a clever application of the difference in the rates of reaction of a chiral derivatizing agent with each of two enantiomers, due to differences in the energy levels of the transition states (Shi et al. 1996). A racemic standard is reacted with a fraction of one equivalent of Mosher's reagent (**93**) at −20°C, producing an unequal mixture of diastereomers. The NMR spectrum of the mixture is taken, using the unequal sizes of peaks from the two diastereomers to assign specific peaks to each diastereomer. Consideration of the most favorable conformation for each of the diastereomers and the anticipated effects on the chemical shifts of nuclei in the derivatized molecules then enables the tentative assignment of absolute configurations. However, final verification of assignments by unambiguous synthesis of one of the two enantiomers is still recommended.

High-field NMR is also useful for determination of the relative stereochemistry of chiral centers, for those compounds containing more than one chiral center. In the cases of the maritime pine bast scale pheromone (**39**, Fig. 8.2; Mori and Harashima 1993a) and matsuone (**40**, Fig. 8.2; Mori & Harashima 1993b) relative stereochemistry was assigned on the basis of a subtle difference in the ^{1}H NMR spectra of substances **39** and **40** with those of their respective anti diastereomers. Furthermore, the fact that the natural pheromone (**87**) is a stereoisomeric mixture (1:1) at C-8 was also clarified by comparison of the NMR spectra of the isolated material with those of the synthetic (8*R*,15*R*)- and (8*S*,15*R*)-stereoisomers (Ernst & Wagner 1989).

8.5. Determination of Absolute Configuration and Enantiomeric Purity by Chromatography with Chiral Phases

In the past decade, there has been a virtual explosion in the development and use of chiral stationary phases in both liquid and gas chromatography, and to a lesser extent, in thin-layer chromatography. Separations that were tedious, difficult, and in many cases virtually impossible (e.g., saturated hydrocarbons) are now routine. Chiral stationary phases work by providing a chiral environment with which each of a pair of enantiomers interacts differently. The differences in interactions, summed over the length of the column, result in resolution of the enantiomers. Unlike most other methods of determination of enantiomeric purity, because the separation is due to the sum of multiple interactions with the stationary phase, the chiral selector on the stationary phase does not have to be chirally or even chemically pure. All that is required is a considerable excess of one of the two enantiomers of the chiral selector.

The chiral selector is normally bonded or adsorbed to, or coated on the stationary phase. Most chiral stationary phases work by one of two general principles, or some combination of the two. First, a phase may contain "chiral cavities," so that separation of enantiomers depends on the comparative "fit" of each enantiomer into the cavity, with the one that fits best being retained longest. This is the principle behind the widely used and highly successful cyclodextrin-based phases, which can resolve even saturated chiral hydrocarbons with no functional groups. Second, phases may contain polar, chiral groups whose electronic interactions with polar functionalities in the analytes, modified by steric interactions, determine resolution. These types of phases require three points of interaction between the solutes and the chiral selector to discriminate between the two enantiomers. This restricts the usefulness of these phases to compounds that contain more than one polar functional group, although this restriction can be circumvented by the addition of further polar functionalities by derivatization.

A limited number of chiral compounds can be resolved on achiral stationary phases using chiral modifiers, such as transition metal ions complexed with chiral ligands, in the mobile phase. However, applications are limited to compounds with two polar functional groups, such as amino acids, which can complex strongly with the metal ions, and are not discussed further here. Interested readers are referred to Davankov (1992).

8.5.1. High-Performance Liquid Chromatography with Chiral Stationary Phases

Although less than 1 mg of sample can be analyzed by HPLC, this technique has not been used extensively for the determination of the absolute configuration and enantiomeric purity of naturally occurring pheromones, usually because pheromones are often obtained in nanogram to microgram quantities. Furthermore, many pheromones lack chromophores with strong absorptions in the ultraviolet region, and so cannot be detected readily (without derivatization) by the UV detectors routinely used with HPLC. However, for those compounds that are nonvolatile (or that cannot be rendered volatile by derivatization), LC is obviously the only chromatographic option.

Chiral stationary phases (CSP) for HPLC can be broadly classified into five types, as follows (Wainer 1988; Stevenson & Wilson 1988).

1. Solute-CSP complexes are formed by attractive interactions, such as H bonding, π-donor and π-acceptor interactions, and dipole stacking. These include the so-called Pirkle phases, which contain amide and/ or carbamate functional groups and one or more aromatic moieties. Derivatization of analytes is often required to improve chromatographic characteristics and/or to provide another polar functional group.

2. Solute-CSP complexes are formed as in type 1, modified by the formation of inclusion complexes. These CSPs include some forms of derivatized cellulose and polymethacrylates.

3. The solute enters chiral cavities within the CSP to form inclusion complexes. Typical examples are the large number of derivatized cyclodextrin CSPs, but also include microcrystalline cellulose.

4. Diastereomeric complex formation is mediated by a metal ion complexed with a chiral ligand on the CSP (chiral ligand exchange). These CSPs are useful only for analytes with two polar functional groups on or adjacent to the asymmetric carbon (e.g., amino acid derivatives).

5. The CSP is a protein, such as bovine serum albumin or a glycoprotein, bound to a support. Solute-CSP complexes form on the basis of mixed hydrophobic and polar interactions.

There are now a plethora of CSP HPLC columns available from a large number of manufacturers, many of whom maintain extensive applications databases (and free brochures) for their products. It is strongly recommended that these be consulted when considering a column purchase. In practice, the most useful phases (i.e., most generally applicable) are those of types 1 and 3, which are now extensively utilized for both analytical and preparative separations of enantiomers.

Historically, a cellulose triacetate column (Fig. 8.7, **103**) was first employed in pheromone chemistry for the optical resolution of the synthetic spiroacetal pheromone (±)-**84** of *Andrena wilkella* (Isaksson et al. 1984). Racemic lactones such as 4-hexanolide (a pheromone component of *Trogoderma glabrum*) were resolved on preparative scale with a cellulose triacetate column (Francotte & Lohmann 1987). HPLC methods also are routinely used in checking the enantiomeric purity of intermediates in pheromone syntheses (e.g., Brevet & Mori 1992). The enantiomeric purity of a synthetic sample of the mosquito pheromone **55** was analyzed with Chiralcel® OF (Fig. 8.7, Henkel et al. 1995).

Among many derivatives of polysaccharides, the following three show high chiral recognition, and they are used as chiral stationary phases for HPLC analysis (Fig. 8.7) (1) 3,5-dimethylphenylcarbamates of cellulose (e.g., Chiralcel® OD [**104**]) and amylose (e.g., Chiralpak® AD [**105**]), and cellulose tris(*p*-methylbenzoate) (e.g., Chiralcel® OJ [**106**]). These chiral stationary phases have enabled the resolution of more than 80% of the racemates tested (Yashima & Okamoto 1995). A very good compilation of examples of enantiomeric resolution with polysaccharide derivatives is available (Daicel Chemical Industries 1995).

Some other chiral stationary phases are also used for enantiomer separation (Fig. 8.7). The representative examples are *N*-3,5-dinitrobenzoylphenylglycine-based stationary phases (e.g., Sumichiral® OA-2000 [**107**]) and triphenylmethyl (+)-polymethacrylate (e.g., Chiralpak® OT [+] [**108**]).

The state of the art of the HPLC enantiomer separation method was thoroughly reviewed recently (Yashima & Okamoto 1995).

8.6. Determination of Absolute Configuration and Enantiomeric Purity by Gas Chromatography

Gas chromatography is the method of choice for the analysis of enantiomers for several reasons. First, only nanogram or smaller amounts of sample are required, depending on the type of detector used, and the sensitivity and dynamic range of most GC detectors is such that enantiomeric purities can be determined with great accuracy (a fraction of 1%). Second, capillary GC has much better inherent resolution than liquid chromatography, so that diastereomers or transitory diastereomeric complexes formed with chiral selectors in the CSP with only small differences in properties will be resolved. Third, the standard flame-ionization detector (FID) used with GC is essentially a universal detector, detecting all compounds regardless of functional group, in contrast to HPLC, where an analyte usually must contain a chromophore to be detected. As with most other methods,

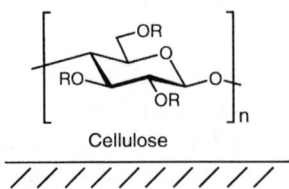

Cellulose

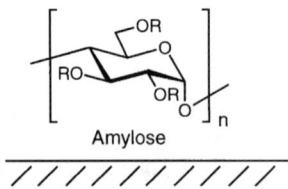

Amylose

///////// Coated on silica gel

///////// Coated on silica gel

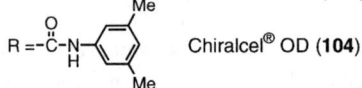

R = Ac   Chiralcel® CA-1 (103)

R = $-\overset{O}{\overset{\|}{C}}$-N   Chiralcel® OD (104)

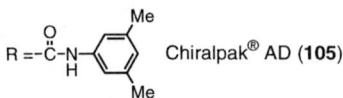

R = $-\overset{O}{\overset{\|}{C}}$-N   Chiralpak® AD (105)

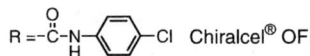

R = $-\overset{O}{\overset{\|}{C}}$-N──Cl   Chiralcel® OF

R = $-\overset{O}{\overset{\|}{C}}$──Me   Chiralcel® OJ (106)

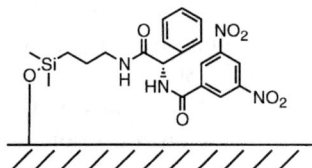

/////////
Bonded to silica gel
Sumichiral® OA-2000 (107)

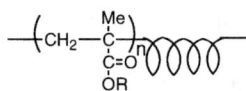

/////////
Coated on silica gel
Chiralpak® OT (+)
R = −C(C₆H₅)₃ (108)

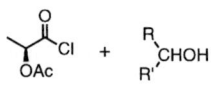

109   + $\overset{R}{\underset{R'}{}}$CHOH $\xrightarrow[\text{Et}_2\text{O}]{\text{C}_5\text{H}_5\text{N}}$ **110**   +   **111**

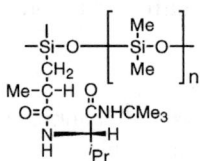

Chirasil L-Val (**112**)

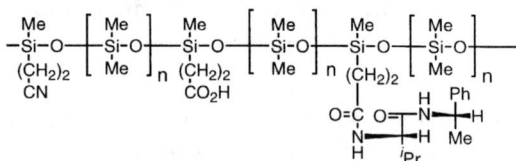

XE-60-L-Val.-(*S*)-α-phenethylamide (**113**)

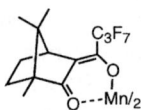

114

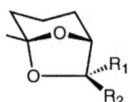

115 R₁ = Et, R₂ = H
116 R₁ = H, R₂ = Et

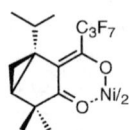

117

Figure 8.7. Structural formulas—2.

synthetic samples of one enantiomer with known absolute configuration, and usually the racemic material as well, are required as references.

There are two methods for GC analysis of enantiomers: (1) derivatization of analytes (e.g., pheromone alcohols or hydroxy esters) with enantiomerically pure derivatizing agents and analysis of the resulting mixture of diastereomers on an achiral GC column and (2) analysis of analytes, with or without derivatization, on chiral GC phases containing chiral peptides, chiral metal chelates, or derivatized cyclodextrins.

8.6.1. Gas Chromatographic Analysis after Derivatization

The derivatization of chiral analytes with chiral reagents, followed by GC analysis of the resulting diastereomeric derivatives is straightforward and widely used. The method has the advantage that the derivatives can be analyzed on standard achiral columns, so that no special equipment is required. Several derivatizing agents can be used (Table 8.1), all of which are commercially available or can be readily made in one or two simple steps. Allenmark (1988) and Souter (1985) provide more extensive compilations. The type of analyte and the list of suitable derivatizing agents are limited by the requirements that the analytes must have a derivatizable functional group, and the resulting derivative must be volatile enough for GC analysis. In practice, derivatization may actually improve the chromatographic properties of analytes with polar functional groups. Numerous examples of these types of analyses in chemical ecology studies were compiled by Mori (1984). More recent examples are discussed below.

Just prior to the advent of cyclodextrin-based chiral stationary phases, Slessor et al. (1985) reported an excellent method for determining the enantiomeric purity of chiral alcohols, lactones, and hydroxy acids in quantities ranging from 25 ng to 10 µg. Before derivatization, lactones and hydroxy acids are converted to the corresponding hydroxy esters by treatment with methanol and BF_3 etherate. For example, treatment of enantiomerically impure alcohols or hydroxy esters with enantiomerically pure 2-acetoxypropanoyl chloride (**109**) derived from (*S*)-(+)-lactic acid gave a diastereomeric mixture of esters (**110**) and (**111**), which were resolved by capillary GC. Slessor's method was used for the determination of the enantiomeric purity of compounds **8, 17, 27, 28, 29, 30, 37, 44, 45, 56, 57, 58,** and **59**.

8.6.2. Gas Chromatographic Analysis with Peptide or Chiral Metal Chelate Phases

Until the late 1980s, peptide phases composed of polysiloxanes modified by α-amino acid derivatives, as represented by Chirasil®-L-Val (**112**) and XE-60-L-Val-(*S*)-α-phenethylamide (**113**) (König 1987) were two of the few CSPs available for GC. Ipsenol (**16**), compounds **20** and **21**, and pityol (**66**) were analyzed with compound **112,** both before and after derivatization with isopropyl isocyanate. Pheromones **71** and **79** were analyzed with compound **113.**

Phases based on chiral metal β-diketonates derived from optically active monoterpenes were also used for enantiomeric separations (Weber & Schurig 1984). For example, homo-rose oxide (**72**) and *endo-* and *exo*-brevicomin (**115 and 116,** respectively) were resolved with Mn(II) bis[3-heptafluorobutanoyl-(1*R*)-camphorate] (**114**) (Schurig et al. 1983). The enantiomers of olean (**83**) were resolved by employing Ni(II) bis[2-heptafluorobutanoyl-(1*S*,5*S*)-4-methylthujonate] (**117**) as the stationary phase (Haniotakis et al. 1986). However, analyses generally had to be conducted at low to moderate temperatures to obtain resolution, limiting the analytes that could be used, and these phases have not been widely commercialized, due in large part to the much greater versatility of the cyclodextrin-based phases described in the next section.

8.6.3. Gas Chromatographic Analysis with Modified Cyclodextrins

In the last decade, experimental and commercial chiral GC stationary phases based on cyclodextrin derivatives have been developed and used with extraordinary success in the resolution of all types of volatile chiral analytes, from nonpolar saturated hydrocarbons to polar, water soluble compounds such as diols. Their versatility, general applicability, ease of use, and relative durability in comparison to many other types of CSPs have resulted in a veritable explosion of applications, and they are now used in the vast majority of GC separations of volatile enantiomers (König 1990, 1992).

The key breakthrough was achieved independently in 1988 by the research groups of Schurig and König, by derivatization of some of the hydroxy groups of cyclodextrins with methyl or pentyl groups followed by further modification by selective alkylation or acylation. The modified cyclodextrins obtained (e.g., Fig. 8.8) are liquid at room temperature, soluble in nonpolar solvents, thermostable up to 400°C, and enantioselective. The rather hydrophilic per-*O*-methylated cyclodextrins could be used at 70–220°C after dilution with a polysiloxane (Schurig et al. 1990).

The cyclodextrin chiral selectors are coated or covalently bonded on inert supports. Selectivity is modified by altering the size of the cyclodextrin ring (6, 7, or 8 glucose units, corresponding to α-, β-, and γ-cyclodextrins, **118–120**), which alters the size of the chiral cavity, and by the number and type of chemical modifications (usually alkylations or acylations with chains of different lengths and functionalities, such as hydroxypropyl, trifluoracetyl, or acyl). Numerous "tailor-made" phases are now available from a number of suppliers. Because the molecular mechanism of chiral discrimination with cyclodextrins is not yet fully understood, it is difficult to predict whether or not a selected phase will solve a specific separation problem. However, a large volume of data concerning various chemical classes has been compiled (König 1992), and the *Journal of High*

α-Cyclodextrin
118 $R_2 = R_3 = R_6 = H$

β-Cyclodextrin
119 $R_2 = R_3 = R_6 = H$

γ-Cyclodextrin
120 $R_2 = R_3 = R_6 = H$

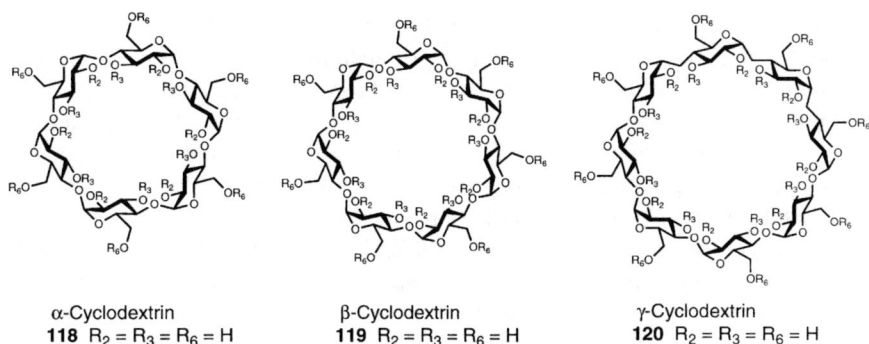

Structure of hydrophobic α-, β-, and γ-cyclodextrin derivatives

$R_2 = R_3 = R_6$ = pentyl	Lipodex® A = per-*O*-pentyl-**118**
$R_2 = R_6$ = pentyl, R_3 = methyl	Lipodex® B = per-*O*-pentyl-**119**
$R_2 = R_3$ = pentyl, R_6 = methyl	Lipodex® C = 3-*O*-acetyl-2,6-di-*O*-pentyl-**118**
$R_2 = R_6$ = pentyl, R_3 = acetyl	Lipodex® D = 3-*O*-acetyl-2,6-di-*O*-pentyl-**119**
$R_2 = R_3$ = pentyl, R_6 = acetyl	α-DEX® = per-*O*-methyl-**118**
$R_2 = R_6$ = pentyl, R_3 = butanoyl	β-DEX® = per-*O*-methyl-**119** = Chirasil Dex®
$R_2 = R_3$ = pentyl, R_6 = butanoyl	γ-DEX® = per-*O*-methyl-**120**

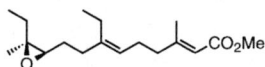

Juvenile hormone I **(121)**

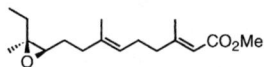

Juvenile hormone II **(122)**

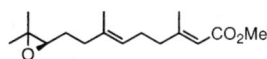

Juvenile hormone III **(123)**

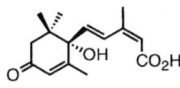

Abscisic acid **(124)**

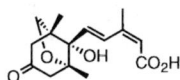

Phaseic acid **(125)**

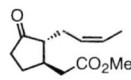

Methyl *trans*-jasmonate **(126)**

Figure 8.8. Structural formulas—3.

Resolution Chromatography maintains a database of separations of chiral compounds, including the phase and conditions used. Stationary phases employed for the separation of pheromone enantiomers are depicted in Figure 8.8.

Cyclodextrin-based chiral stationary phases were successfully employed for the resolution of numerous chiral compounds other than pheromones, including epoxy alcohols (König et al. 1989), insect juvenile hormones I–III **(121–123)** (König et al. 1993), abscisic and phaseic acids **(124** and **125)** (Balsevich et al. 1994), and sesquiterpene hydrocarbons (Hardt et al. 1995). The analysis of commercial samples of one of the most widely used chiral derivatizing agents, Mosher's acid (α-methoxy-α-trifluoromethylphenylacetic acid) as its methyl ester (König et al. 1990) showed them to be of 97.98–99.80% enantiomeric excess.

This analysis dramatically illustrated the limitations imposed by the chiral purity of the derivatizing agent on the accuracy of enantiomeric purity determinations based on the formation and analysis of diastereomeric derivatives.

Preparative scale separation of the enantiomers of methyl *trans*-jasmonate (**126**) in 2-mg scale yielded both the (+)-isomer (93% e.e.) and its (−)-antipode (91% e.e.) by employing 2,6-di-*O*-methyl-3-*O*-pentyl-β-cyclodextrin in OV-1701 as the chiral stationary phase (Hardt & König 1994).

In practice, there are several points to keep in mind when using capillary GC with chiral stationary phases. First, the phases are less durable than many bonded GC phases, requiring rigorous exclusion of air and moisture, including sealing the column ends when not in use. Furthermore, their upper temperature limits may be comparatively low, restricting the types of analytes that can be studied. Second, resolution usually decreases with increasing temperature because increasing random thermal motion of molecules disrupts the relatively weak interactions with the chiral selectors. Thus, the best resolutions are obtained at lower temperatures, even at the expense of long retention times. Analyses are often conducted isothermally, or with slow temperature programs ($1-2°C/min$), and retention times of >1 h are not uncommon. This may again restrict the molecular weights of analytes. It may be possible to compensate to some extent by using H_2 as a carrier gas instead of He or N_2 (retention time is proportional to carrier gas molecular weight, see chapter 3), or by increasing the carrier gas flow rate. Third, retention times may shift slightly with sample size and concentration, so test samples and standards should be of similar concentration, and as final confirmation, the isolated compound and the standard should be coinjected.

The practical procedure for the determination of the absolute configuration of a natural product such as a pheromone can include the following steps.

1. Take a synthetic racemate of the compound under investigation, and test whether the chosen chiral stationary phase will separate the enantiomers. Note that the flow rate and temperature conditions can be very important parameters, so these should be optimized carefully by a series of iterative steps.

2. Take a pure synthetic enantiomer of the compound, with known absolute configuration, and add it to the racemate. In the gas chromatogram obtained, enhancement of one of the peaks will provide knowledge as to the order of elution of the two enantiomers. Steps 1 and 2 can be combined in so far as a sample resulting from a moderately enantioselective synthesis can be used, provided the absolute configuration of the predominating enantiomer is known.

3. Run the natural sample, and determine its enantiomeric composition.

4. Add an aliquot of the natural sample to the racemate of step 1 or to the enriched sample of step 2. An enantiomerically pure natural product

will augment one of the two peaks produced by the racemate or by the enriched sample, conclusively identifying the absolute configuration of the natural product.

The last step is particularly crucial with respect to relative proportions of the synthetic and natural samples. The synthetic sample and the natural product should have similar concentrations. That is, it is useless to mix, for example, 100:1 or 1:100, since the differences in enantiomeric compositions will be too low to be reliable. Coinjection is always preferable to simple comparison of retention times, since small changes in the oven temperatures between runs or concentration effects may cause misattributions. Even an internal standard eluting close to the target compound is less reliable than coinjection.

GC separation of enantiomers also can be particularly useful in the determination of the biological activity of the pure enantiomers of pheromones. The on-line combination of chiral stationary phases and electroantennograms has been reported (Perez et al. 1994; Leal et al. 1995; Zhang et al. 1997).

8.6.3.1. Examples of Enantiomeric Separations on Chiral Cyclodextrin Phases

Five examples of the separation of pheromone enantiomers are illustrated below. The experimental details were contributed by five researchers in chemical ecology.

Example 1

GC analysis with a chiral stationary phase was used for the determination of the absolute configuration of 2-ethyl-1,5-dimethyl-6,8-dioxabicyclo[3.2.1]octane (bicolorin,**77**) in frass of the males of the European beech bark beetle, *Taphrorychus bicolor* (W. Francke & F. Schröder, unpublished). The experimental data were supplied by Prof. W. Francke. As shown in Figure 8.9, the absolute configuration of bicolorin was 1*S*,2*R*,5*R*. Note that the number of stereoisomers of compound **77** is restricted by the bicyclic ring structure.

Example 2

Gas chromatographic-electroantennographic detection (GC-EAD) with a chiral stationary phase was used for the determination of the absolute configuration of cruentatol (**13**), the aggregation pheromone of the palmetto weevil (*Rhynchophorus cruentatus*), and compound **14,** that of *R. ferrugineus, R. bilineatus, R. vulneratus,* and *Metamasius hemipterus* (Perez et al. 1994; Oehlschlager et al. 1995). The experimental data were supplied by Prof. G. Gries. As shown in Figures 8.10 and 8.11, the absolute configurations of compounds **13** and **14** were both 4*S*,5*S*.

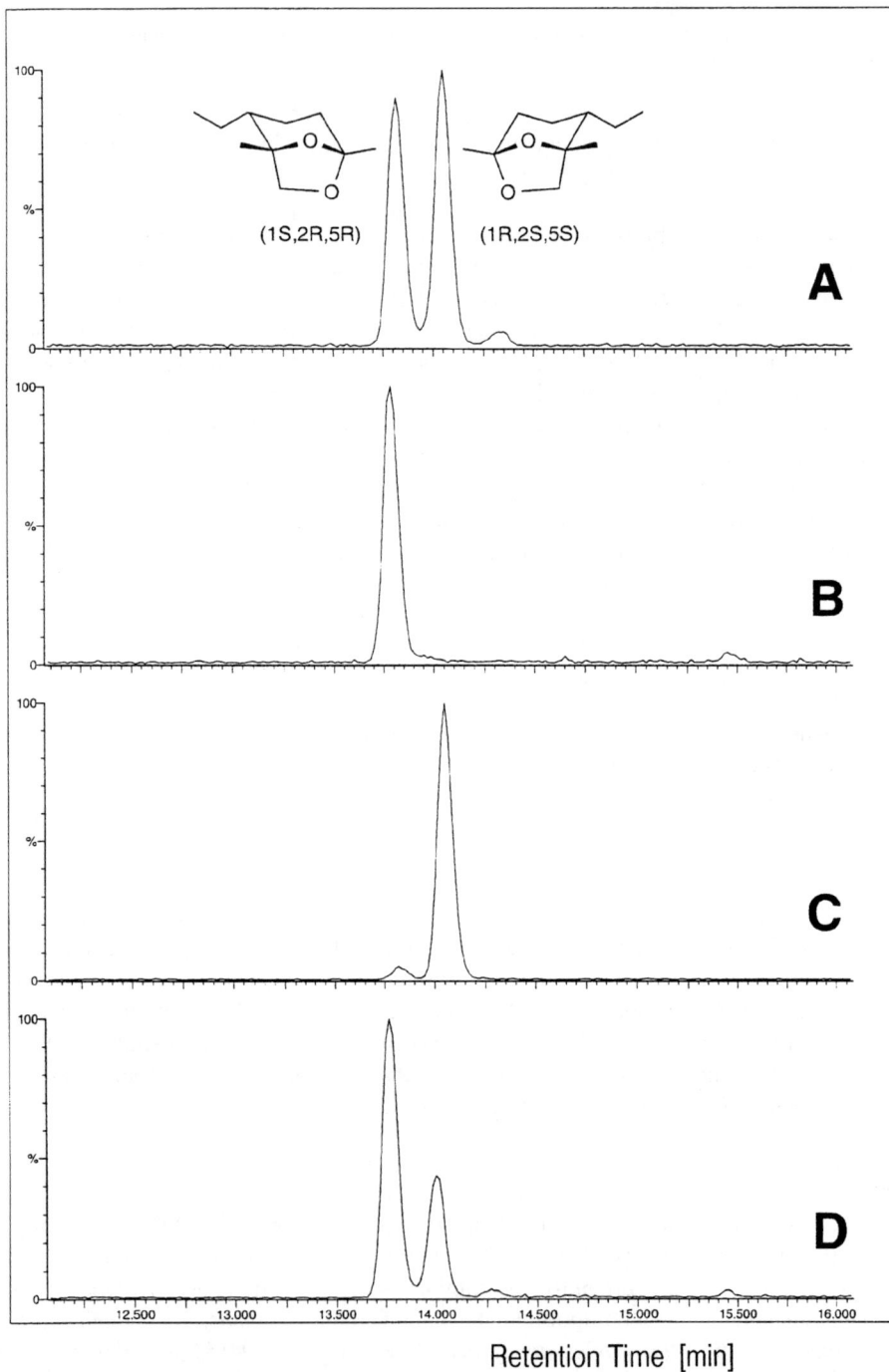

Figure 8.9. Determination of the absolute configuration of bicolorin (**77**): (**A**) synthetic (±)-**77**, (**B**) natural extract, (**C**) synthetic (1*R*,2*S*,5*S*)-**77**, and (**D**) a combination of the substances in (**B**) and (**C**). Conditions: fused silica column (25 m × 0.25 mm i.d.) coated with a 1:1 mixture of OV 1701 and heptakis(2,6-di-*O*-methyl-3-*O*-pentyl)-β-cyclodextrin; 5 min at 60°C, 5°C/min to 170°C.

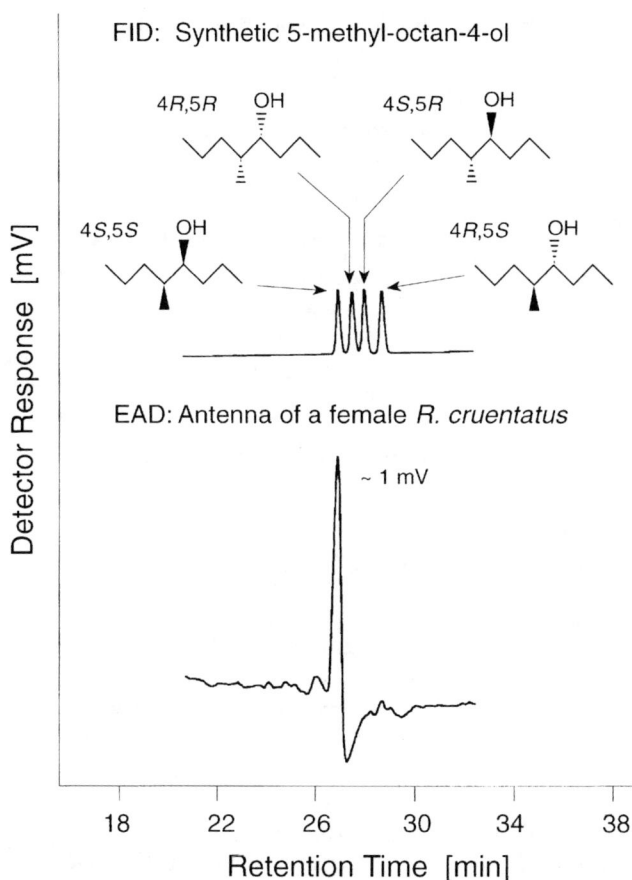

Figure 8.10. Flame-ionization and electroantennographic detector (male *Rhynchophorus cruentatus* antenna) responses to the stereoisomers of cruentatol (**13**), the pheromone of *R. cruentatus*. Conditions: split injection, Cyclodex®-B column, 30 m × 0.25 mm i.d.; 90°C isothermal; linear flow velocity of carrier gas, 35 cm/s; injection temperature, 220°C.

A critical prerequisite for optimal GC-EAD analyses is optimal treatment of test animals during shipment and laboratory maintenance. A weevil antenna was carefully dislodged (not cut!) from the weevil's head and suspended between two saline-filled (Staddon & Everton 1980) glass capillary electrodes (1 mm o.d. × 0.58 mm i.d., A-M Systems, Everett, WA) with inserted, chloridized silver wire, mounted on a micromanipulator. The antennal flagellum was pierced with a drawn-out recording electrode. To accommodate the antennal base, the opening of the indifferent, L-shaped electrode was slightly widened. During GC-EAD recordings, nonhumidified, column effluent-laden medical air (20°C) passed the

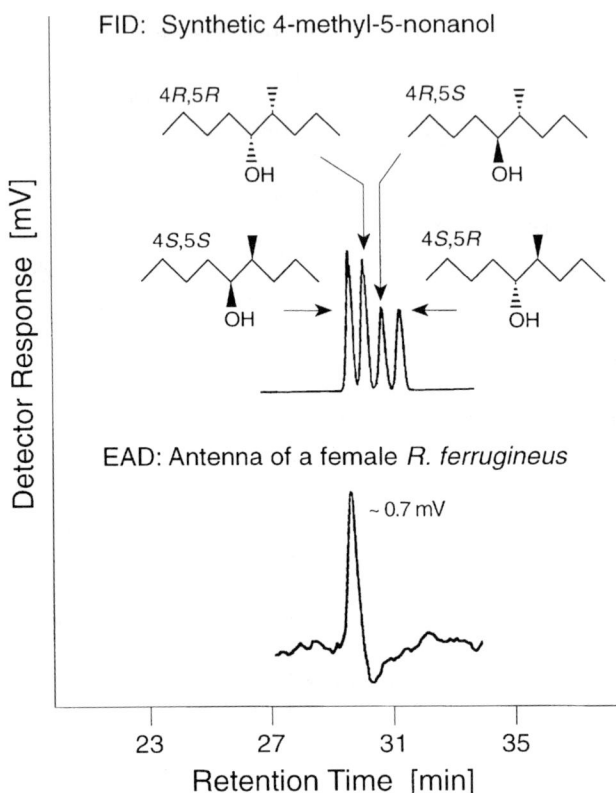

Figure 8.11. FID and EAD (male *Rhynchophorus ferrugineus* antenna) responses to the stereoisomers of 4-methyl-5-nonanol (**14**), the pheromone of *R. ferrugineus, R. bilineatus, R. vulneratus,* and *Metamasius hemipterus.* Conditions: split injection, Cyclodex®-B column; 100°C isothermal; linear flow velocity of carrier gas, 35 cm/s; injection temperature, 220°C.

antennal preparation. Antennal responses were amplified by a custom-built amplifier with a passive low-pass filter and a cutoff frequency of 10 KHz.

Example 3

GC analysis with a chiral stationary phase was used for the confirmation of the high enantiomeric purity of a synthetic sample of japonilure (**53**) prepared by a known (Senda & Mori 1983) method. The experimental data shown in Figure 8.12 were supplied by Dr. W. S. Leal.

Resolution was achieved on a trifluoroacetylated γ-cyclodextrin-based capillary column, Chiraldex® GTA (Astec, Whippany, NJ). Chiraldex® GTA is a coated phase, and the film integrity is adversely affected by high temperature. Column

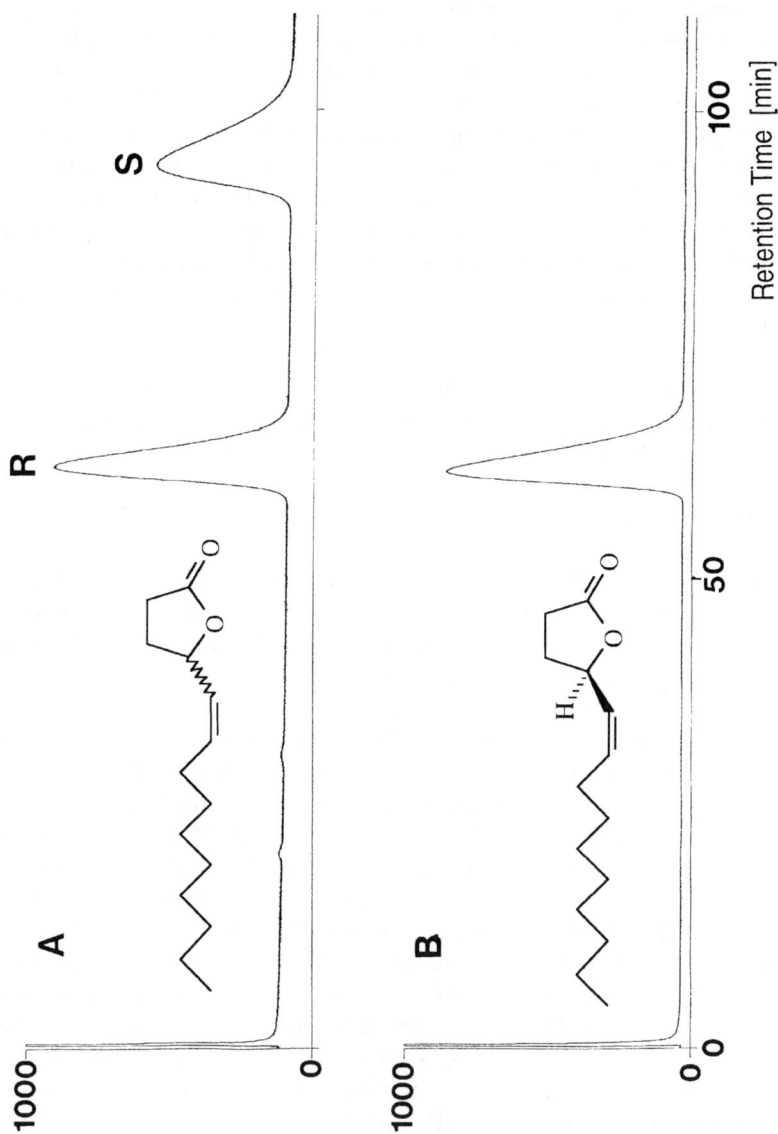

Figure 8.12. Gas chromatograms obtained from (**A**) (±)-japonilure (**53**) and (**B**) (*R*)-**53** prepared according to a known method (Senda & Mori 1983). Conditions: Chiraldex® GTA (20 m × 0.25 mm i.d.; 0.125-mm film thickness), splitless mode; 125°C; carrier gas helium; linear flow velocity, 146 cm/s; (*R*)-Japonilure **53**, Retention time (Rt) = 61.9 min, (*S*)-**53**, Rt = 93.8 min.

life will be extended by operating it at a temperature as low as possible (maximum operating temperature is 150°C). In the resolution of japonilure, high head pressure was used to compensate for the low temperature. Unlike the bonded-phase cyclodextrin columns, splitless or on-column injections may permanently damage the coating at the front of the column, but this can be minimized by use of a retention gap (see chapter 2). Chiraldex® GTA is also very sensitive to moisture; even methylene chloride extracts of an aqueous sample contain sufficient water to cause the hydrolysis of the trifluoroacetyl groups. Evaporation to near dryness in the presence of 2,2-dimethoxypropane will adequately dry the sample. Because moisture may alter the selectivity of the column during the storage, it should be flame-sealed after use. The manufacturer describes a procedure for regeneration of the phase by treatment with trifluoroacetic anhydride vapor in an applications brochure.

With Chirasil-Dex® CB (Chrompack, Middelburg, The Netherlands), various chiral and polar compounds were resolved, but no resolution of japonilure (**53**), and other lactones with a double bond adjacent to the lactone ring was obtained (W.S. Leal, unpublished).

Example 4

GC analysis with a chiral stationary phase was used for the determination of the enantiomeric composition of ipsdienol (**17**) isolated from two populations of the pine engraver beetle (*Ips pini*), a pest of pines in North America. The experimental data and Figure 8.13 were supplied by Prof. S. J. Seybold.

Ipsdienol-containing volatiles were trapped on Porapak Q® (chapter 1) from populations of male *I. pini* from Suffolk County, New York (Fig. 8.13A) (Seybold et al. 1995a) and Lassen County, California (Fig. 8.13B) (Seybold et al. 1995b). During the collection period, males fed for 168 h on logs of Eastern white pine (Fig. 8.13A) and Jeffrey pine (Fig. 8.13B). The Porapak® was extracted with pentane, the crude extract was concentrated from 350 to 10 ml and ipsdienol (**17**) was isolated from a 1-ml aliquot of the extract by HPLC (column: Nucleosil® 50-5, 50 cm × 10 mm; Alltech, Deerfield, IL; mobile phase: hexane/acetone [94:6, vol/vol]; flow rate: 1.6 ml/min; UV detector 235 nm). Following concentration of the relevant HPLC fraction from ~3 ml to ~200 µl, the enantiomeric composition of the isolated compound **17** was determined by split injection (split ratio ~90:1) of 0.5 µl (Fig. 8.13A) or 1.4 µl (Fig. 8.18B) onto a Cydex®-B column (Scientific Glass Engineering, Austin, TX). Approximately 0.5–1.0 µg of compound **17** was analyzed in Figures 8.13A and 8.13B. Under the conditions used, the separation factor (α) for the enantiomers of compound **17** was 1.025 and the peak resolution (Rs) was 2.32 (Poole & Schuette 1984). Based on five injections of compound **17** from the New York population and three injections of compound **17** from the California population, the mean enantiomeric compositions were 34.9%-(*R*) and 97.0%-(*R*), respectively (Seybold et al. 1995a, 1995b). The elution order of

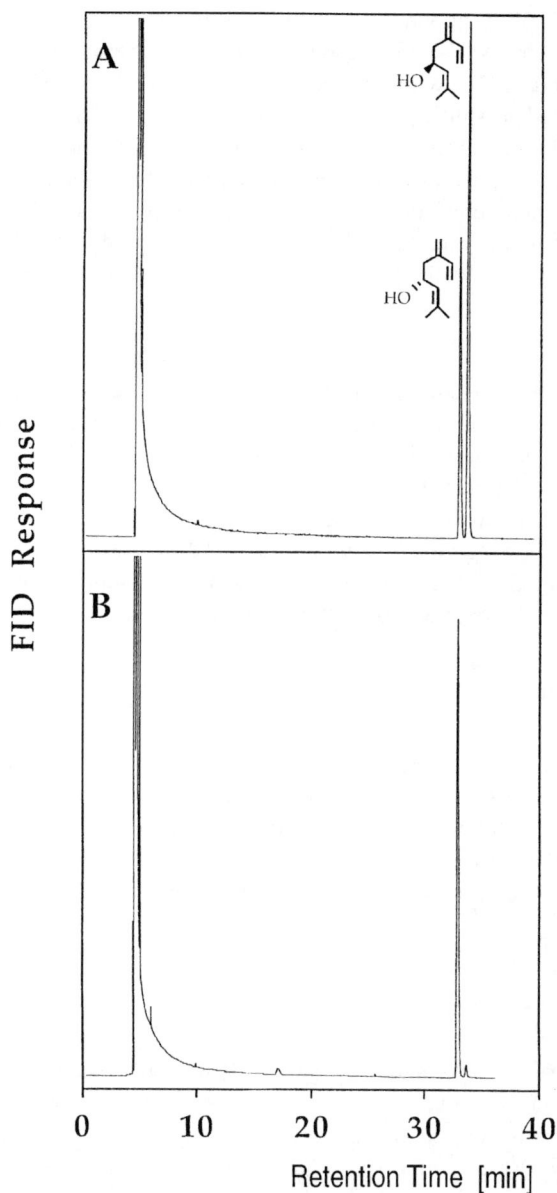

Figure 8.13. Gas chromatograms of ipsdienol (**17**) obtained from (**A**) male *Ips pini* from Suffolk County, NY, and (**B**) male *I. pini* from Lassen County, CA. Conditions: Cydex®-B column (50 m × 0.22 mm i.d., 0.25-mm film; SGE, Austin, TX), carrier gas helium; linear velocity 20.4 cm/s; 110°C isothermal; injector and detector temperature 225°C.

the enantiomers of compound **17** on this column had previously been verified as (*R*)-(−)-**17** before (*S*)-(+)-**17** by co-injection of the racemate (#P407, Bedoukian Research, Danbury, CT) with authentic (*R*)- and (*S*)-**17** (Seybold 1992).

When operated in splitless mode, purified samples of compound **17** isolated from individual insects can also be analyzed for enantiomeric composition. However, without prior purification, coeluting compounds in the crude volatile or hindgut extracts may confound GC analysis of compound **17**. When preliminary purification is not possible, the samples can be analyzed by GC-MS with selected ion monitoring (SIM) of the base peak of compound **17** (m/z = 85).

Example 5

GC analysis with a chiral stationary phase was used for the determination of the absolute configuration of phoenicol (**10**), the aggregation pheromone of *Rhynchophorus phoenicis*. The natural pheromone was (3*S*,4*S*)-**10**. The experimental data and Figure 8.14 were supplied by Dr. C. Malosse. A racemic and diastereomeric mixture of phoenicol was analyzed first, using a Cyclodex®-B column. Analysis of synthetic samples of (3*R*,4*R*)- and (3*S*,4*S*)-**10** determined that (3*S*,4*S*) eluted first, followed by (3*R*,4*R*). The natural pheromone gave a single peak, which coincided with the (3*S*,4*S*) isomer.

8.7. Conclusion

Recent developments in analytical methods, particularly in GC with chiral stationary phases, have made it possible to determine the enantiomeric composition of a compound in nanogram quantities. It must be remembered, however, that synthetic samples of known absolute configuration are a necessity, emphasizing the importance of collaborations between chemical ecologists and synthetic chemists.

8.8. Acknowledgments

I thank Profs. W. Francke and W. König of Universität Hamburg, Germany, and Prof. V. Schurig of Universität Tübingen, Germany, for their suggestions and help. The experimental examples were provided by Prof. W. Francke; Prof. G. Gries of Simon Fraser University, Canada; Dr. W. Soares Leal of the National Institute of Sericultural and Entomological Science, Japan; Dr. C. Malosse of the Institut National de la Recherche Agronomique, France; and Dr. S.J. Seybold of the University of Nevada, Reno, to whom I express my sincere thanks. I thank Dr. H. Takikawa at Science University of Tokyo for preparing the typescript and Figs. 8.1–8.8. My special thanks are due to Prof. J.G. Millar for his editorial efforts.

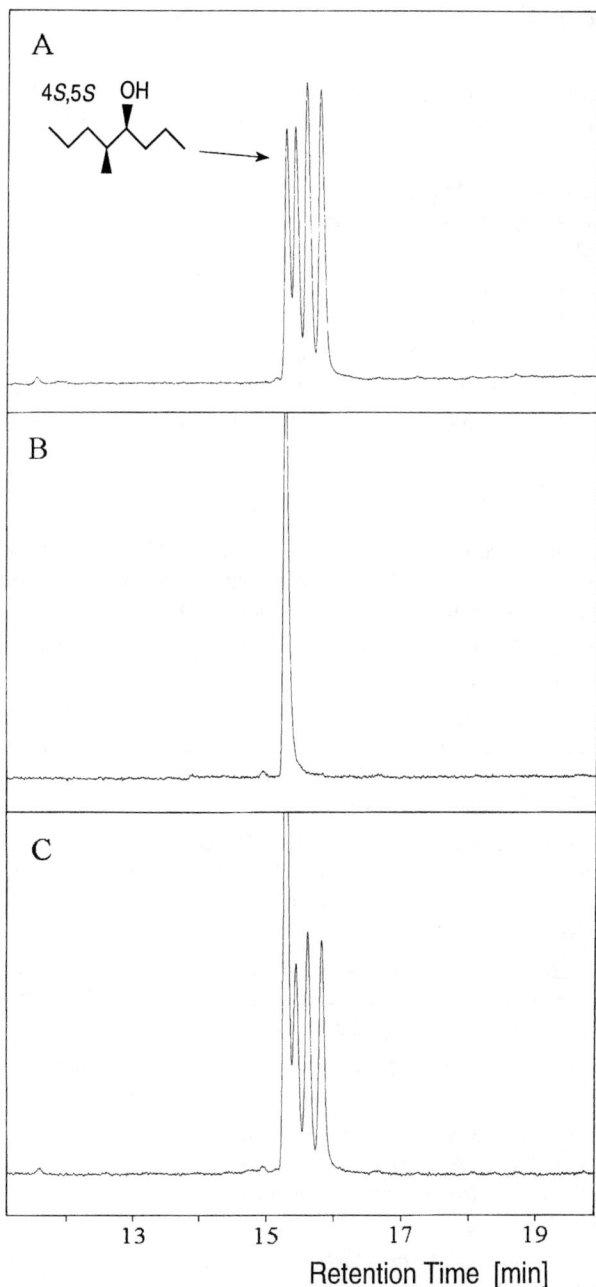

Figure 8.14. Determination of the absolute configuration of phoenicol (**13**): (**A**) synthetic mixture of the four stereoisomers, (**B**) natural pheromone, and (**C**) a combination of the substances in (**A**) and (**B**). Conditions: Cydex®-B (25 m × 0.22 mm i.d., 0.25 mm film), carrier gas helium, 15 psi; 80°C/1 min, 2°C/min to 240°C; injector 230°C, detector 240°C.

8.9. References

Aldrich, J.R., G.K. Waite, C. Moore, J.A. Payne, W.R. Lusby & J.P. Kochansky. 1993. Male-specific volatiles from nearctic and Australian true bugs. J. Chem. Ecol. **19**:2767–2781.

Alexakis, A., S. Mutti & P. Mangeney. 1992. A new reagent for the determination of the optical purity of primary, secondary, and tertiary chiral alcohols and of thiols. J. Org. Chem. **57**:1224–1237.

Allenmark, S.G. 1988. Chromatographic Enantioseparation: Methods and Applications. John Wiley & Sons, New York.

Avgerinos, A. & R.J. Hutt. 1987. Determination of the enantiomeric composition of ibuprofen in human plasma by HPLC. J. Chromatog. **415**:75–83.

Baeckström, P., G. Bergström, F. Björkling, H.-Z. He, H.-E. Högberg, U. Jacobsson, G.-Q. Lin, J. Löfqvist, T. Norin & A.-B. Wassgren. 1989. Structures, absolute configurations, and syntheses of volatile signals from three sympatric anti-lion species, *Euroleon nostras, Grocus bore,* and *Myrmeleon formicarius.* J. Chem. Ecol. **15**:61–80.

Balsevich, J.J., S.R. Abrams, N. Lamb & W.A. König. 1994. Identification of unnatural phaseic acid as a metabolite derived from exogenously added (–)-abscisic acid in a maize cell suspension culture. Phytochem. **36**:647–650.

Bartelt, R.J., A.M. Schaner & L.L. Jackson. 1989. Aggregation pheromone components in *Drosophila mulleri.* J. Chem. Ecol. **15**:399–411.

Bergström, G., A.-B. Wassgren, H.-E. Högberg, E. Hedenström, A. Hefetz, D. Simon, T. Ohlsson & J. Löfqvist. 1992. Species-specific two-component, volatile signals in two sympatric anti-lion species: *Synclysis baetica* and *Acanthaclisis occitanica* (Neuroptera: Myrmeleontidae). J. Chem. Ecol. **18**:1177–1188.

Birgersson, G., G.L. Debarr, P. de Groot, M.J. Dalusky, H.D. Pierce, Jr., J.H. Borden, H. Meyer, W. Francke, K.E. Espelie & C.W. Berisford. 1995. Pheromones in white pine cone beetle, *Conophthorus coniperda* (Schwarz) (Coleoptera: Scolytidae). J. Chem. Ecol. **21**:143–167.

Brevet, J.-L. & K. Mori. 1992. Pheromone synthesis 139: enzymatic preparation of (2*S*,3*R*)-4-acetoxy-2,3-epoxybutan-1-ol and its conversion to the epoxy pheromones of the gypsy moth and the ruby tiger moth. Synthesis pp. 1007–1012.

Cammaerts, M.-C. & K. Mori. 1987. Behavioral activity of pure chiral 3-octanol for the ants *Myrmica scabrinoides* Nyl. and *Myrmica rubra* L. Physiol. Entomol. **12**:381–385.

Cossé, A.A., R. Cyjon, I. Moore, M. Wysoki & D. Becker. 1992. Sex pheromone components of the giant looper, *Boarmia selenaria* Schiff. (Lepidoptera: Geometridae): identification, synthesis, electrophysiological evaluation, and behavioral activity. J. Chem. Ecol. **18**:165–181.

Daicel Chemical Industries. 1995. Application Guide for Chiral Column Selection—Crownpak-Chiralcel-Chiralpak, 2nd ed. Daicel Chemical Industries, Tokyo.

Dale, J.A. & H.S. Mosher. 1973. Nuclear magnetic resonance enantiomer reagents. Configurational correlations via nuclear magnetic resonance chemical shifts of diastereomeric mandelate, O-methylmandelate, and α-methoxy-α-trifluoromethylphenyl acetate (MTPA) esters. J. Am. Chem. Soc. **95**:512–519.

Davankov, V.A. 1992. Ligand-exchange chromatography of chiral compounds. *In:* Complexation Chromatography, ed. D. Cagniant, Marcel Dekker, New York. pp. 197–245.

Doolittle, R.E. & R.R. Heath. 1984. (*S*)-Tetrahydro-5-oxo-2-furancarboxylic acid: a chiral derivatizing agent for asymmetric alcohols. J. Org. Chem. **49**:5041–5050.

Dunkelblum, E., R. Gries, G. Gries, K. Mori & Z. Mendel. 1995. Chirality of Israeli pine bast scale, *Matsucoccus josephi* (Homoptera: Matsucoccidae), sex pheromone. J. Chem. Ecol. **21**:849–858.

Eliel, E.L., S.H. Wilen & L.N. Mander. 1994. Stereochemistry of Organic Compounds. John Wiley & Sons, New York.

Ernst, B. & B. Wagner. 1989. Synthesis of the oviposition-deterring pheromone (ODP) in *Rhagoletis cerasi* L. Helv. Chim. Acta **72**:165–171.

Feringa, B.L., A. Smaardijk & H. Wynberg. 1985. Simple ^{31}P NMR method for the determination of enantiomeric purity of alcohols not requiring chiral auxiliary compounds. J. Am. Chem. Soc. **107**:4798–4799.

Fletcher, M.T., J.A. Wells, M.F. Jacobs, S. Krohn, W. Kitching, R.A.J. Drew, C.J. Moore & W. Francke. 1992. Chemistry of fruit-flies, spiroacetal-rich secretions in several *Bactrocera* species from the south-west Pacific region. J. Chem. Soc., Perkin Trans. pp. 2827–2832.

Francke, W., M.-L. Pan, W.A. König, K. Mori, P. Puapoomcharoen, H. Heuer & J.P. Vité. 1987. Identification of "pityol" and "grandisol" as pheromone components of the bark beetle *Pityophthorus pityographus*. Naturwissenschaften **74**:343–345.

Francke, W., J. Bartels, H. Meyer, F. Schröder, U. Kohnle, E. Baader & J.P. Vité. 1995. Semiochemicals from bark beetles: new results, remarks, and reflections. J. Chem. Ecol. **21**:1043–1063.

Francotte, E. & D. Lohmann. 1987. Analytic and preparative resolution of racemic γ- and δ-lactones by chromatography on cellulose triacetate. Relationship between elution order and absolute configuration. Helv. Chim. Acta **70**:1569–1582.

Hallett, R.H., A.L. Perez, G. Gries, R. Gries, H.D. Pierce, Jr., J. Yue, A.C. Oehlschlager, L.M. Gonzalez & J.H. Borden. 1995. Aggregation pheromone of coconut rhinoceros beetle, *Oryctes rhinoceros* (L.) (Coleoptera: Scarabaeidae). J. Chem. Ecol. **21**:1549–1570.

Haniotakis, G., W. Francke, K. Mori, H. Redlich & V. Schurig. 1986. Sex-specific activity of (*R*)-(−)- and (*S*)-(+)-1,7-dioxaspiro[5,5]undecane, the major pheromone of *Dacus oleae*. J. Chem. Ecol. **12**:1559–1568.

Hardt, I.H. & W.A. König. 1994. Preparative enantiomer separation with modified cyclodextrins as chiral stationary phases. J. Chromatog. A **666**:611–615.

Hardt, I.H., A. Rieck, C. Fricke & W.A. König. 1995. Enantiomeric composition of sesquiterpene hydrocarbons of the essential oil of *Cedrela odorata* L. Flavor Fragrance J. **10**:165–171.

Hashimoto, S. & E.J. Corey. 1981. A practical process for large-scale synthesis of (*S*)-

5-hydroxy-6-trans-8,11,14-cis-eicosatetraenoic acid (5-HETE). Tetrahedron Let. **22**:299–302.

Henkel, B., A. Kunath & H. Schick. 1995. Enzymes in organic synthesis 23: chemo-enzymatic synthesis of (5*R*,6*S*)-6-acetoxyhexadecan-5-olide-the major component of the mosquito oviposition attractant pheromone. Liebigs Ann. pp. 921–923.

Hibbard, B.E. & F.X. Webster. 1993. Enantiomeric composition of grandisol and grandisal produced by *Pissodes strobi* and *P. nemorensis* and their electroantennogram response to pure enantiomers. J. Chem. Ecol. **19**:2129–2141.

Ichikawa, A., T. Yasuda & S. Wakamura. 1995. Absolute configuration of sex pheromone for tea tussock moth, *Euproctis pseudoconspersa* (Strand) via synthesis of (*R*)- and (*S*)-10,14-dimethyl-1-pentadecyl isobutyrates. J. Chem. Ecol. **21**:627–634.

Isaksson, R., T. Liljefors & P. Reinholdsson. 1984. Preparative separation of the enantiomers of *trans,trans*- and *cis,trans*-2,8-dimethyl-1,7-dioxaspiro[5.5]undecane, main pheromone components of *Andrena* bees, by liquid chromatography on triacetylcellulose. J. Chem. Soc., Chem. Commun. pp. 137–138.

Jactel, H., P. Menassieu, M. Lettere, K. Mori & J. Einhorn. 1994. Field response of maritime pine scale, *Matsucoccus feytaudi* Duc. (Homoptera, Margarodidae), to synthetic sex pheromone stereoisomers. J. Chem. Ecol. **20**:2159–2170.

James, D.G. & K. Mori. 1995. Spined citrus bugs, *Biprorulus bibax* Breddin (Hemiptera, Pentatomidae), do not discriminate between enantiomers in their aggregation pheromone. J. Chem. Ecol. **21**:403–406.

James, D.G., C.J. Moore & J.R. Aldrich. 1994. Identification, synthesis, and bioactivity of a male-produced aggregation pheromone in assassin bug, *Pristhesancus plagipennis* (Hemiptera: Reduviidae). J. Chem. Ecol. **20**:3281–3295.

Johnson, C.R., R.C. Elliott & T.D. Penning. 1984. Determination of enantiomeric purities of alcohols and amines by a ^{31}P NMR technique. J. Am. Chem. Soc. **106**:5019–5020.

Kitching, W., J.A. Lewis, M.V. Perkins, R. Drew, C.J. Moore, V. Schurig, W.A. König & W. Francke. 1989. Chemistry of fruit flies. Composition of the rectal gland secretion of (male) *Dacus cucumins* (cucumber fly) and *Dacus halfordiae*. Characterization of (*Z,Z*)-2,8-dimethyl-1,7-dioxaspiro[5.5]undecane. J. Org. Chem. **54**:3893–3902.

Klimetzek, D., J. Bartels & W. Francke. 1989. Das Pheromon-System des Bunten Ulmen-bastkäfers *Pteleobius vittatus* (F.) (Coleoptera: Scolytidae). J. Appl. Entomol. **107**:518–523.

Knierzinger, A., W. Walther, B. Weber, R.K. Muller & T. Netscher. 1990. A new method for the stereochemical analysis of acyclic terpenoid carbonyl compounds. Helv. Chim Acta **73**:1087–1107.

König, W.A. 1987. The Practice of Enantiomer Separation by Capillary Gas Chromatography. Hüthig, Heidelberg.

König, W.A. 1990. Modifizierte Cyclodextrine als chirale Trennphasen in der Gaschroma-tographie. Kontakte **1990**(2):3–14.

König, W.A. 1992. Gas Chromatographic Enantiomer Separation with Modified Cyclodex-trins. Hüthig, Heidelberg.

König, W.A., S. Lutz, G. Wenz, G. Görgen, C. Neumann, A. Gäbler & W. Boland. 1989. Gas chromatographic determination of the enantiomeric composition of epoxy alcohols. Angew. Chem. Int. Ed. Engl. **28**:178–179.

König, W.A., K.-S. Nippe & P. Mischnick. 1990. Enantiomeric purity of Mosher's acid. Tetrahedron Lett. **31**:6867–6868.

König, W.A., B. Gehrcke, M.G. Peter & G.D. Prestwich. 1993. Direct gas chromatographic enantiomeric resolution of juvenile hormones I-III. Tetrahedron: Asymmetry **4**:165–168.

Krohn, S., M.T. Fletcher, W. Kitching, R.A.I. Drews, C.J. Moore & W. Francke. 1991. Chemistry of fruit flies: nature of glandular secretion and volatile emission of *Bactrocera cacuminatus* (Héring). J. Chem. Ecol. **17**:485–495.

Kruse, K., W. Francke & W.A. König. 1979. Gas chromatographic separation of chiral alcohols, amine alcohols, and amines. J. Chromatog. **170**:423–429.

Kubo, I., T. Matsumoto, D.L. Wagner & J.N. Shoolery. 1985. Isolation and structure of hepialone; principal component from male sex scales of *Hepialus californicus*. Tetrahedron Lett. **26**:563–566.

Kusumi, T., H. Takahashi, T. Hashimoto, Y. Kan & Y. Asakawa. 1994. Determination of the absolute configuration of ginnol, a long-chain aliphatic alcohol, by use of a new chiral anisotropic reagent. Chem. Lett. pp. 1093–1094.

Kuwahara, S. & K. Mori. 1990. Pheromone synthesis 124: synthesis of both the enantiomers of Hauptmann's periplanone-A and clarification of the structure of Persoons's periplanone-A. Tetrahedron **46**:8083–8092.

Kuwahara, Y., K. Matsumoto, Y. Wada & T. Suzuki. 1991. Chemical ecology of astigmatid mites 29: aggregation pheromone and kairomone activity of synthetic lardolure (1*R*,3*R*,5*R*,7*R*)-1,3,5,7-tetramethyldecyl formate and its optical isomers to *Lardoglyphus konoi* and *Carpoglyphus lactis*. Appl. Entomol. Zool. **26**:85–89.

Langer, W. & D. Seebach. 1979. Enantioselective 1,4 additions of organometallic compounds to conjugated systems in the chiral medium DDB. Helv. Chim. Acta **62**:1710–1722.

Lanier, G.N., A. Classon, T. Stewart, J.J. Piston & R.M. Silverstein. 1980. *Ips pini:* the basis for interpopulational differences in pheromone biology. J. Chem. Ecol. **6**:677–687.

Laurence, B.R., K. Mori, T. Otsuka, J.A. Pickett & L.J. Wadhams. 1985. Absolute configuration of mosquito oviposition attractant pheromone, 6-acetoxy-5-hexadecanolide. J. Chem. Ecol. **11**:643–648.

Leal, W.S., S. Matsuyama, Y. Kuwahura, S. Wakamura & M. Hasegawa. 1992. An amino acid derivative as the sex pheromone of a scarab beetle. Naturwissenschaften **79**:184–185.

Leal, W.S., M. Sawada & M. Hasegawa. 1993a. The scarab beetle *Anomala cuprea* utilizes the sex pheromone of *Popillia japonica* as a minor component. J. Chem. Ecol. **19**:1303–1313.

Leal, W.S., M. Sawada, S. Matsuyama, Y. Kuwahara & M. Hasegawa. 1993b. Unusual periodicity of sex pheromone production in the large black chafer *Holotrichia parallela*. J. Chem. Ecol. **19**:1381–1391.

Leal, W.S., M. Hasegawa, M. Sawada, M. Ono & Y. Ueda. 1994a. Identification and field evaluation of *Anomala octiescostata* (Coleoptera: Scarabaeidae) sex pheromone. J. Chem. Ecol. **20**:1643–1655.

Leal, W.S., F. Kawamura & M. Ono. 1994b. The scarab beetle *Anomala albopilosa sakishimana* utilizes the same sex pheromone blend as a closely related and geographically isolated species *Anomala cuprea*. J. Chem. Ecol. **20**:1667–1676.

Leal, W.S., X.W. Shi, D.S. Liang, C. Schal & J. Meinwald. 1995. Application of chiral gas chromatography with electroantennographic detection to the determination of the stereochemistry of a cockroach sex pheromone. Proc. Natl. Acad. Sci. USA **92**:1033–1037.

Levinson, H.Z. & K. Mori. 1983. Chirality determines pheromone activity. Naturwissenschaften **70**:190–192.

Levinson, H.Z., A. Levinson, Z. Ren & K. Mori. 1990. Comparative olfactory perception of the aggregation pheromones of *Sitophilus oryzae* (L.), *S. zeamais* (Motsch.) and *S. granarius* (L.), as well as stereoisomers of these pheromones. J. Appl. Entomol. **110**:203–213.

Li, J., G. Gries, R. Gries, J. Bikic & K.N. Slessor. 1993a. Chirality of synergistic sex pheromone components of the western hemlock looper, *Lambdina fiscellaria lugubrosa*. J. Chem. Ecol. **19**:2547–2561.

Li, J., R. Gries, G. Gries, K.N. Slessor, G.G.S. King, W.W. Bowers & R.J. West. 1993b. Chirality of 5,11-dimethylheptadecane, the major sex pheromone component of the hemlock looper, *Lambdina fiscellaria* (Lepidoptera: Geometridae). J. Chem. Ecol. **19**:1057–1062.

Lindgren, B.S., G. Gries, H.D. Pierce, Jr. & K. Mori. 1992. *Dendroctonus pseudotsugae* Hopkins (Coleoptera, Scolytidae): production of and response to enantiomers of 1-methylcyclohex-2-en-1-ol. J. Chem. Ecol. **18**:1201–1208.

Millar, J.G., S.P. Foster & M.O. Harris. 1991. Synthesis of the stereoisomers of the female sex pheromone of the Hessian fly, *Mayetiola destructor*. J. Chem. Ecol. **17**:2437–2447.

Miller, D.R., J.H. Borden & K.N. Slessor. 1989. Inter- and intrapopulation variation of the pheromone, ipsdienol, produced by male pine engravers *Ips pini* (Say). J. Chem. Ecol. **15**:233–247.

Moore, C.J., A. Hübener, Y.Q. Tu, W. Kitching, J.R. Aldrich, G.K. Waite, S. Schulz & W. Francke. 1994. A new spiroketal type from the insect kingdom. J. Org. Chem. **59**:6136–6138.

Mori, K. 1973. Absolute configuration of (−)-14-methyl-*cis*-8-hexadecen-1-ol and methyl (−)-14-methyl-*cis*-8-hexadecenoate, the sex attractant of female dermestid beetle, *Trogoderma inclusum* Le Conte. Tetrahedron Lett. pp. 3869–3872.

Mori, K. 1984. The significance of chirality: methods for determining absolute configuration and optical purity of pheromone and related compounds. *In:* Techniques in Pheromone Research, eds. H.E. Hummel and T.A. Miller, pp. 323–370, Springer-Verlag, New York.

Mori, K. 1992. The synthesis of insect pheromones, 1979–1989. *In:* The Total Synthesis of Natural Products, Vol. 9, ed. J. ApSimon, pp. 1–532, John Wiley & Sons, New York.

Mori, K. & N.P. Argade. 1994. Pheromone synthesis 163: synthesis of (9Z,25S,26R,43Z)-25,26-epoxy-9,43-henpentacontadiene and its antipode, components of the nymph recognition pheromone produced by nymphs of the cockroach *Nauphoeta cinerea.* Liebigs Ann. Chem. pp. 695–700.

Mori, K. & T. Ebata. 1986. Synthesis of optically active pheromone with an epoxy ring, (+)-disparlure and both the enantiomers of (3Z,6Z)-*cis*-9,10-epoxy-3,6-heneicosadiene. Tetrahedron **42**:3471–3478.

Mori, K. & S. Harashima. 1993a. Synthesis of (2E,4E,6R,10S)-4,6,10-trimethyl-2,4-dodecadien-7-one-the major component of the sex pheromone of the maritime pine scale (*Matsucoccus feytaudi*)-and its stereoisomers. Liebigs Ann. Chem. pp. 391–401.

Mori, K. & S. Harashima. 1993b. Pheromone Synthesis 155: Synthesis of (2E,3E,6R,10R)-4,6,10,12-tetramethyl-2,4-tridecadien-7-one (matsuone)—the primary component of the sex pheromone of three *Matsucoccus* pine bast scales—and its antipode. Liebigs Ann. Chem. pp. 993–1001.

Mori, K. & M. Ikunaka. 1987. Optical rotatory dispersion of spiroacetals. Liebigs Ann. Chem. pp. 333–336.

Mori, K. & S. Kuwahara. 1986. Synthesis of both the enantiomers of lardolure, the aggregation pheromone of the acarid mite, *Lardoglyphus konoi.* Tetrahedron **42**:5539–5544.

Mori, K. & Y. Nakazono. 1988. Synthesis of lactone components of the pheromone of *Anastrepha suspensa,* suspensolide, and the enantiomers of anastrephin and epianastrephin. Liebigs Ann. Chem. pp. 167–174.

Mori, K. & S. Senda. 1985. Synthesis of the enantiomers of *cis*-2-methyl-5-hexanolide, the major component of the sex pheromone of the carpenter bee. Tetrahedron **41**:541–546.

Mori, K. & H. Takikawa. 1990. A new synthesis of the four stereoisomers of 3,11-dimethyl-2-nonacosanone, the female-produced sex pheromone of the German cockroach. Tetrahedron **46**:4473–4486.

Mori, K., T. Uematsu, M. Minobe & K. Yanagi. 1983. Synthesis and absolute configuration of both the enantiomers of lineatin. Tetrahedron **39**:1735–1743.

Mori, K., T. Uematsu, K. Yanagi & M. Minobe. 1985. Synthesis of the optically active forms of 4,10-dihydroxy-1,7-dioxaspiro[5.5]undecane and their conversion to the enantiomers of 1,7-dioxaspiro[5.5]undecane, the olive fly pheromone. Tetrahedron **41**:2751–2758.

Mori, K., H. Harada, P. Zagatti, A. Cork & D.R. Hall. 1991. Pheromone Synthesis 126: Synthesis and biological activity of four stereoisomers of 6,10,14-trimethyl-2-pentadecanol, the female-produced sex pheromone of rice moth (*Corcyra cephalonica*). Liebigs Ann. Chem. pp. 259–267.

Mori, K., M. Amaike & J.E. Oliver. 1992. Pheromone synthesis 163: synthesis and absolute configuration of the hemiacetal pheromone of the spined citrus bug *Biprorulus bibax.* Liebigs Ann. Chem. pp. 1185–1190.

Mori, K., K. Fukamatsu & M. Kodo. 1993a. Pheromone synthesis 152: synthesis of blattellastanosides A and B, chlorinated steroid glucosides isolated as the aggregation pheromone of the German cockroach, *Blattella germanica* L. Liebigs Ann. Chem. pp. 665–670.

Mori, K., H. Kiyota & D. Rochat. 1993b. Pheromone synthesis 154: synthesis of the stereoisomers of 3-methyl-4-octanol to determine the absolute configuration of the naturally occurring (3*S*,4*S*)-isomer isolated as the male-produced aggregation pheromone of the African palm weevil, *Rhynchophorus phoenicis*. Liebigs Ann. Chem. pp. 865–870.

Mori, K., H. Kiyota, C. Malosse & D. Rochat. 1993c. Synthesis of the enantiomers of *syn* 4-methyl-5-nonanol to determine the absolute configuration of the naturally occurring (4*S*,5*S*)-isomer isolated as the male-produced pheromone compound of *Rhynchophorus vulneratus* and *Metamasius hemipterus*. Liebigs Ann. Chem. pp. 1201–1204.

Morrison, J.D. ed. 1983. Asymmetric Synthesis, Vol. 1, Analytical Methods. Academic Press, New York.

Nichols, D., C.F. Barfknecht & D.B. Rusterholz. 1973. Asymmetric synthesis of psychotomimetic phenylisopropylamines. J. Med. Chem. **16**:480–483.

Novotny, M.V., T.M. Xie, S. Harvey & D. Wiesler. 1995. Stereoselectivity in mammalian chemical communication. Experientia **51**:738–743.

Oehlschlager, A.C., G.G.S. King, H.D. Pierce, Jr., A.M. Pierce, K.N. Slessor, J.G. Millar & J.H. Borden. 1987. Chirality of macrolide pheromones of grain beetles in the genera *Oryzaephilus* and *Cryptolestes* and its implication for species specificity. J. Chem. Ecol. **13**:1543–1554.

Oehlschlager, A.C., H.D. Pierce, Jr., B. Morgan, P.D.C. Wimalaratne, K.N. Slessor, G.G.S. King, G. Gries, R. Gries, J.H. Borden, L.F. Jiron, C.M. Chinchilla & R.G. Mexzan. 1992. Chirality and field activity of rhynchophorol, the aggregation pheromone of the American palm weevil. Naturwissenschaften **79**:134–135.

Oehlschlager, A.C., P.N.B. Prior, A.L. Perez, R. Gries, G. Griers, H.D. Pierce, Jr. & S. Laup. 1995. Structure, chirality and field testing of a male-produced aggregation pheromone of Asian palm weevil *Rhynchophorus bilineatus* (Montr). J. Chem. Ecol. **21**:1619–1629.

Oertle, K., H. Beyeler, R.O. Duthaler, W. Lottenbach, M. Riediker & E. Steiner. 1990. A facile synthesis of optically pure (−)-(*S*)-ipsenol using a chiral titanium complex. Helv. Chim. Acta **73**:353–358.

Ohtani, I., T. Kusumi, Y. Kashman & H. Kakisawa. 1991. High-field FT NMR application of Mosher's method. The absolute configuration of marine terpenoids. J. Am. Chem. Soc. **113**:4092–4096.

Ôi, N., M. Horiba & H. Kitahara. 1981. Gas chromatographic separation of chrysanthemic acids on an optically active stationary phase. Agric. Biol. Chem. **45**:1509–1510.

Olaifa, J.I., F. Matsumura, T. Kikukawa & H.C. Coppel. 1988. Pheromone-dependent species recognition mechanisms between *Neodiprion pinetum* and *Diprion similis* on white pine. J. Chem. Ecol. **14**:1131–1144.

Oliver, J.E. & R.M. Waters. 1995. Determining enantiomeric composition of disparlure. J. Chem. Ecol. **21**:199–211.

Park, S.C., A.J. Wi & H.S. Kim. 1994. Response of *Matsucoccus thumbergianae* males to synthetic sex pheromone and its utilization for monitoring the spread of infestation. J. Chem. Ecol. **20**:2185–2196.

Perez, A.L., G. Gries, R. Gries, R.M. Giblin-Davis & A.C. Oehlschlager. 1994. Pheromone chirality of African palm weevil, *Rhynchophorus phoenicis* (F.) and palmetto weevil, *Rhynchophorus cruentatus* (F.). J. Chem. Ecol. **20**:2653–2671.

Perkins, M.V. & W. Kitching. 1990. An (*S*)-(+)-lactic acid route to (2*S*,6*R*,8*S*)-2,8-di-methyl-1,7-dioxaspiro[5.5]undecane and (2*S*,6*R*,8*S*)-2-ethyl-8-methyl-1,7-dioxaspiro [5.5]undecane and demonstration of their presence in the rectal glandular secretion of *Bactrocera nigrotibialis*. J. Chem. Soc., Perkin Trans. 1 pp. 2501–2506.

Pierce, A.M., H.D. Pierce, Jr. A.C. Oehlschlager & J.H. Borden. 1991. 1-Octen-3-ol, attractive semiochemical for foreign grain beetle, *Ahasversus advena* (Waltl). J. Chem. Ecol. **17**:567–580.

Pierce, Jr., H.D., A.M. Pierce, B.D. Johnston, A.C. Oehlschlager & J.H. Borden. 1988. Aggregation pheromone of square-necked grain beetle, *Cathartus quadricollis* (Guér). J. Chem. Ecol. **14**:2169–2184.

Pierce, Jr., H.D., P. de Groot, J.H. Borden, S. Ramaswamy & A.C. Oehlschlager. 1995. Pheromones in red pine cone beetle, *Conophthorus resinosae* Hopkins, and its synonym, *C. banksianae* McPherson. J. Chem. Ecol. **21**:169–185.

Pirkle, W.H. & M.S. Hockston. 1974. Synthesis of a broad-spectrum resolving agent and resolution of 1-(1-napthyl)-2,2,2-trifluoroethanol. J. Org. Chem. **39**:3904–3905.

Pirkle, W.H., D.L. Sikkenga & M.S. Pavlin. 1977. Nuclear magnetic resonance determina-tion of enantiomeric composition and absolute configuration of γ-lactone using chiral 2,2,2-trifluoro-1-(9-anthryl)ethanol. J. Org. Chem. **42**:384–387.

Poole, C.F. & S.A. Schuette. 1984. Contemporary Practice of Chromatography. Elsevier Science Publishers, Amsterdam.

Rhainds, M., G. Gries, J. Li, R. Gries, K.N. Slessor, C.M. Chinchilla & A.C. Oehlschlager. 1994. Chiral esters: sex pheromone of the bagworm, *Oiketicus kirbyi*. J. Chem. Ecol. **20**:3083–3096.

Rochat, D., F. Akamou, A. Sangaré, D. Mariau & K. Mori. 1995. Field trapping of *Rhynchophorus phoenicis* with stereoisomers of the synthetic aggregation pheromone. C.R. Acad. Sci. Paris, Sciences de la Vie. **318**:183–190.

Rudmann, A.A. & J.R. Aldrich. 1987. Chirality determinations for a tertiary alcohol. Ratios of linalool enantiomers in insects and plants. J. Chromatog. **407**:324–329.

Sato, R., N. Abe, H. Sugie, M. Kato, K. Mori & Y. Tamaki. 1986. Biological activity of the chiral sex pheromone of the peach leafminer moth, *Lyonetia clerkella* Linné. Appl. Entomol. Zool. **21**:478–480.

Schaner, A.M., L.D. Tanico-Hogan & L.L. Jackson. 1989. (*S*)-2-Pentadecyl acetate and 2-pentadecanone, components of aggregation pheromone of *Drosophila busckii*. J. Chem. Ecol. **15**:2577–2588.

Schreier, P., A. Bernreuter & M. Huffer. 1995. Analysis of Chiral Organic Molecules: Methodology and Applications. Walter de Gruyter, Berlin.

Schulz, S., W. Francke, W.A. König, V. Schurig, K. Mori, R. Kittmann & D. Schneider. 1990. Male pheromone of swift moth, *Hepialus hecta* L. J. Chem. Ecol. **16**:3511–3521.

Schurig, V. 1985–1986. Current methods for the determination of enantiomeric compositions (Part 1–3). Kontakte **1985**(1):54–60, (2):22–36; **1986**(1):3–22.

Schurig, V., R. Weber, D. Klimetzek, U. Kohnle & K. Mori. 1982. Enantiomeric composition of lineatin in three sympatric ambrosia beetles. Naturwissenschaften **69**:602.

Schurig, V., R. Weber, G.J. Nicholson, A.C. Oehlschlager, H.D. Pierce, Jr., A.M. Pierce, J.H. Borden & L.C. Ryker. 1983. Enantiomer composition of natural *exo-* and *endo-* brevicomin by complexation gas chromatography/selected ion mass spectrometry. Naturwissenschaften **70**:92–93.

Schurig, V., D. Schmalzing, U. Mühleck, M. Jung, M. Schleimer, P. Mussche, C. Duvekot & J.C. Bruyten. 1990. Gas chromatographic enantiomer separation on polysiloxane-anchored permethyl-β-cyclodextrin (Chirasil-Dex). J. High Res. Chromatog. **13**:713–717.

Senda, S. & K. Mori. 1983. Asymmetric synthesis of (*R,Z*)-(−)-(1-decenyl)oxacyclopentan-2-one, the pheromone of the Japanese beetle. Agric. Biol. Chem. **47**:2595–2598.

Seybold, S.J. 1992. The role of chirality in the olfactory-directed aggregation behavior of pine engraver beetles in the genus *Ips*. Ph.D. Thesis. University of California at Berkeley.

Seybold, S.J., T. Ohtsuka, D.L. Wood & I. Kubo. 1995a. Enantiomeric composition of ipsdienol: a chemotaxonomic character for North American populations of *Ips* spp. in the *pini* subgeneric group. J. Chem. Ecol. **21**:995–1016.

Seybold, S.J., D.R. Quilici, J.A. Tillman, D. Venderwel, D.L. Wood & G.J. Blomquist. 1995b. *De novo* biosynthesis of the aggregation pheromone components ipsenol and ipsdienol by the pine bark beetles *Ips paraconfusus* Lanier and *Ips pini* (Say). Proc. Natl. Acad. Sci. USA **92**:8393–8397.

Shi, X., W.S. Leal & J. Meinwald. 1996. Assignment of absolute stereochemistry to an insect pheromone by chiral amplification. Bioorg. Med. Chem. **4**:297–303.

Silverstein, R.M., R.F. Cassidy, W.E. Burkholder, T.J. Shapas, H.Z. Levinson, A.R. Levinson & K. Mori. 1980. Perception by *Trogoderma* species of chirality and methyl branching at a site removed from a functional group in a pheromone component. J. Chem. Ecol. **6**:911–917.

Slessor, K.N., G.G.S. King, D.R. Miller, M.L. Winston & T.L. Cutforth. 1985. Determination of chirality of alcohols or latent alcohol semiochemicals in individual insects. J. Chem. Ecol. **11**:1659–1667.

Slessor, K.N., L.-A. Kaminski, G.G.S. King & M.L. Winston. 1990. Semiochemicals of the honeybee queen mandibular glands. J. Chem. Ecol. **16**:851–860.

Souter, R.W. 1985. Chromatographic Separations of Stereoisomers. CRC Press, Boca Raton, FL.

Staddon, B.W. & I.J. Everton. 1980. Haemolymph of the milkweed bug *Oncopeltus fasciatus:* inorganic constituents and amino acids. Comp. Biochem. Physiol. **65A**:371–374.

Steghaus-Kovâc, S., U. Maschwitz, A.B. Attygale, R.T.S. Frighetto, N. Frighetto, O. Vostrowsky & H.J. Bestmann. 1992. Trail-following responses of ant *Leptogenys diminuta* to stereoisomers of 4-methyl-3-heptanol. Experientia **48**:691–694.

Stevenson, D. & I.D. Wilson. 1988. Chiral Separations. Plenum, New York.

Swedenborg, P.D., R.L. Jones, H.-Q. Zhou, J. Shin & H.-W. Liu. 1994. Biological activity of (3*R*,5*S*,6*R*)- and (3*S*,5*R*,6*S*)-3,5-dimethyl-6-(methylethyl)-3,4,5,6-tetrahydropyran-2-one, a pheromone of wasp *Macrocentrus grandii* (Goidanich), J. Chem. Ecol. **20**:3373–3380.

Szöcs, G., M. Tóth, W. Francke, F. Schmidt, P. Philipp, W.A. König, K. Mori, B.S. Hansson & C. Löfstedt. 1993. Species discrimination in five species of winter-flying geometrids (Lepidoptera) based on chirality of semiochemicals and flight season. J. Chem. Ecol. **19**:2721–2735.

Tengö, J., L. Ågren, B. Baur, R. Isaksson, T. Lilijefors, K. Mori, W. König & W. Francke. 1990. *Andrena wilkella* male bees discriminate between enantiomers of cephalic secretion components. J. Chem. Ecol. **16**:429–441.

Tóth, M., H.R. Buser, A. Peña, H. Arn, K. Mori, T. Takeuchi, L.N. Nikolaeva & B.G. Kovalev. 1989a. Identification of (3Z,6Z)-1,3,6–9,10-epoxyheneicosatriene and (3Z,6Z)-1,3,6–9,10-epoxyeicosatriene in the sex pheromone of *Hyphantria cunea*.Tetrahedron Lett. **30**:3405–3408.

Tóth, M., G. Helmchen, U. Leikauf, Gy. Sziráki & G. Szöcs. 1989b. Behavioral activity of optical isomers of 5,9-dimethylheptadecane, the sex pheromone of *Leucoptera scitella* L. J. Chem. Ecol. **15**:1535–1543.

Tóth, M., G. Szöcs, E.J. van Nieukerken, P. Philipp, F. Schmidt & W. Francke. 1995. Novel type of sex pheromone structure identified form *Stigmella malella* (Stainton), J. Chem. Ecol. **21**:13–27.

Trost, B.M., J.L. Belletire, S. Godleski, P.G. McDougal, J.M. Balkovec, J.J. Baldwin, M.E. Christy & G.S. Ponticello. 1986. On the use of *O*-methylmandelate ester for establishment of absolute configuration of secondary alcohols. J. Org. Chem. **51**:2370–2374.

Tumlinson, J.H., M.G. Klein, R.E. Doolittle, T.L. Ladd & A.T. Proveaux. 1977. Identification of the female Japanese beetle sex pheromone: inhibition of male response by an enantiomer. Science **197**:789–792.

Vigneron, J.P., R. Méric, M. Larchevêque, A. Debal, G. Kunesch, P. Zagatti & M. Gallois. 1982. Absolute configuration of eldanolide, the wing gland pheromone of the male African sugar cane borer, *Eldana saccharina* (Wlk.), Synthesis of its (+) and (–) enantiomers. Tetrahedron Lett. **23**:5051–5054.

Wainer, I.W. 1988. A Practical Guide to the Selection and Use of HPLC Chiral Stationary Phases. J.T. Baker, Phillipsburg, NJ.

Walgenbach, C.A., J.K. Phillips, W.E. Burkholder, G.G.S. King, K.N. Slessor & K. Mori. 1987. Determination of chirality in 5-hydroxy-4-methyl-3-heptanone, the aggregation pheromone of *Sitophilus oryzae* (L.) and *S. zeamais* Matschulsky. J. Chem. Ecol. **13**:2159–2169.

Wassgren, A.-B. & G. Bergström. 1995. Quantitative high-resolution gas chromatographic determination of stereoisomeric composition of chiral volatile compounds in the picogram range by EC-detection. J. Chem. Ecol. **21**:987–994.

Weber, R. & V. Schurig. 1984. Complexation gas chromatography, a valuable tool for the stereochemical analysis of pheromones. Naturwissenschaften **71**:408–413.

Yashima, E. & Y. Okamoto. 1995. Chiral discrimination on polysaccharide derivatives. Bull. Chem. Soc. Jpn. **68**:3289–3307.

Zhang, A., P.S. Robbins, W.S. Leal, C.E. Linn, Jr., M.G. Villani, and W.L. Roelofs. 1997. Essential amino acid methyl esters: major sex pheromone components of the cranberry white grub, *Phyllophaga anxia* (Coleoptera: Scarabaeidae). J. Chem. Ecol. **23**:231–245.

9

Electrophysiological Methods

Louis B. Bjostad

9.1. Introduction

Electrophysiological techniques have proven useful in the isolation, identification, and elucidation of the behavioral and physiological roles of semiochemicals, particularly for insects, whose chemoreceptors are readily accessible. These methods include the electroantennogram (EAG) technique for recording from whole insect antennae (Schneider 1957a, 1957b; Roelofs 1977; Roelofs 1984), the cut-sensillum technique for single-cell recording (SCR) from olfactory chemoreceptors (Kaissling 1974; Van der Pers & Den Otter (1978), the tungsten microelectrode technique for SCR from olfactory chemoreceptors (Boeckh 1962), and the tip-recording technique for contact chemoreceptive sensilla (Hodgson et al. 1955; Hodgson and Roeder 1956; Dethier 1976; Städler 1984). Details of these techniques are discussed below. Reviews are available on insect chemosensory electrophysiology (Boeckh 1984; Frazier and Hanson 1986; De Kramer and Hemberger 1987) and insect neurophysiological techniques in general (Miller 1979).

The most important cautionary note with respect to electrophysiologically active compounds is that they are not always behaviorally active (or not in the context that an investigator may expect). For example, SCR with the black bean aphid, *Aphis fabae,* seemed to implicate (*E*2)-hexenal as a host-derived component of the alarm pheromone, but this compound proved to have no activity in behavioral bioassays (Wadhams 1990). As a second example, of the compounds that are EAG-active for tsetse flies, *Glossina morsitans,* some are attractants (1-octen-3-ol, 4-methylphenol, and 3-propylphenol), some are repellents (acetophenone, 2-methoxyphenol), and some have no apparent behavioral activity (2,6,10,10-tetramethyl-1-oxaspiro[4.5]dec-2-en-8-one) (Gough et al. 1987; Bursell et al. 1988).

9.2. Electroantennogram Recording

The EAG technique was originally developed by Schneider (Schneider 1957a, 1957b) to measure electrophysiological responses from antennae of male silk moths (*Bombyx mori*) to volatile compounds from the female sex pheromone gland. An excised male antenna was suspended between two electrodes connected to a voltage amplifier, which was in turn connected to an oscilloscope. When the male antenna was given a puff of the sex pheromone, a sharp negative deflection in voltage (corresponding to 1–2 mV) was observed, followed by a slow return of the signal to baseline after 1 or 2 s. The EAG technique was subsequently shown to work with a number of other insect taxa, including Lepidoptera, Coleoptera, Hymenoptera, Homoptera, Orthoptera, and Trichoptera (Roelofs 1984; Jewett et al. 1996).

This technique was later adapted as an analytical tool in the isolation of sex pheromone components (Roelofs 1977, 1984). Behavioral bioassays are required to evaluate the entire complement of compounds in a sex pheromone blend, but EAG recording is usually helpful in detecting the most important components in the blend. Fractions from the gas chromatographic (GC) separation of a sex pheromone blend can be tested by EAG within an hour or two, whereas comparable tests with behavioral bioassays with flight tunnels (Baker & Linn 1984) or other techniques (Baker & Cardé 1984) may require days or weeks to complete. The EAG technique is particularly useful in rapidly evaluating a set of GC fractions prior to behavioral bioassays.

Most workers have found that the simplest EAG systems work best in practice (Roelofs 1984). Electrodes, an amplifier (with high input impedance, typically 10^{12} Ω), and a display are the essential elements (Fig. 9.1). Electrodes are easily made (less than $1). The amplifier is simple and can be constructed for about $50 (Bjostad & Roelofs 1980; Bjostad 1988). Amplifiers are also available commercially (Section 9.2.3). The display is usually the most expensive component. An oscilloscope, a chart recorder, or an integrator can be connected directly to the amplifier. Personal computers and graphing calculators require a hardware-software data acquisition package (ranging widely in price from $350 to $3500. Voltmeters require construction of an additional circuit ($20) for signal processing (Bjostad & Roelofs 1980).

9.2.1. Electrodes

The electrodes must provide excellent electrical conductivity with the antennal tissue. Either saline-filled glass electrodes with gold wires or silver wires coated

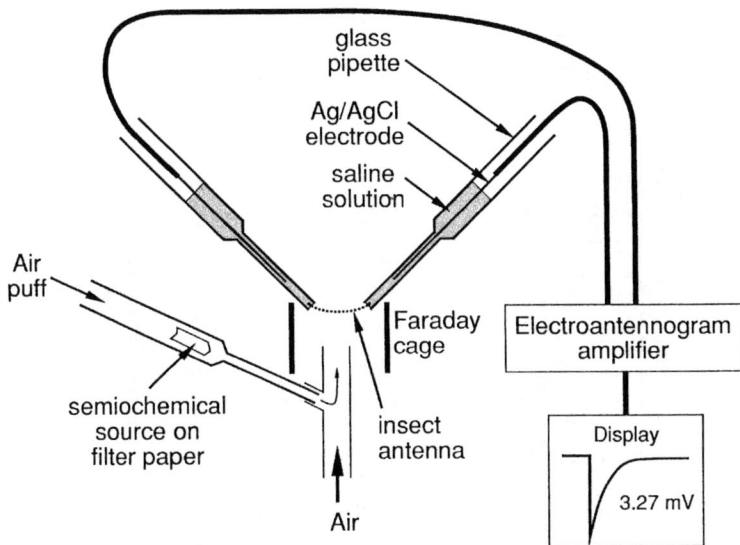

Figure 9.1. Electroantennogram system.

Table 9.1. Composition (in millimoles/liter) of Saline Solutions Used in Electroantennogram Recordings

	Hemolymph[a]	Receptor lymph[a]	Drosophila Ringer	Beadle-Ephrussi-Ringer	TES Ringer[b]	Leal[c]	Hidoh et al. (1992)	Wadhams (1990)
Glucose	354	22.5						
NaCl	12	25	46	129	140	241	120	129
KCl	6.4	172	182	4.7	5	2.7	10	8.6
CaCl$_2$	1	1	3	1.9	7	3.6	4	2
MgCl$_2$	12	3			1		8	8.5
KH$_2$PO$_4$	20	9.2						
K$_2$HPO$_4$		10.8						
Other	9.6 KOH	1.5 HCl	10 Tris-HCl		Footnote b	2 NaHCO$_3$	5 HEPES	10.2 NaHCO$_3$, 4.3 NaH$_2$PO$_4$
pH	6.5	6.5	7.2			7.8-8.2	7.0	

[a]Kaissling and Thorson (1980) in Roelofs (1984).

[b]Other components: 5-mM 2-{[tris(hydroxymethyl)methyl]amino}-1-ethanesulfonic acid (TES), 4-mM NaHCO$_3$, 5-mM trehalose, 100-mM sucrose (Strausfeld et al. 1983).

[c]W. S. Leal, personal communication (1996).

with silver chloride or microsharpened tungsten electrodes are used for most EAG work, and both are simple and inexpensive to make.

9.2.1.1. Glass Electrodes

Glass electrodes of many different configurations have been used. One convenient approach is to use disposable glass Pasteur pipettes and short lengths (4 cm) of gold wire (Aldrich Chemical). An older, less convenient approach involves the use of silver wire, which requires a coating of AgCl at one end to provide adequate electrical conductivity with the saline in the electrode. To apply the AgCl coating, the entire surface of each piece of silver wire is first cleaned by scraping with fine steel wool until the wire is shiny. Small fragments of steel wool that may adhere to the wire and ruin the AgCl coating must be removed. One end of each clean silver wire is dipped briefly (<1 s) into a few grams of molten AgCl (melting point 455°C) in a crucible to provide a stabilized coating about 1 cm long. The AgCl-coated end of the silver wire is a flat gray color, easily distinguished from the shiny, noncoated end of the wire. The AgCl must melt completely, because partially melted AgCl makes a thick coating that provides poor conductivity. The silver wire must be dipped and removed from the molten AgCl so that the very tip of the wire is the last point to leave the surface of the molten liquid to prevent a large drop of AgCl cooling at the end of the wire. The cooled AgCl is stored in the dark. Relatively pure AgCl cools to a light gray solid; if the cooled salt is black, it should be replaced.

Alternatively, the silver wire electrode can be coated by electrolysis. The electrode is scraped clean as described above, then connected to the negative terminal of two 1.5-V batteries in series (i.e., 3 V), with the other terminal connected to a spare piece of silver wire. The first ~1 cm of the electrode and the silver wire are dipped into 0.1-M HCl, and the current is run for a few seconds until gas bubbles (hydrogen) appear at the electrode. This process strips a thin layer of silver off the electrode to provide a clean surface. The polarity is then reversed by connecting the electrode to the positive terminal. In a few seconds, the electrode turns pale gray as a layer of AgCl is deposited. The coated electrode is rinsed with distilled water and is ready for use. This method is more convenient than using molten silver chloride, and recoating can be done easily and frequently.

For the reference electrode attached to the base of the antenna, the noncoated end of the silver wire slides into a Molex Soldercon® IC terminal (DigiKey part no. 1938-4) that is attached to a piece of insulated copper wire (10 cm, stripped at both ends). The Soldercon® terminal allows the silver wire to be removed easily for cleaning (about weekly) and recoating with AgCl, as described above. If a small-parts jig is used to hold the electrodes (section 9.2.3), the reference electrode is kept at ground potential by connecting the other end of the copper wire to a screw on the arm of the small-parts jig that holds the reference electrode.

This arrangement minimizes the wiring required by using the jig itself to provide the connection to ground.

The recording electrode is constructed similarly, but the Soldercon® IC terminal is attached (with a wire-wrapping kit, Radio Shack part no. 276-1570) to the center wire of a 50-cm length of coaxial cable (3 mm diameter) that carries the antennal signal to the amplifier. One end of a piece of insulated copper wire (10 cm, stripped at both ends) is attached to the corresponding screw on the other arm of the jig, and the other end of this wire pierces the outer insulation of the coaxial cable of the recording electrode, contacting the braided wire sheath that is connected to ground in the amplifier. The recording electrode must be shielded from electromagnetic radiation (e.g., from building wiring), and an effective Faraday cage can be made by wrapping aluminum foil around the Pasteur pipette of the recording electrode.

For many insects, disposable Pasteur pipettes pulled to a fine point with a glassworking torch can be used. For finer work, a microelectrode puller (Sutter Instrument Co., Novato, CA) can be used with glass capillary tubing (50×2 mm), and tubing with an inner filament (Sutter Instrument Co.) may facilitate filling with saline (Cork et al. 1990).

An alternative to using pipette electrodes is to construct a cell with channels and wells to hold the antenna, an approach that is used particularly for EAG recordings from GC effluents (section 9.2.9). A perspex holder was constructed for use with a portable EAG apparatus (Suckling et al. 1994), and a similar acrylic stage was used in EAG characterization of sex pheromones of the brownbanded cockroach, *Supella longipalpa* (Leal et al. 1995).

The composition of the saline solution in the electrodes is apparently not critical for some applications (Roelofs 1984), and KCl solutions (usually 0.1–0.3 M) have been used with excellent success by a number of laboratories for recording from different insect species (e.g., Vet et al. 1990; Gonzalez et al. 1994). However, special attention to saline composition may be crucial for species in which EAG responses decline rapidly. Careful matching of the ion concentration, osmolarity, and pH in the chemoreceptor cells (Kaissling & Thorson 1980) may be required, and is often well worth the effort, particularly for single-cell recording (section 9.3.1.1). Hemolymph compositions have been determined for a number of insect species (e.g., Usherwood 1969; Florkin & Jeuniaux 1974; Bindokas & Adams 1988; Pelletier & Clark 1992). A number of saline "recipes" for EAG are listed in Table 9.1.

Evaporation of water from electrode tips can be a problem, particularly in work with microelectrodes as it can leave a deposit of highly concentrated salt at the tip of an electrode. This problem can be mitigated by the addition of 0.2% agar to the saline solution (Dethier & Hanson 1965; Cork et al. 1990), or by the addition of a 5% aqueous solution of polyvinylpyrrolidone K90 (Fluka AG, Buchs, Switzerland) to the saline (1:10 parts vol/vol). The tip of the pipette where the antenna enters has also been coated with a film of Vaseline™ (Van der Pers & Den Otter 1978; Ramachandran et al. 1990; Gonzalez et al. 1994).

9.2.1.2. Tungsten Electrodes

Tungsten electrodes for EAG work have largely been replaced with saline-filled glass electrodes, although tungsten electrodes were used successfully in early work (Payne 1970). Tungsten electrodes must be sharpened before use, and details of electrode preparation were presented by Hubel (1957) and Miller (1979). Sharpened tungsten wires are also used to punch a hole into hard antennal cuticles into which a glass electrode is then inserted (White & Birch 1987; White & Chambers 1989; White et al. 1989; Chinta & Dickens 1994).

9.2.2. Micromanipulators

The electrodes are commonly held and positioned with micromanipulators, such as the small joystick type (model MN-151, Narishige Inc., Greenvale, NY). However, many insect antennae are sufficiently large that the fine control achievable with a micromanipulator is unnecessary for EAG. A far less expensive method is to attach each glass electrode to one arm of a small-parts jig (Helping Hands® part no. 64-2093, Radio Shack), a device that enables the distance between the electrode tips to be adjusted easily. The small-parts jig includes a horizontal bar with a jointed arm at each end, and the jointed arms are used to hold the two electrodes. The most convenient working arrangement is to disconnect the horizontal arm of the small-parts jig from the base unit, and attach the horizontal bar instead to a laboratory ring stand, which allows the entire antennal preparation to slide up and down for selection of a comfortable working height. Each of the two electrode attachments to the jointed arms of the jig is made from a piece of thin aluminum (5 × 5 cm, cut from a soft drink can or a baking pan and thoroughly scraped with steel wool to remove coatings). This is wrapped around the electrode pipette, with the ends of the strip forming a stiff tab that can be clipped into the alligator clip at the end of one of the jointed arms.

9.2.3. Amplifier

The key consideration for an EAG amplifier is high input impedance (1 TΩ, or 10^{12} Ω). The amplifier is generally of the direct current (DC) type, which amplifies all frequency components of the signal, rather than the alternating current (AC) type, which passes only the high-frequency components. Appropriate DC amplifiers with at least 1 TΩ input impedance (e.g., Grass P-16 amplifier) are commercially available (Table 9.2).

When an insect antenna is puffed with a volatile chemical stimulus, the voltage produced is in the range 0.1–10 mV, but the current generated is extremely weak.

Table 9.2.   Manufacturers of Electrophysiological Amplifiers

Syntech	George Johnson Electronics
P.O. Box 1547	506 Woodside Road
NL-1200 BM	Baltimore, MD 21229
Hilversum, The Netherlands	Phone: 410-362-6841
E-mail: syntech@knoware.nl	
Astro-Med/Grass Instruments	World Precision Instruments
600 E. Greenwich Avenue	175 Sarasota Center Boulevard
West Warwick, RI 02893	Sarasota, FL 31240-9258
Phone: 401-828-4000	Phone: 941-371-1003
Fax: 401-822-2430	Fax: 941-377-5428
E-mail: astro-med@astro-med.com	E-mail: wpi@wpiinc.com
Web: http://www.astro-med.com	Web: http://www.wpiinc.com

The most important function of the amplifier is actually to buffer the input signal from the insect antenna, in addition to its role in amplifying the signal. For this reason, many commercial amplifiers are unsuitable for EAG work because their small input impedance (typically about 1 $M\Omega$) drains the current from the insect antenna so badly that the signal from the antenna may be completely eliminated. The desired input impedance of 1 $T\Omega$ is easily achievable with an input stage based on the junction field-effect transistor (JFET) family of operational amplifiers (Berlin 1991). The time course of an EAG signal is typically slow (usually 0.1–1 s), and the amplifier usually does not require high-frequency response, which allows the use of a fairly abbreviated amplifier design. In fact, many of the subtleties of conventional neurophysiological recording techniques that are critical for work with single nerve cells (such as capacitance compensation, notch filtering, etc.) can be safely ignored for routine EAG work, due to the slow, robust nature of the EAG signal.

Although commercial amplifiers for EAG work are available, the circuitry is simple and inexpensive enough that many workers prefer to build their own. A suitable amplifier can be built for about $50, with ~8 h work, and key considerations for a homebuilt amplifier are described in detail below. The main advantage of a laboratory-built amplifier is that it is an inexpensive interface that can be used with many different types of common display devices.

A simple EAG amplifier (Fig. 9.2) and power supply (Fig. 9.3) can be constructed with inexpensive parts available from several retail suppliers (Table 9.3). Both these circuits can be housed in the same project case, but each must be built on a separate circuit board, and the EAG amplifier must be screened from the power supply with a Faraday cage constructed from a thin aluminum sheet, with the Faraday cage placed over the amplifier circuit board inside the main project case. As the figures indicate, one convenient approach is to make printed circuits in the laboratory that are made specifically for 44-contact, solder-eye

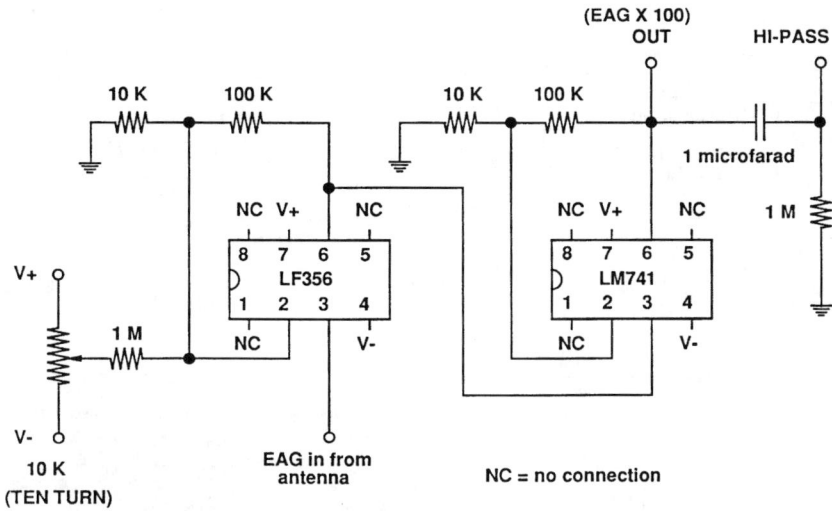

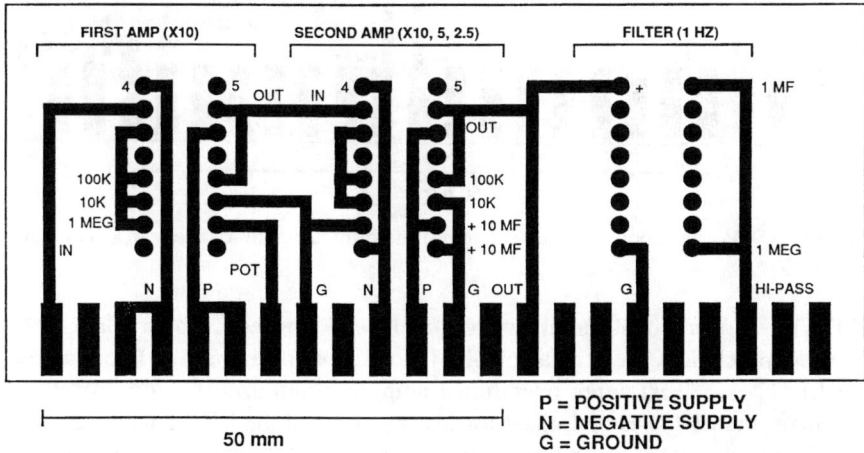

Figure 9.2. Electroantennogram amplifier schematic diagram and printed circuit.

edgeboard connectors (section 9.6). Two edgeboard connectors can be bolted to the floor of the project case, parallel with the front panel, with the transformer and the power supply board to the rear.

A control knob on the front of the project case, attached to a potentiometer (see Fig. 9.2) inside the case, is used to adjust the output from the amplifier to ~0V on the display. The EAG signal tends to wander as the electrolytic environment of the insect antenna changes, and this potentiometer is adjusted to position the EAG baseline near the top of the display.

A general-purpose choice for the first stage of the EAG amplifier is the LF356

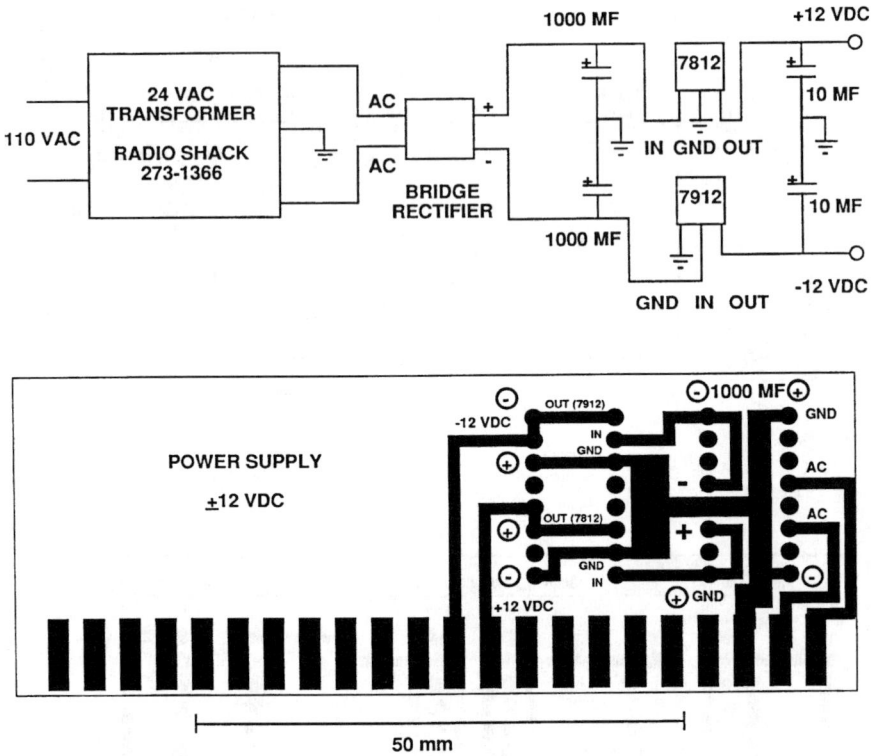

Figure 9.3. Power supply schematic diagram and printed circuit for use with an EAG amplifier.

JFET input operational amplifier (National Semiconductor, Santa Clara, CA). Other good choices are now available, tailored for special needs. For example, the LF411 is a JFET input operational amplifier optimized for low offset and low drift; the LF441 is optimized for low power; and the LF351 requires a low supply current, yet has low noise and offset voltage drift (although for applications where these requirements are critical, the LF356 is recommended). Data sheets for all these devices are available on-line.

A convenient total amplification to use for EAG work is 100 ×, but problems with oscillation sometimes occur if this is attempted with a single operational amplifier. Using a second stage of amplification tends to avoid this problem, and a convenient combination is 10 × amplification with a first stage (such as the LF356) and 10 × amplification with a second stage (such as the LM741). The LM741 is a good choice for a second amplification stage because it is less expensive than the LF356, and its lack of high-input impedance is not a drawback for use in the second stage.

Miniaturization of the EAG amplifier would be desirable for recording from

Table 9.3.   Electronics Parts and Data Systems Suppliers

Digi-Key
P.O. Box 677
Thief River Falls, MN 56701-0677
Phone: 218-681-6674
Fax: 218-681-3380
E-mail: webmaster@digikey.com
Web: http://www.digikey.com

Radio Shack
One Tandy Center
Fort Worth, TX 76102
Phone: 817-390-3200
Fax: 817-390-3292
E-mail: rs.customer.relations@tandy.com
Web: http://www.radioshack.com

Jameco
1355 Shoreway Road
Belmont, CA 94002
Phone: 415-592-8097
Fax: 415-592-2503
E-mail: info@jameco.com
Web: http://www.jameco.com

DC Electronics
P.O. Box 3203
Scottsdale, AZ 85271-3203
Phone: 602-945-7736
Fax: 602-994-1707

National Semiconductor
2900 Semiconductor Drive
P.O. Box 58090
Santa Clara, CA 95052-8090
Phone: 800-272-9959
Web: http://www.national.com

a freely moving small insect. A promising effort along these lines is the recent development of a technique for recording from a direct junction between an insect antenna and a field effect transistor (Schütz et al. 1997).

9.2.4. Faraday Cage

Stray electromagnetic radiation in the working environment adds considerable noise to the antennal signal. The coaxial cable in the recording electrode provides grounded shielding of the EAG signal up to the EAG amplifier, and the grounded metal project case surrounding the EAG amplifier provides shielding for the critical input stage. The grounded aluminum foil wrapped around the recording electrode, as mentioned above, is often the only external Faraday cage that is needed. However, the noise in the EAG baseline can often be improved considerably by adding one additional Faraday cage around an insect antenna after it is attached to the electrodes. If a ring stand is used to hold the electrodes (by means of a small-parts jig, as described above), an external Faraday cage (4 × 4 × 2 cm, made from a thin aluminum sheet) can be mounted on a grounded clamp that slides up and down the bar of the ring stand to surround the antennal preparation, protecting the preparation from stray air currents as well as electrical noise.

9.2.5. High-Pass Filters and Signal Processing

The output of a simple amplifier, as described above, is often the best all-around choice to observe on a display device. A disadvantage, however, is that the baseline tends to wander, requiring frequent adjustment to keep the baseline positioned near the top of the display. The wandering can be eliminated by using a high-pass filter with a cutoff frequency of about 1 Hz (Bjostad & Roelofs 1980; Struble & Arn 1984; Leal et al. 1995). There are two disadvantages. First, high-pass filters have a tendency to lose information about the shape of the EAG signal, which is often some of the most useful information in the EAG response. For many lepidopteran species, for example, the EAG response to the main pheromone components is not only larger than it is to nonpheromone components, but it also has a characteristic slow return to baseline (Roelofs 1984). Second, during initial setup of an antenna, it is desirable to use the potentiometer to set the EAG baseline near 0 V to ensure that the amplified signal is not too close to the upper or lower limit of the power supply, which would distort or truncate the signal. The EAG baseline voltage is determined in part by the internal electrolytic environment of the insect antenna, and this varies among species and among individuals. By observing a filtered output alone, it is not easily possible to tell if the baseline is near the power supply limits. A simple way around this problem is to build the amplifier with a switch between the filtered and unfiltered outputs, so that the unfiltered baseline can be adjusted near 0 V, and then the amplifier can be switched to the high-pass-filtered output for recording.

9.2.6. Display

This is the component of the EAG system for which there is the widest variation in available choices and cost. Choices include

1. personal computer (data-acquisition hardware and software also required),
2. storage oscilloscope,
3. graphing calculator (data-acquisition hardware and software also required),
4. strip chart recorder,
5. integrator,
6. voltmeter (construction of signal-processing circuit also required).

Personal computers, in conjunction with data-acquisition hardware and software, are now commonly used for data collection and display (Bjostad 1988; Karg & Sauer 1995; Rumbo et al. 1995; Marion-Poll 1996).

The cost of the hardware-software data-acquisition package can vary widely.

Most data-acquisition packages range in price from $1300 to $3000. The Grams32® package offered by Galactic Industries includes capabilities for chromatographic analysis, and their chromatographic software for analysis of peak height and peak area can be adapted easily to the analysis of EAG peaks. One of the least expensive packages ($350) is offered by Vernier Software (Portland, OR), with versions for Macintosh® and Windows®. Most manufacturers now have complete descriptions of their hardware and software on the World Wide Web (Table 9.4). It is also possible to construct data-acquisition systems in the laboratory, at relatively low cost (Bjostad 1988; Johnson 1993; McCombs 1997).

Digital oscilloscopes are used for extended display of the EAG signal, and can also be programmed to measure EAG amplitudes automatically. An excellent choice is the HP54603B ($2,000, Hewlett-Packard, Wilmington, DE). However, given the cost of an oscilloscope and its limited capabilities, a computer-based hardware-software system may be a better choice.

A strip chart recorder connected to the amplifier can provide a permanent record of the EAG signal. Disadvantages are that the limited mechanical response time of the pen may cause very fast signals to be reported inaccurately and that EAG responses must be measured manually. Electronic integrators intended for chromatographic use are an improvement, and they also calculate peak heights and areas, the exact features that are of interest in EAG work. A suitable instrument is the HP3396 Series III integrator (Hewlett-Packard Corp.), which also allows BASIC programming.

A graphing calculator ($100–$200) in conjunction with a suitable hardware-software data-acquisition package ($99–$350, Vernier Software, Portland, OR) offers perhaps the best combination of versatility and economy. This system is very lightweight, an advantage for using an EAG apparatus in the field.

At rock-bottom cost, it is possible to display the EAG signal using only a voltmeter (Bjostad & Roelofs 1980), but additional circuitry ($20) must be built to process the EAG signal, and detailed information about the time course of the EAG signal is not available. Such information is helpful when something has gone wrong and the source of the problem must be determined (dirty electrodes, damaged operational amplifier, etc.). Moreover, Baker and Roelofs (1976) showed that parameters such as the recovery time to baseline can be important, and additional circuitry would be required to capture this information.

9.2.7. Antennal Preparation

Fresh disposable glass pipettes should be used for each EAG session, because electrodes become dirty with use, resulting in poor electrical connections. Each pipette is positioned at a 45° angle from vertical, and the tip of each pipette is filled by dipping briefly into the appropriate saline (section 9.2.1.1). The gold wire (or AgCl-coated tip of each silver wire) must extend into the saline in each pipette, but must not contact the wire-Soldercon® connector connection. The tips

Table 9.4.   Data-Acquisition Systems for Personal Computers

Data Translation
100 Locke Drive
Marlboro, MA 01752-1192
Phone: 508-481-3700
Fax: 508-481-8620
E-mail: Info@datx.com
Web: http://www.datx.com

Galactic Industries
395 Main Street
Salem, NH 03079
Phone: 603-898-7600
Fax: 603-898-6228
E-mail: support@galactic.com
Web: http://www.galactic.com

GW Instruments, Inc.
35 Medford Street
Somerville, MA 02143-4237
Phone: 617-625-4096
Fax: 617-625-1322
E-mail: info@gwinst.com
Web: http://www.gwinst.com

IOtech, Inc.
25971 Cannon Road
Cleveland, OH 44146
Phone: 216-439-4091
Fax: 216-439-4093
E-mail: sales@iotech.com
Web: http://www.iotech.com

Keithley MetraByte
440 Myles Standish Boulevard
Taunton, MA 02780
Phone: 508-880-3000
Fax: 508-880-0179
E-mail: info@metrabyte.com
Web: http://www.metrabyte.com

Microstar Laboratories, Inc.
2265 116th Avenue NE
Bellevue, WA 98004
Phone: 206-453-2345
Fax: 206-453-3199
E-mail: info@mstarlabs.com
Web: http://www.mstarlabs.com

National Instruments
6504 Bridge Point Parkway
Austin, TX 78730
Phone: 512-794-0100
Fax: 512-794-8411
E-mail: info@natinst.com
Web: http://www.natinst.com

OMEGA Engineering
One Omega Drive
P.O. Box 4047
Stamford, CT 06907-0047
Phone: 203-359-1660
Fax: 203-359-7700
E-mail: info@omega.com
Web: http://www.omega.com

Vernier Software
8565 SW Beaverton-Hillsdale Highway
Portland, OR 97225-2429
Phone: 503-297-5317
Fax: 503-297-1760
E-mail: info@vernier.com
Web: http://www.vernier.com

of the pipettes are brought into contact, completing the circuit, and the amplifier is zeroed.

If this does not work, the most likely problems are that the pipette tips are not in electrical contact or that the AgCl-coated silver wires are not contacting the saline. If difficulties persist, there may be a problem with the amplifier itself. This can be tested by pulling the wires out of the Pasteur pipettes and touching the silver electrodes to one another on the upper clean regions (not the AgCl-coated tips). If it is now possible to move the baseline, the amplifier is working, and dirty electrodes are the likely problem (replace the pipettes and saline, clean and recoat the silver wires). If the EAG amplifier still does not work, the usual problem is that the integrated circuit in the input stage (typically a LF356 op-amp) has been damaged by static electricity, and must be replaced (this costs about $5).

If the amplifier and display function properly when the electrodes are touching, then the tips should be moved a short distance apart, ready for the insect antenna to be connected. As soon as electrical contact is broken, the display baseline jumps to the top or bottom of the screen (if the baseline does not move, it means that there is a problem, usually a short circuit in the electrodes or a problem with the amplifier).

An antenna is plucked from the head of the insect, by gripping (but not crushing!) the base of the antenna with a pair of fine forceps. The base of the antenna is quickly mounted in the tip of the ground electrode, to prevent dessica-tion. The very tip of the antenna is then snipped off, and the cut tip is inserted into the saline in the tip of the recording electrode. Thread-trimmers available from fabric stores or a sharp razor blade are much superior to scissors for making a clean cut and minimizing crushing of the tissues. The movable arms of the small-parts jig allow the distance between the pipette tips to be adjusted quickly, and the excised antenna hangs by surface tension between the saline-filled tips. By convention, the base of the antenna is connected with the ground electrode and the tip with the recording electrode (reversing this arrangement allows per-fectly good recordings, but the observed voltage deflection is in the reverse direction). If the antenna has been connected properly, it is now possible to bring the baseline on scale by turning the control knob. Once a good antennal connection has been obtained, the clamp holding the Faraday cage is slid up the pole on the ring stand until the Faraday cage surrounds the antennal preparation, and a reduction in EAG baseline noise on the display is usually observed.

The use of a small-parts jig for positioning the electrodes is simpler than a method described previously (Bjostad & Roelofs 1980; Roelofs 1984), which entailed pressing the base of the insect antenna into a dish of Tackiwax® (often resulting in a broken antenna) and using a micromanipulator to place the tip of the recording electrode over the tip of the antenna.

The cut-tip technique has been used successfully with many types of insects (Visser 1979; Roelofs 1984; Gonzalez et al. 1994), but it is essential to cut off only the very tip cleanly without crushing it. Similarly, when removing the antenna, the base must not be crushed with the forceps. Many variations have also been used successfully. For example, for EAG recordings from the tarnished plant bug, *Lygus lineolaris,* glass-saline electrodes were placed into the antenna after puncturing the scape and the terminal antennal segment with a sharpened tungsten wire (Chinta & Dickens 1994). Longer-lived preparations can sometimes be obtained by excising the entire head and mounting the cervix on the tip of the saline-filled ground electrode (Averill et al. 1988; Hidoh et al. 1992; Okada et al. 1992), and the recording electrode then contacts the tip of the antenna.

For some species (such as *Diabrotica virgifera virgifera*, the western corn rootworm), it is not necessary to cut off the tip of the antenna, and excellent electrical connections are obtained when the uncut tip of the insect antenna is brought directly into contact with the saline solution in the tip of the recording electrode. Successful EAG recordings from the European corn borer, *Ostrinia nubilalis*, have been made by wrapping thin strips of filter paper saturated with saline solution at intervals along the antennae (Nagai 1981). EAG recordings from tsetse flies, *Glossina morsitans*, have been made by touching the tip of the recording electrode to the uncut distal end of the funiculus of the antenna (Den Otter & Saini 1985; Bursell et al. 1988). When experimenting with these alternative methods, if it is possible to zero the baseline by adjusting the potentiometer on the EAG amplifier, a successful connection has been made. Experimentation is worthwhile when a new species is tested because an antennal preparation usually lasts longer if the tip is not cut off.

It is more difficult to obtain EAG responses from the antennae of some insect species than those of others, and recordings from "problem" species can often be improved dramatically by careful selection of the type of electrode and saline, and by using more than just the antennae for the preparation. At one extreme, for example, excellent EAG responses are usually obtained from male moths to sex pheromones by using an excised antenna with a cut tip, which can often be used for an hour or more. At the other extreme, it is notoriously difficult to obtain EAG responses from the antennae of many hemipteran species, and it may only be possible to obtain usable recordings from an antenna for a few minutes. For these problem insects, it is often best to leave the antennae intact on the live insect, puncturing the antenna near the base and near the tip with glass electrodes pulled to a fine point. Alternatively, preparations can be made from part of the insect, such as the head or head and thorax.

9.2.8. Volatile Sample Delivery

The air delivery system for introduction of volatile compounds to the antenna is controlled with a needle valve. Air is humidified by bubbling air (flow ~300 ml/min) through an Erlenmeyer flask filled with water. Alternatively, the air can be passed through a tube loosely packed with wet glass wool. The humidified air is passed into a glass or metal tube that has a 3-mm hole in the side to allow a volatile sample to be puffed into the airstream from a sample pipette. If necessary, a guide tube (0.5 cm long, 3 mm diameter) can be permanently mounted in the small hole in the side of the air delivery tube to shield the tip of a sample pipette from the airstream and prevent the small deflection of the EAG trace that can occur due to brief disruption of the airstream and consequent mechanical stimulation of the antenna. The antenna is centered in the orifice of the air delivery tube. A variation is the use of suction to pull air at a constant rate past the antenna, for example, in a portable EAG apparatus (Suckling et al. 1994). The antenna was mounted in a Perspex™ holder, held between saline-filled wells cut into the Perspex.™

The simplest (and usually the best) way to test a volatile compound is the manual "puff" technique. A solution of the volatile compound is placed on a small piece of filter paper, and the solvent is allowed to evaporate. The filter paper is then placed inside a disposable glass Pasteur pipette. The tip of the pipette is placed into the hole in the side of the air delivery system described above. A 5-ml syringe is then used to provide a puff of air (~1 ml) through the pipette and into the stimulus tube. If the stimulus is EAG-active, a sudden downward deflection in voltage is observed on the display at the instant the puff is made (Fig. 9.4). Slight differences in the speed or volume of the puff have little effect on the amplitude of the EAG response, which depends largely on the structure and dose of the stimulus. The syringe must make a good seal with the pipette, accomplished by mounting a silicone rubber GC septum over the syringe tip. Air must not be drawn into the syringe through the sample pipettes to prevent contamination and artifact production.

The puff technique can also be used with volatiles collected by preparative GC in glass capillaries (chapter 3) by fitting the 5-ml syringe with a short length of Teflon® tubing that fits snugly over the end of each collection tube. Each collection tube can be tested many times because only a fraction of the collected volatiles are evaporated with each puff.

It is important to puff the stimulus into an air delivery system and not directly onto the antenna because mechanical stimulation of the antenna can also generate an electrophysiological response, which obscures any response to the test stimuli. A control puff typically elicits little or no response at all from the antenna if the air delivery system is properly configured.

In an interesting application, EAG has been used to show that changes in behavioral responses to various compounds are not entirely due to complex

Figure 9.4. Electroantennogram traces from a female mosquito, *Culex quinquefasciatus*, antenna stimulated by (from left to right) nonanal (1000 μg), nonanal (100 μg), nonanal (10 μg), 3-methylindole (100 μg), and alpha-terpineol (100 μg). Compounds were loaded on 25- × 8-mm filter papers, and a puff of air (3 ml) was passed through the stimulus delivery tube holding the filter paper, and over the antenna. Figure courtesy of Y.-J. Du.

processes in the brain. Instead, an observed shift in odor preference in the wasp *Leptopilina heterotoma* was accompanied by a corresponding shift in the relative sensitivity of olfactory receptor neurons (Vet et al. 1990).

9.2.9. Coupled Gas Chromatography-Electroantennographic Detection

The power of the EAG technique is substantially increased by using an EAG as a detector for the effluent from a GC system (electroantennographic detection, or EAD; Moorhouse et al. 1969; Arn et al. 1975; Struble & Arn 1984; Cork et al. 1990; Wadhams 1990), a technique that has also been adapted for single-cell recording (Wadhams 1984). The EAD approach (Fig. 9.5) has numerous

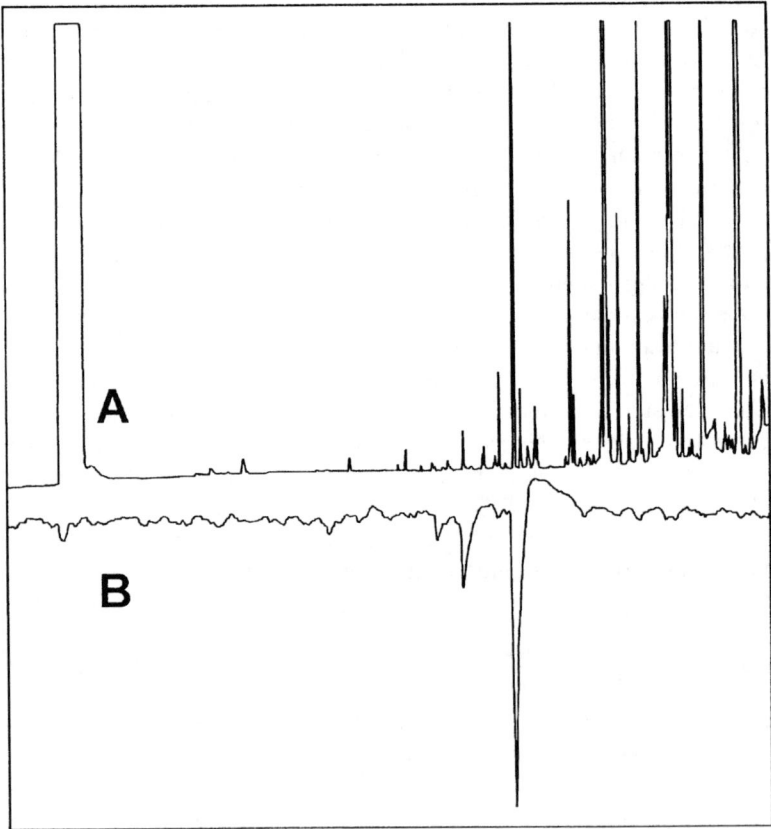

Figure 9.5. Coupled GC-electroantennogram detection traces: *A*, chromatogram from an extract of pheromone glands of the saturniid moth *Coloradia velda*, and *B*, simultaneously recorded responses of a male *C. velda* antenna to compounds in the pheromone gland extract. Note the medium to strong responses to the three pheromone components, and the lack of response to most other compounds in the extract. Figure courtesy of J.S. McElfresh.

applications, including characterization of responses to pheromones (Wadhams 1990; Burger et al. 1991; Leal et al. 1992; Leal et al. 1995) and host odors (Thiery et al. 1990; Henning & Teuber 1992; Wadhams et al. 1994; Cossé et al. 1995). Unlike the traditional method of collecting and testing individual GC fractions in capillary tubes (Roelofs 1984), coupled GC-EAD takes full advantage of the GC resolution capabilities (Wadhams 1990), so that antennal responses to compounds eluting from the GC within a couple of seconds of each other can be readily distinguished. This is particularly useful for isomers, which often separate only slightly by GC. For example, stereoisomers of supellapyrone, the sex pheromone of the brownbanded cockroach, *Supella longipalpa,* were resolved with a chiral column GC and concurrent EAD responses established the natural configuration as $2'R,4'R$ (Leal et al. 1995). Continuous monitoring of the GC effluent by both GC and EAD is invaluable in making distinctions of this sort. The retention times on both detectors should be identical for a given GC run.

In practice, the column effluent is split (equally or unequally) with a fused silica "Y" connector (e.g., J&W Scientific, Folsom, CA; SGE, Austin, TX; Struble and Arn 1984; Wadhams 1990; Burger et al. 1991; also see chapter 3) into two uncoated fused silica capillaries, with part of the effluent going to the GC detector and the other part directed out of the oven through a heated outlet port and through a small hole in the stimulus delivery tube, which delivers the stimulus to the insect antenna. The lengths of the two capillaries should be the same so that compounds reach the antenna and the GC detector at the same time. The internal cross-sectional area of the two capillaries determines what proportion of the effluent is directed to the GC detector and the EAD. As for EAG, the airflow (~250–500 ml/min) through the stimulus delivery tube should be humidified to minimize dessication of the antenna. Depending on the diameter of the delivery tube (normally ~1 cm), the airflow should be adjusted to sweep through the tube fairly quickly, to maintain the integrity of the GC peaks, but not so fast that the mechanoreceptors on the antenna are stimulated, creating high background noise levels. Higher airflows also result in greater dilution of the stimulus.

Some researchers do not split the effluent. Instead, the sample is run first with the column connected to the GC detector to determine retention times. The column end is then transferred to the heated EAD port, and the sample is run again, over the insect antenna. This maximizes the amount of material passing over the antenna per run, and minimizes differences in measured retention times between the GC detector and the EAD, but it means that EAD runs are made essentially "blind."

Several useful operational tips are as follows. First, it is crucial to prevent the eluting compounds from condensing out on cold spots at the end of the capillary where it enters the stimulus tube, which leads to broad, smeared-out responses. To minimize this problem, the capillary should be heated right up to the point where it enters the tube, and the end of the capillary should protrude only ~1 mm into the tube. Furthermore, instead of using a Y connector, a four-way X-cross can be used instead, which allows makeup gas to be added to increase the flow through each of the capillary arms and thus sweep them out more quickly, hindering condensation.

Second, the distance between where the capillary enters the stimulus delivery tube and the antennal preparation should be kept fairly short (a few centimeters) to minimize dilution of the sample and to minimize absorption of the sample on the walls of the delivery tube.

Third, the air temperature at the end of the delivery tube should be measured under normal operating conditions. If the air temperature is >25°C, it may be necessary to cool the airstream to minimize dessication of the antenna, particularly for small, fragile antennae. This can be done by cooling the incoming air (e.g., by running the air through a coil of copper tubing in a cooling bath or a bucket of ice). Even better, the stimulus delivery tube can incorporate a water jacket, cooled by a cooling recirculator. This method both cools the humidified air and supersaturates it with moisture, hindering antennal dessication. However, a cooling-jacketed tube may also result in greater likelihood of a cold spot where the capillary enters the side of the stimulus delivery tube.

A clever alternative arrangement (Cork et al. 1990) involves two capillary columns (one polar, one nonpolar) whose outlets are joined with a Y connector, with a short length of capillary tubing leading to a second Y connector that splits the effluent between the GC detector and an EAD. In this arrangement, one of the two columns is used for separation at 1 ml/min, and the other is used to provide makeup gas at 2 ml/min, to minimize peak broadening. By switching the flows and injecting alternately into each column, the retention characteristics of EAD-active compounds are determined without having to switch columns. The EAD in this case was also set up differently, with the column effluent emptying into a collection chamber whose contents were puffed across the antenna at 15-s intervals, so that the EAD output produced a series of short spikes, due to the combined response from chemoreceptors and mechanoreceptors. This approach provided better sensitivity than a continuous delivery system, and also provided a useful indicator of the condition of the EAG preparation, but at the expense of exploiting the full resolution of the capillary column.

Unlike EAG, GC-EAD typically involves continuous recordings for 30–60 min. Baseline drift is therefore a serious problem, correctable by high-pass filters. Burger et al. (1991) used a Murphy Developments (Syntech, Hilversum, The Netherlands) AMS-025 amplifier with a time constant of 12 s to remove the drift. High-pass filters distort the shape of the EAD signal by removing low-frequency components, but this is not normally a disadvantage when recording responses to closely spaced GC peaks.

9.2.10. Portable EAG

Portable EAG systems have been developed, primarily to measure pheromone concentrations in fields treated with pheromones for mating disruption (Sauer et al. 1992; Karg & Sauer 1995; Rumbo et al. 1995). A turnkey system is available (Syntech).

9.2.11. Troubleshooting EAG and EAD

The integration of biological and electronic components creates possibilities for a number of problems that can be frustrating to pinpoint. In practice, many EAG problems with home-built systems can be solved more quickly, effectively, and inexpensively by disposal-and-replacement rather than diagnosis-and-repair. The following short list of remedies solves most problems:

1. Replace the old glass electrodes, even if they appear to be clean.
2. Clean the gold wires (or clean and recoat the silver wires).
3. Replace the operational amplifier chip that receives the signal from the insect antenna. This chip is vulnerable to damage from static electricity discharge through the wire that receives the EAG signal from the insect antenna. The chip is inexpensive, and replacement is usually necessary only once or twice a year.

9.3. Single-Cell Recording from Olfactory Receptors

Boeckh introduced the use of microelectrodes for recording from individual receptor neurons (1962; Schneider & Boeckh 1962). Although it is more difficult to set up an SCR than an EAG preparation, SCR offers advantages for some applications. The EAG response is an integration of potentials from a large number of chemoreceptors (Schneider 1957a; De Kramer & Hemberger 1987), and is normally useful for detecting responses to the most important behaviorally active compounds under investigation. However, behaviorally important compounds do not always elicit large EAG responses. The SCR technique can be important in such instances because insect chemoreceptors are often highly specialized for particular compounds, and even if a specialized chemoreceptor type for a given compound is uncommon, its responses will be easily observed once the receptor has been located (Fig. 9.6). For example, recordings from 545 cells showed that the aggregation pheromone of *Scolytus scolytus* includes (−)-α-cubebene, a host compound that was detected by SCR but not EAG (Wadhams 1982, 1990). In a study of antennal perception of volatiles from oilseed rape, *Brassica napus,* by the cabbage seed weevil, *Ceutorhynchus assimilis,* EAG and SCR techniques were compared (Blight et al. 1995). Specific olfactory cells were located by SCR for five compounds that did not elicit significant EAG responses. A second point in favor of the SCR technique is that some insects, such as aphid species, have such limited numbers of receptors on their antennae that EAG

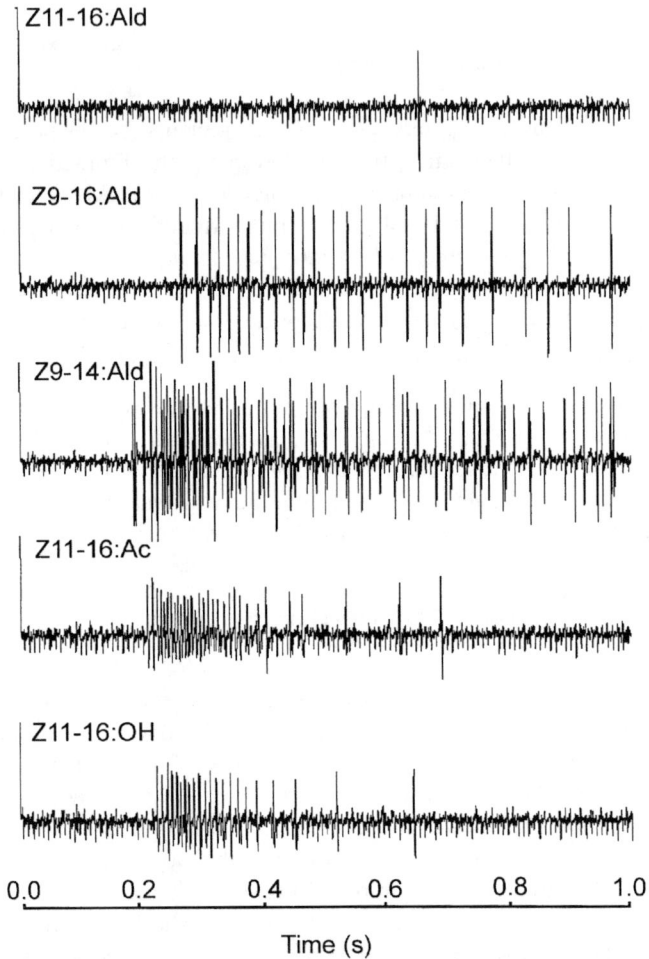

Figure 9.6. Single-cell responses of two receptor neurons present in one type of trichoid sensilla on male *Helicoverpa zea* antennae, after 20-ms stimulations with two sex phero-mone components (Z9-16; Ald, Z11-16:Ald) and three behavioral antagonists (Z9-14:Ald, Z11–16:Ac, Z11-16:OH). Note that Z9-16:Ald stimulates only a large-spiking receptor neuron, Z11-16:Ac and Z11-16:OH stimulate only a small-spiking neuron, and Z9-14:Ald stimulates both the large- and the small-spiking neuron. Figure courtesy of T.C. Baker and J.L. Todd.

responses are small despite the summation they provide, and SCR is therefore competitive in ease of detection (Wadhams 1990).

9.3.1. Differences between EAG and Single-Cell Recordings

Different techniques for electrophysiological recording generate qualitatively different kinds of information. In EAG recordings, a single, large depolarization is observed after olfactory stimulation, due to the summation of relatively low frequency generator potentials from thousands of neurons. DC amplifiers are therefore of greatest interest for EAG recording. In contrast, in single-cell recordings from olfactory receptors or tip recordings from contact chemoreceptors, the main feature of interest is the burst of action potentials that follows stimulation. Action potentials from a given nerve cell are all the same height, elicited in an all-or-none fashion. More active compounds elicit more rapid spiking, i.e., higher frequencies of action potential responses, but do not result in action potentials with higher amplitudes. AC amplifiers are therefore of greatest interest for single-cell recordings, because action potentials consist principally of high-frequency components, and the low-frequency components of the recorded signal are usually a nuisance, causing the baseline to wander off scale.

Action potential frequencies can be calculated automatically by adding circuitry or computer software (Marion-Poll & Van der Pers 1996; Marion-Poll 1995; Marion-Poll & Tobin 1991) that counts action potentials over discrete time intervals. Action potentials of different amplitudes are sometimes observed in single-cell recordings or tip recordings, and each distinct amplitude is due to a different cell or cell type. It is also possible to use additional circuitry or computer software to discriminate among the different amplitudes and thereby obtain simultaneous assessments of action potential frequencies from several different cells (Frazier & Hanson 1986).

9.3.2. Electrodes for SCR

The main techniques for SCR from olfactory receptors are analogous to those described above for EAG recording, involving saline-filled glass microelectrodes (cut-sensillum SCR) or tungsten microelectrode SCR. The principal advantage of the cut-sensillum technique is that a very low impedance electrical path is created between the antennal hemolymph and the saline in the electrode, allowing low-noise recordings (De Kramer & Hemberger 1987). The principal advantage of the tungsten microelectrode technique is that connections are more easily achieved with sensilla that are densely packed or very small (Mustaparta 1975; Hansen 1983; De Kramer & Hemberger 1987).

9.3.2.1. Saline-Filled Glass Microelectrodes

The cut-sensillum technique has been used with particular success for studies of sex pheromone receptors of male moths (Kaissling 1974; Todd et al. 1992,

1995; Todd & Baker 1993, 1996; Van der Pers & Minks 1993). In a typical approach (Todd et al. 1992), one antenna is excised from the head of a male moth, and the antennal base is placed in a saline-filled ground electrode. The antenna is maneuvered with a micromanipulator until a single sensillum tricho-deum rests on the sharpened blade of a stationary, vertically positioned glass microknife, with its tip hanging over the edge. The sensillar tip is cut off using a mobile glass knife placed in a Leitz joystick type micromanipulator that permits movement in three dimensions, and the cut end is then immediately brought into contact with a saline-filled Ag-AgCl recording electrode. Glass microknives are sharpened with a rotating carborundum stone (no. 1200) (Van der Pers & Den Otter 1978). Tungsten microknives can also be used, and are stronger and less vulnerable to damage than glass (Van der Pers & Minks 1993).

Single-cell recording with the cut-sensillum technique has even been success-fully conducted in the field, to study responses to synthetic sex pheromone dispensers by the moths *Aegeria myopaeformis* and *Adoxophyes orana* (Van der Pers & Minks 1993). A portable module was constructed with two small joystick micromanipulators (Narishige, model MN-151) mounted on a vertical column, with a tungsten microknife and a high-impedance input amplifier for the recording electrode mounted on one of the micromanipulators. With the base of the insect antenna mounted on the indifferent electrode, the tip was cut from a single sensory hair by using the microknife on the micromanipulator to wipe against a rotatable vertical microknife mounted on the vertical column, and a swivel joint then allowed the recording electrode to be brought immediately into contact with the cut tip. Total preparation time was less than 2 min.

Saline composition may be more important for SCR than for EAG work, but the evidence for this is conflicting (as noted in section 9.2.1.1). In SCR with ermine moths, preparations lasted only 3 min when Beadle-Ephrussi saline was used in the electrodes, but preparations lasted 30 min when more specialized saline solutions were used in the electrodes (Van der Pers & Den Otter 1978). Other workers have found that a simple saline solution was as effective as "insect hemolymph saline" (Roessingh et al. 1992). Physiological saline solutions for many insect species have been developed by matching hemolymph plasma compo-sitions (Usherwood 1969; Florkin & Jeuniaux 1974; Bindokas & Adams 1988; Pelletier & Clark 1992; Table 9.1).

9.3.2.2. Tungsten Microelectrodes

Techniques for tungsten microelectrode preparation have been described in detail (Hubel 1957). The details of optimum electrode placement, usually deter-mined empirically, are somewhat variable from species to species. For example, single-cell recordings were made from the cabbage seed weevil, *Ceutorhynchus assimilis,* by placing the indifferent electrode in the antennal scape, and the recording electrode was brought into contact with the surface of the antenna until impulses were recorded (Blight et al. 1995), amplifying signals with a UN-03b amplifier (Syntech). Single-cell responses from the ambrosia beetle, *Trypodendron lineatum,* were made with tungsten microelectrodes inserted into the base of the hairformed sensilla, in the area of the antennal club where the density of olfactory sensilla is highest (Tømmerås and Mustaparta 1989).

9.3.3. Amplifier and Display

AC amplifiers are usually used with SCR because bursts of action potentials in response to candidate volatile compounds are often of greatest interest in characterizing specialized receptors that may be uncommon on the antenna. However, both low-frequency and high-frequency components of the signal may be useful in SCR interpretation, and both can be recorded at the same time by using high-pass and low-pass filters in conjunction with a DC amplifier of the same sort used in EAG recording, with at least 1 TΩ input impedance. Capacitance compensation and notch-filtering may be more useful in SCR work than in EAG work, due to the small size and concomitantly different electrical characteristics of the olfactory receptors, and DC amplifiers with these features are commercially available (Table 9.2). An amplifier with small bias current (<10 pA) may be advisable, since chemoreceptors are sensitive to small electric currents (Städler 1984). The most popular display for SCR due to the rapid time course of action potentials is a personal computer with a data acquisition system (section 9.2.6).

9.3.4. Sample Delivery

Stimulus delivery to single chemoreceptors is analogous to that used in EAG work. It may not be necessary to construct small delivery tubes for introducing candidate volatile compounds, because a larger airstream bathing the entire antenna with odor is simpler to construct and stimulates individual receptors just as effectively.

9.4. Tip-Recording from Contact Chemoreceptors

The tip-recording technique for contact chemoreceptive sensilla (Fig. 9.7) was introduced by Hodgson et al. (1955; Hodgson & Roeder 1956), with later modifications devised by other workers (Dethier 1976; Städler 1984). The conventional tip-recording technique for contact chemoreceptors is somewhat different from the SCR technique for olfactory receptors. For SCR from olfactory receptors, receptor connections are first established with the two electrodes, and the volatile stimulus is then puffed across the receptor. In contrast, for tip-recordings from contact chemoreceptors, a water-soluble test compound (a sugar, salt, amino acid, etc.) is dissolved in the saline solution in the recording electrode. When the electrode is brought into contact with the tip of the sensory hair, two things

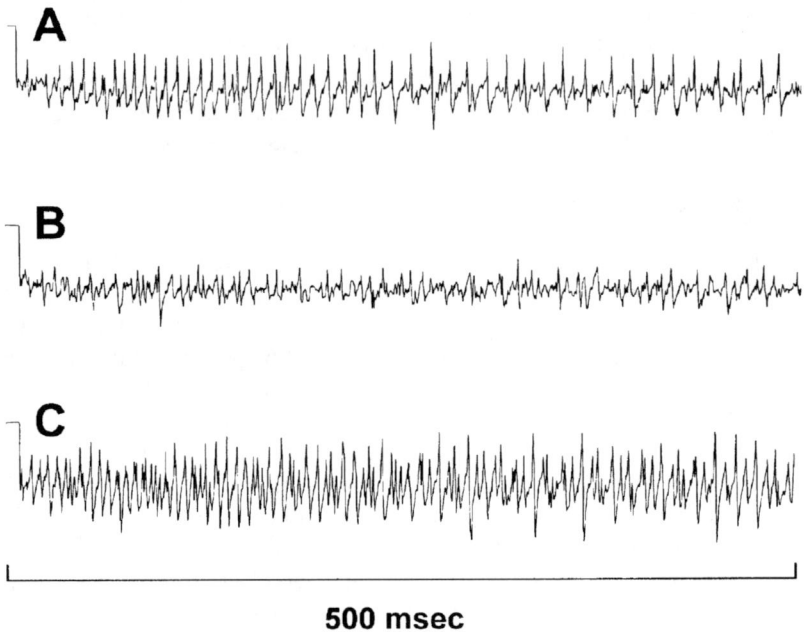

500 msec

Figure 9.7. Contact chemoreceptor tip-recording from *Trichoplusia ni* larva styloconic sensillum stimulated with *A*, 0.1-mM sinigrin; *B*, 0.1-mM sinigrin + 1-mM sucrose + 100-mM inositol; and *C*, 0.1-mM sinigrin + 1 mM-sucrose. Figure courtesy of B.K. Mitchell.

happen simultaneously. First, the test compound elicits a burst of action potentials, if the compound is perceived by one or more of the chemosensory cells in the sensory hair. Second, the saline solution completes an electrical connection with the antenna, allowing the action potentials to be observed on the display. Typical responses are shown in Fig. 9.7.

9.4.1. Electrodes and Sample Delivery

The saline composition for tip-recordings is often matched carefully to the hemolymph composition, but much simpler saline solutions may prove adequate as well. In phytochemical contact chemoreception tests with the tarsal B hairs of the black swallowtail butterfly, *Papilio polyxenes,* the indifferent electrode was filled with "insect hemolymph saline" (Kaissling & Thorson 1980), and the recording electrode was filled with 10-mM KCl plus the test phytochemical (Roessingh et al. 1991). This arrangement allowed stable recordings for many hours with a Johnson nonblocking amplifier. In similar work with the cabbage root fly, *Delia radicum,* 100-mM NaCl or "insect hemolymph saline" were used with equal success in the indifferent electrode, and with 30-mM KCl (plus the test phytochemical) in the recording electrode (Roessingh et al. 1992).

Because the electrode saline is also the sample delivery system, modifications have been made to accommodate the introduction of a variety of compounds. Water-soluble compounds such as sugars, salts, and amino acids are no problem, but nonpolar compounds require special approaches. Blaney (1975) used a suspension of leaf surface waxes in 50-mM NaCl to deal with such compounds, while Städler (1984) obtained better results using 10% methanol extracts of leaf surfaces in 100-mM NaCl. No apparent damage or stimulation due to methanol was observed in different fly species tested.

9.4.2. Amplifier and Display

Contact chemoreceptor recordings begin as soon as the recording electrode contacts the receptor because the stimulus compounds are present in the saline solution and initiate action potentials on contact. The main difficulty is that the sensillum has a large offset potential, typically about 100 mV (Marion-Poll & Van der Pers 1996) with respect to the indifferent electrode, and this may cause the amplified signal to exceed the limits of the amplifier power supply. This does not damage the amplifier, but the output is useless since it merely shows the maximum output voltage possible for the amplifier. Most investigators have avoided this problem by using an AC instead of a DC amplifier for these recordings. Use of an appropriate nonblocking filter amplifier avoids exceeding the limits of the amplifier (Maes 1977), but it distorts the waveform of the action potentials, which may interfere with spike discrimination (Marion-Poll & Van der Pers 1996). One way around this is to use an amplifier having two filters and automatic selection between them (Frazier & Hanson 1986), such as the Johnson "clamping preamplifier" (George Johnson Electronics, Baltimore, MD), similar to one described in the literature (De Kramer & van der Molen 1980).

An innovative alternative to high-pass filtering has recently been developed, involving automatic baseline compensation that takes place when the electrode contacts the chemoreceptor (Marion-Poll & Van der Pers 1996). This is accomplished by a circuit (marketed as TastePROBE® from Syntech) that automatically detects contact between a saline electrode and a taste receptor, stores the sensillum offset voltage (typically 100 mV) in a capacitor at the beginning of the recording, and then subtracts the stored value from the subsequent incoming signal. The main advantage of this approach over high-pass filtering is that it does not continuously remove the low-frequency components of the signal from the insect, which often contain important information. As with SCR, a personal computer (section 9.2.6) is used for data acquisition and processing.

9.5. Restraints for Difficult Insects

Insect restraints are sometimes essential for effective electrode placement, particularly if whole insects are used for recording; detached antennae are not satisfactory for all insect species. Common problems are limited preparation life and awkward geometry with respect to electrode placement or chemoreceptor exposure. For example, the pine engraver, *Ips pini,* was mounted in a Plexiglass® block by inserting the beetle into a hole that narrowed at the top, such that only the head and a portion of the thorax protruded. The antenna was attached with two-sided sticky tape to a metal sheet on the top of the block, with the recording electrode inserted into the antennal club and the indifferent electrode inserted into the mouth, using glass electrodes (Angst & Lanier 1979). The flagellate antenna of the scarab beetle, *Anomala cuprea,* required construction of a special acrylic chamber to spread the antennal segments open for access of volatile compounds (Leal et al. 1992).

For tip-recordings from the medial sensillum on the galea of the red turnip beetle, *Entomoscelis americana,* an elaborate system of restraints was required to expose the galea properly (Sutcliffe & Mitchell 1982). A sharpened silver reference electrode was inserted into the metathorax and anchored with a beeswax-resin mixture. The beetle was then inserted headfirst into a tapering glass tube that allowed only the head and prothorax to protrude from the top, and the prothorax was waxed to the top of the glass tube. To expose the galea, a fine human hair was tied around the maxillary palp to pull it out of the way, and the free end of the hair was waxed to the glass tube.

For EAG recording from the olive beetle, *Phloeotribus scarabaeoides,* the insect was introduced posteriorly under pressure into a conical plastic tube, the head was wedged against the thorax with a segment of the tube, and adhesive tape was used to bind the legs (Gonzalez et al. 1994). The indifferent electrode was introduced into the head capsule via the mouth, and the glass recording electrode was brought into contact with the cut tip of the antenna.

With the common furniture beetle, *Anobium punctatum,* the head was excised and mounted on a cork stage using double-sided adhesive tape (White & Birch 1987). The indifferent glass electrode was inserted into the base of the head, and the opening was sealed with petroleum jelly to inhibit dessication. The recording microelectrode was inserted into a small hole punched in the terminal antennal segment with a tungsten needle.

In the weevils *Pissodes strobi* and *P. nemorensis,* the thorax and abdomens of live individuals were wrapped in Teflon® tape to immobilize the legs, and a specially modified pair of forceps with a semicircle ground into each tong was used to hold the snout and prevent the antennae from retracting into the grooves on the snout (Hibbard & Webster 1993). The insect was secured to the forceps with tape, and the forceps were connected to a micromanipulator. The indifferent glass saline electrode was placed over the snout of the insect, and the recording electrode was connected to the tip of one antenna.

Single-cell recordings from caterpillars can be hampered by excessive move-ments by the insect, and one way around this is to use excised heads mounted on the reference electrode (Frazier & Hanson 1986). Longer recording times can be obtained if whole caterpillars are used, but special immobilization procedures are required (Devitt 1983; Gothilf & Hanson 1994). One approach is to introduce larvae head first into tapered glass tubes, restraining the mouthparts with paraffin film, beeswax, or dental floss. This approach was not adequate with fifth-instar tobacco hornworm, *Manduca sexta,* and an effective alternative was to immerse the posterior of the caterpillar in 100-mM KCl that served to anesthetize the insect without drowning it (Gothilf & Hanson 1994).

Adaptations for recording from very tiny insects have also been developed, suitable for recording from aphids (van Giessen et al. 1992, 1994) and their parasitoids (Vet et al. 1990; Vaughn et al. 1995).

9.6. Electronic Construction

Texts are available for the more academic aspects of basic electronics, including the theoretical basis for modern electronics (Horowitz & Hill 1995), the conceptual foundation for prototype development (Hayes & Horowitz 1989), and the practical elements of electronic construction (Reis 1995). A few guidelines can be helpful:

1. Use integrated circuit (IC) sockets, which allows ICs to be plugged in and removed as often as necessary. They can also be used for nearly all other circuit elements, including resistors, capacitors, potentiometers, transformers, fuses, switches, and display devices. A small piece of double-sided tape can be used to hold each socket to the perfboard, or one edge of the socket can be tacked down with a tiny drop of 5-min epoxy glue or a tiny drop of hot-melt glue (from glue guns for hobby use, with small nozzles and cool-melting glue). IC "breadboards" (cat. no. 923253, Digi-Key), usually used for building circuit prototypes, can also be used but are more expensive.

2. Use wire-wrapping, if few revisions or copies are anticipated. Novices assume that soldering is the only good way to make electrical connections, but wire-wrapping is an excellent choice for single copies of a circuit. A wire-wrapping tool (Radio Shack 276-1570) can be purchased for about $8. Very slender insulated wire (30-gauge) with different colors of insulation is manufactured specifically for wire-wrapping. Instead of soldering, a highly reliable electrical connection is created by stripping the insulation from the end of the wire, and then using the wire-wrapping tool to wrap the bare wire for several turns around a socket pin. Large numbers of alterations are easily made by unwrapping or cutting the wire from a pin. Wire-wrapped connections are highly reliable for many years, and can be soldered at a later time if desired.

3. A computer can be used easily to design, fabricate, revise, and copy printed circuits. Conventional graphics software is used to create the circuit layout, and a laser printer can be used to print the reversed design on special film (TEC-200 image film, DC Electronics, Scottsdale, AZ). A hot iron is used to transfer the reversed circuit design to a copperclad board (single-sided is best), and the board is etched with saturated $FeCl_3$. The computer-designed printed circuit board approach is less risky than wire-wrapping because the old board is not physically altered and can be used again immediately if the new board is unsuitable for some reason.

4. Use edgeboards, which slot into connectors bolted to the project case. Each subcircuit is built on a separate edgeboard, allowing a subcircuit to be removed with no trailing wires, facilitating wiring alterations within the subcircuit. This approach is particularly advantageous with computer-designed printed circuits, in which each board is designed with a row of metal tabs etched along one edge to fit into the edgeboard connector. The most useful edgeboard connectors are the 44-contact, solder-eye type (cat. no. V1062-ND, Digi-Key).

5. Divide the project into subcircuits, with only a few inputs and outputs, and test each subcircuit before building the next. For example, the power supply converts a high AC voltage from a wall socket to one or more low DC voltages, and typically has only three inputs (including ground) and three outputs. The power supply is built and tested on one board as the first subcircuit, and the small number of inputs and outputs facilitates troubleshooting. The high-input-impedance amplifier is then built as a second subcircuit on a separate board (it is essential to build a Faraday cage around this amplifier, and the subcircuit approach facilitates this). The amplifier has only four inputs (the signal from the insect antenna and the three power supply connections) and one output (the buffered and amplified signal from the insect antenna). If the amplifier subcircuit does not work on the first try, you know that the pretested power supply circuit cannot be the problem. Because the number of inputs and outputs for each subcircuit is small, building and testing each subcircuit before going on to the next greatly facilitates troubleshooting. If the EAG system fails because of a single damaged component, disconnecting the subcircuits makes it easier to pinpoint the problem (by removing all edgeboards, for example, and replacing them one by one by one with subsequent testing). The subcircuit approach also facilitates adding capabilities at a later time, such as active filters, integrators, sample-and-hold capability, or automatic offset compensation.

9.7. Acknowledgments

I thank J. G. Millar, Y.-J. Du, J. S. McElfresh, T. C. Baker, and B. K. Mitchell for providing figures of electrophysiological recordings. Support was provided by the Colorado Agricultural Experiment Station.

9.8. References

Angst, M.E. & G.N. Lanier. 1979. Electroantennogram responses of two populations of *Ips pini* (Coleoptera: Scolytidae) to insect-produced and host tree compounds. J. Chem. Ecol. **5**:131–140.

Arn, H., E. Städler & S. Rauscher. 1975. The electroantennographic detector—a selective and sensitive tool in the gas chromatographic analysis of insect pheromones. Z. Naturforsch. **30c**:722–725.

Averill, A.L., W.H. Reissig & W.L. Roelofs. 1988. Specificity of olfactory responses in the tephritid fruit fly, *Rhagoletis pomonella.* Entomol. Exp. Appl. **47**:211–222.

Baker, T.C. & R.T. Cardé. 1984. Techniques for behavioral bioassays. *In:* Techniques in Pheromone Research, eds. H.E. Hummel & T.A. Miller, Springer-Verlag, New York. pp. 45–73.

Baker, T.C. & C.E. Linn Jr. 1984. Wind tunnels in pheromone research. *In:* Techniques in Pheromone Research, eds. H.E. Hummel & T.A. Miller, Springer-Verlag, New York. pp. 75–110.

Baker, T.C. & W.L. Roelofs. 1976. Electroantennogram responses of the male moth *Argyrotaenia velutinana* to mixtures of sex pheromone components of the female. J. Insect. Physiol. **22**:1357–1364.

Berlin, H.M. 1991. Op-amp circuits and principles. SAMS Macmillan, Cormel, IN.

Bindokas, V.P. & M.E. Adams. 1988. Hemolymph composition of the tobacco budworm, *Heliothis virescens* F. (Lepidoptera: Noctuidae). Comp. Biochem. Physiol. A. **90**:151–155.

Bjostad, L.B. 1988. Insect electroantennogram responses to semiochemicals recorded with an inexpensive personal computer. Physiol. Entomol. **13**:139–145.

Bjostad, L.B. & W.L. Roelofs. 1980. An inexpensive electronic device for measuring electroantennogram responses to sex pheromone components with a voltmeter. Physiol. Entomol. **5**:309–314.

Blaney, W.M. 1975. Behavioural and electrophysiological studies of taste discrimination by the maxillary palps of larvae of *Locusta migratoria* (L.). J. Exp. Biol. **62**:555–569.

Blight, M.M., J.A. Pickett, L.J. Wadhams & C.M. Woodcock. 1995. Antennal perception of oilseed rape, *Brassica napus* (Brassicaceae), volatiles by the cabbage seed weevil *Ceutorhynchus assimilis* (Coleoptera: Curculionidae). J. Chem. Ecol. **21**:1649–1664.

Boeckh, J. 1962. Electrophysiologische Untersuchungen an einzelnen Geruchs-Rezeptoren auf den Antennen des Totengräbers (Necrophorus: Coleoptera). Z. Vergl. Physiol. **46**:212–248.

Boeckh, J. 1984. Neurophysiological aspects of insect olfaction. *In*: Insect Communication, ed. T. Lewis, pp. 83–104, Academic Press, New York.

Burger, B.V., A.E. Nell & W.G.B. Petersen. 1991. Analysis of pheromones: enrichment on thick film capillary traps and GC detection with a living detector (EAD). J. High Res. Chromatog. **14**:718–725.

Bursell, E., J.E. Gough, P.S. Beevor, A. Cork, D.R. Hall & G.A. Vale. 1988. Identification of components of cattle urine attractive to tsetse flies, *Glossina* spp. (Diptera: Glossinidae). Bull. Ent. Res. **78**:281–291.

Chinta, S. & J.C. Dickens. 1994. Olfactory reception of potential pheromones and plant odors by tarnished plant bug, *Lygus lineolaris* (Hemiptera: Miridae). J. Chem. Ecol. **20**:3251–3267.

Cork, A., P.S. Beevor, A.J.E. Gough & D.R. Hall. 1990. Gas chromatography linked to electroantennography: a verstaile technique for identifying insect semiochemicals. *In:* Chromatography and Isolation of Insect Hormones and Pheromones, eds. A.R. McCaffery & I.D. Wilson, Plenum Press, New York. pp. 271–274.

Cossé, A.A., J.L. Todd, J.G. Millar, L.A. Martinez & T.C. Baker. 1995. Electroantennographic and coupled gas-chromatographic electroantennographic responses of the Mediterranean fruit-fly, *Ceratitis capitata,* to male-produced volatiles and mango odor. J. Chem. Ecol. **21**:1823–1836.

De Kramer, J.J. & J. Hemberger. 1987. The neurobiology of pheromone reception. *In:* Pheromone Biochemistry, eds. G.D. Prestwich & G.J. Blomquist, Academic Press, New York. pp. 433–472.

De Kramer, J.J. & J.N. van der Molen. 1980. Special purpose amplifier to record spike trains of insect taste cells. Med. Biol. Eng. Comput. **18**:371–374.

Den Otter, C.J. & R.K. Saini. 1985. Pheromone perception in the tsetse fly, *Glossina morsitans morsitans.* Entomol. Exp. Appl. **39**:155–161.

Dethier, V.G. 1976. The Hungry Fly. Harvard University Press, Cambridge, MA.

Dethier, V.G. & F.E. Hanson. 1965. Taste papillae of the blowfly. J. Cell. Comp. Physiol. **65**:93–100.

Devitt, B. 1983. The contact chemosensory system of the dark-sided cutworm *Euxoa messoria* (Harris) (Lepidoptera: Noctuidae). University of Toronto.

Florkin, M. & C. Jeuniaux. 1974. Haemolymph: composition (Chapter 6). *In:* The Physiology of Insecta, ed. M. Rockstein, Academic Press, New York.

Frazier, J.L. & F.E. Hanson. 1986. Electrophysiological recording and analysis of insect chemosensory responses. *In:* Insect-Plant Interactions, eds. J.R. Miller & T.A. Miller, Springer-Verlag, New York. pp. 285–330.

Gonzalez, R., A. Alvarez & M. Campos. 1994. An electroantennogram apparatus for testing the activity of semiochemicals on the olive beetle, *Phloeotribus scarabaeoides* (Coleoptera: Scolytidae)—first recordings of the response to ethylene. Physiol. Entomol. **19**:301–306.

Gothilf, S. & F.E. Hanson. 1994. A technique for electrophysiologically recording from chemosensory organs of intact caterpillars. Entomol. Exp. Appl. **72**:305–310.

Gough, A.J.E., D.R. Hall, P.S. Beevor, A. Cork, E. Bursell & G.A. Vale. 1987. Attractants for tsetse from cattle urine. Proc. 19th Meeting International Scientific Council for Trypanosomiasis Research Control, Lome, Togo, 27 March–4 April.

Hansen, K. 1983. Reception of bark beetle pheromone in predaceous clerid beetle, *Thanasius formicarius* (Coleoptera: Cleridae). J. Comp. Physiol. **150**:371–378.

Hayes, T.C. & P. Horowitz. 1989. Student Manual for the Art of Electronics. Cambridge University Press, New York.

Henning, J.A. & L.R. Teuber. 1992. Combined gas-chromatography-electroantennogram characterization of alfalfa floral volatiles recognized by honey-bees (Hymenoptera: Apidae). J. Econ. Entomol. **85**:226–232.

Hibbard, B.E. & F.X. Webster. 1993. Enantiomeric composition of grandisol and grandisal produced by *Pissodes strobi* and *P. nemorensis* and their electroantennogram response to pure enantiomers. J. Chem. Ecol. **19**:2129–2141.

Hidoh, O., T. Kawashima, J.I. Fukami & Y. Kainoh. 1992. EAG responses of parasitoids, *Ascogaster reticulatus* Watanabe (Hymenoptera, Braconidae), to the female sex-pheromone. Appl. Entomol. Zool. **27**:587–589.

Hodgson, E.S. & K.D. Roeder. 1956. Electrophysiological studies of arthropod chemoreception. I. General properties of the labellar chemoreceptors of Diptera. J. Cell. Comp. Physiol. **48**:51–75.

Hodgson, E.S., J.Y. Lettvin & K.D. Roeder. 1955. Physiology of a primary chemoreceptor unit. Science **122**:417–418.

Horowitz, P. & W. Hill. 1995. The Art of Electronics, 2nd ed. Cambridge University Press, Cambridge, MA.

Hubel, D.H. 1957. Tungsten microelectrode for recording from single units. Science **125**:549–550.

Jewett, D., D.L. Brigham & L.B. Bjostad. 1996. *Hesperophylax occidentalis* (Banks) (Trichoptera: Limnephilidae) sex pheromone structure-activity study with electroantennograms. J. Chem. Ecol. **22**:123–138.

Johnson, J.H. 1993. Build your own low-cost data acquisition and display devices. Tab Books, Summit, PA.

Kaissling, K.-E. 1974. Sensory transduction in insect olfactory receptors. *In:* Biochemistry of Sensory Functions, ed. L. Jaenicke, Springer, Berlin. pp. 243–273.

Kaissling, K.E. & J. Thorson. 1980. Insect olfactory sensilla: structural, chemical and electrical aspects of the functional organization. *In:* Receptors for Neurotransmitters, Hormones and Pheromones in Insects, eds. D. B. Sattelle, L. M. Hall & J. G. Hildebrand, Elsevier/North-Holland Biomedical Press, New York. pp. 261–282.

Karg, G. & A.E. Sauer. 1995. Spatial distribution of pheromones in fields treated for mating disruption of the European grape wine moth *Lobesia botrana* measured with electroantennograms. J. Chem. Ecol. **21**:1299–1314.

Leal, W.S., F. Mochizuki, S. Wakamura & T. Yasuda. 1992. Electroantennographic detection of *Anomala cuprea* Hope (Coleoptera: Scarabaeidae) sex pheromone. Appl. Entomol. Zool. **27**:289–291.

Leal, W.S., X.W. Shi, D.S. Liang, C. Schal & J. Meinwald. 1995. Application of chiral gas chromatography with electroantennographic detection to the determination of the stereochemistry of a cockroach sex pheromone. Proc. Natl. Acad. Sci. USA **92**:1033–1037.

Maes, F.W. 1977. Non-blocking AC preamplifier for tip recording from insect hairs. Med. Biol. Eng. Comput. **15**:470–471.

Marion-Poll, F. 1995. Object-oriented approach to fast display of electrophysiological data under MS-Windows. J. Neuroscience Meth. **63**:197–204.

Marion-Poll, F. 1996. Display and analysis of electrophysiological data under Windows™. Entomol. Exp. App. **80**:116–119.

Marion-Poll, F. & Tobin, T.R. 1991. Software filter for detecting spikes superimposed on a fluctuating baseline. J. Neuroscience Meth. **37**:1–6.

Marion-Poll, F. & J. Van der Pers. 1996. Unfiltered recordings from insect taste sensilla. Entomol. Exp. Appl. **80**:113–115.

McCombs, D. 1997. PC Data Acquisition with C: Cost-Effective Data Recording and Analysis. Miller-Freeman, San Francisco, CA.

Miller, T.A. 1979. Insect Neurophysiological Techniques. Springer-Verlag, New York.

Moorhouse, J.E., R. Yeadon, P.S. Beevor & B.F. Nesbitt. 1969. Method for use in studies of insect chemical communication. Nature **223**:1174–1175.

Mustaparta, H. 1975. Responses of single olfactory cells to insect- and host-produced volatiles in the bark beetle *Ips pini* (Say). J. Comp. Physiol. **79**:271–290.

Nagai, T. 1981. Electroantennogram response gradient on the antenna of the European corn borer *Ostrinia nubilalis*. J. Insect. Physiol. **27**:889–894.

Okada, K., A. Watanabe, M. Mori, K. Shimazaki, T. Chuman, F. Mochizuki & T. Shibuya. 1992. Olfactory responses to the sex pheromone component and its behavioural inhibitor in the male cigarette beetle, *Lasioderma serricorne*. J. Insect Physiol. **38**:705–709.

Payne, T.L. 1970. Sex pheromones of noctuid moths: factors influencing antennal responsiveness in males of *Trichoplusia ni*. J. Insect Physiol. **16**:1043–1055.

Pelletier, Y. & C.L. Clark. 1992. The haemolymph plasma composition of adults, pupae, and larvae of the Colorado potato beetle *Leptinotarsa decemlineata* (Say), and development of physiological saline solutions. Can. Entomol. **124**:945–949.

Ramachandran, R., Z.R. Khan, P. Caballero & B.O. Juliano. 1990. Olfactory sensitivity of two sympatric species of rice leaf folders (Lepidoptera: Pyralidae) to plant volatiles. J. Chem. Ecol. **16**:2647–2666.

Reis, R.A. 1995. Electronic Project Design and Fabrication. Prentice-Hall, Englewood Cliffs, NJ.

Roelofs, W.L. 1977. The scope and limitations of the electroantennogram technique in identifying pheromone components. *In:* Crop Protection Agents: Their Biological Evaluation, ed. N.R. McFarlane, Academic Press, New York. pp. 147–165.

Roelofs, W.L. 1984. Electroantennogram assays: rapid and convenient screening procedures for pheromones. *In:* Techniques in Pheromone Research, eds. H.E. Hummel & T.A. Miller, Springer-Verlag, New York. pp. 131–159.

Roessingh, P., E. Stadler, R. Schoni & P. Fenny. 1991. Tarsal contact chemoreceptors of the black swallowtail butterfly *Papilio polyxenes*—responses to phytochemicals from host plants and non-host plants. Physiol. Entomol. **16**:485–495.

Roessingh, P., E. Stadler, G.R. Fenwick, J.A. Lewis, J.K. Nielsen, J. Hurter & T. Ramp. 1992. Oviposition and tarsal chemoreceptors of the cabbage root fly are stimulated by glucosinolates and host plant extracts. Entomol. Exp. Appl. **65**:267–282.

Rumbo, E.R., D.M. Suckling & G. Karg. 1995. Measurements of airborne pheromone concentrations using EAG equipment: Interactions between environmental volatiles and pheromone. J. Insect Physiol. **41**:465–471.

Sauer, A.E., G. Karg, U.T. Koch, R. Milli & J.J. de Kramer. 1992. A portable EAG system for the measurement of pheromone concentrations in the field. Chem. Senses **17**:543–588.

Schneider, D. 1957a. Electrophysiological investigation on the antennal receptors of the silk moth during chemical and mechanical stimulation. Experientia **13**:89–91.

Schneider, D. 1957b. Elektrophysiologische Untesuchungen von Chemo- und Mechanotezeptorin der Antenne des Seidenspinners *Bombyx mori* L. Z. Vergl. Physiol. **40**:8–41.

Schneider, D. & J. Boeckh. 1962. Rezeptorpotential und Nervenimpulse einzelner olfaktorischer Sensillen der Insektenantenne. Z. Vergl. Physiol. **45**:405–412.

Schütz, S., B. Weisbecker, H.E. Hummel, M.J. Schönig, A. Riemer, P. Kordos & H. Lüth. 1997. Field effect transistor-insect antenna junction. Naturwissenschaften. **84**:86–88.

Städler, E. 1984. Contact Chemoreception. *In:* Chemical Ecology of Insects, eds. W.J. Bell & R.T. Cardé, Sinauer, Sunderland, MA. pp. 3–35.

Strausfeld, N.J., H.S. Seyan, D. Wohlers & J.P. Bacon. 1983. Lucifer yellow histology. *In:* Functional Neuroanatomy, ed. N.J. Strausfeld, Springer-Verlag, New York. pp. 132–155.

Struble, D.L. & H. Arn. 1984. Combined gas chromatography and electroantenogram recording of insect olfactory responses. *In:* Techniques in Pheromone Research, eds. H.E. Hummel & T.A. Miller, Springer-Verlag, New York. pp. 161–178.

Suckling, D.M., G. Karg, S.J. Bradley & C.R. Howard. 1994. Field electroantennogram and behavioral responses of *Epiphyas postvittana* (Lepidoptera, Tortricidae) under low pheromone and inhibitor concentrations. J. Econ. Entomol. **87**:1477–1487.

Sutcliffe, J.F. & B.K. Mitchell. 1982. Characterization of galeal sugar and glucosinolate-sensitive cells in *Entomoscelis americana* adults. J. Comp. Physiol. A **146**:393–399.

Thiery, D., J.M. Bluet, M.H. Phamdelegue, P. Etievant & C. Masson. 1990. Sunflower aroma detection by the honeybee—study by coupling gas-chromatography and electroantennography. J. Chem. Ecol. **16**:701–711.

Todd, J.L. & T.C. Baker. 1993. Response of single antennal neurons of female cabbage loopers to behaviorally active attractants. Naturwissenschaften **80**:183–186.

Todd, J.L. & T.C. Baker. 1996. Antennal lobe partitioning of behaviorally active odors in female cabbage-looper moths. Naturwissenschaften **83**:324–326.

Todd, J.L., K.F. Haynes & T.C. Baker. 1992. Antennal neurons specific for redundant pheromone components in normal and mutant *Trichoplusia ni* males. Physiol. Entomol. **17**:183–192.

Todd, J.L., S. Anton, B.S. Hansson & T.C. Baker. 1995. Functional-organization of the macroglomerular complex related to behaviorally expressed olfactory redundancy in male cabbage looper moths. Physiol. Entomol. **20**:349–361.

Tømmerås, B.Å & H. Mustaparta. 1989. Single cell responses to pheromones, host and non-host volatiles in the ambrosia beetle *Trypodendron lineatum*. Entomol. Exp. Appl. **52**:141–148.

Usherwood, P.N.R. 1969. Electrochemistry of insect muscle. *In:* Advances in Insect Physiology, eds. J.W.L. Beament, J.E. Treherne & V.B. Wigglesworth, Academic Press, New York. pp. 205–278.

Van der Pers, J.N.C. & C.J. Den Otter. 1978. Single cell responses from olfactory receptors of small ermine moths to sex attractants. J. Insect Physiol. **24**:337–343.

Van der Pers, J.N.C. & A.K. Minks. 1993. Pheromone monitoring in the field using single sensillum recording. Entomol. Exp. Appl. **68**:237–245.

van Giessen, W.A., J.K. Peterson, & O.W. Barnett. 1992. Electroantennogram responses of aphids to plant volatiles and alarm pheromones. *In:* Proc. 8th Int. Symp. Insect-Plant Relationships, Wageningen. Kluwer Academic, Dordrecht.

van Giessen, W.A., H.W. Fescemyer, P.M. Burrows, J.K. Peterson & O.W. Barnett. 1994. Quantification of electroantennogram responses of the primary rhinaria of *Acyrthosiphon pisum* (Harris) to C4–C8 primary alcohols and aldehydes. J. Chem. Ecol. **20**:909–927.

Vaughn, T.T., M.F. Antolin & L.B. Bjostad. 1995. Behavioral and physiological responses of *Diaeretiella rapae* to semiochemicals in its environment. Entomol. Exp. Appl. **78**:187–196.

Vet, L.E.M., R. De Jong, W.A. van Giessen & J.H. Visser. 1990. A learning-related variation in electroantennogram responses of a parasitic wasp. J. Insect Physiol. **15**:243–247.

Visser J.H. 1979. Electroantennogram responses of the Colorado potato beetle, *Leptinotarsa decemlineata,* to plant volatiles. Entomol. Exp. Appl. **25**:86–97.

Wadhams, L.J. 1982. Coupled gas chromatography—single cell recording: a new technique for use in the analysis of insect pheromones. Z. für Naturforschung **37C**:947–952.

Wadhams, L.J. 1984. The coupled gas chromatography-single cell recording technique. *In:* Techniques in Pheromone Research, eds. H.E. Hummel & T.A. Miller, Springer-Verlag, New York. pp. 179–189.

Wadhams, L.J. 1990. The use of coupled gas chromatography: electrophysiological techniques in the identification of insect pheromones. *In:* Chromatography and Isolation of Insect Hormones and Pheromones, eds. A.R. McCaffery & I.D. Wilson, Plenum Press, New York. pp. 289–298.

Wadhams, L.J., M.M. Blight, V. Kerguelen, M. Le Métayer, F. Marion-Poll, C. Masson, M.H. Pham-Delègue & C.M. Woodcock. 1994. Discrimination of oilseed rape volatiles by honey bee: novel combined gas chromatographic-electrophysiological behavioral assay. J. Chem. Ecol. **20**:3221–3231.

White, P.R. & M.C. Birch. 1987. Female sex pheromone of the common furniture beetle *Anobium punctatum* (Coleoptera: Anobiidae): extraction, identification, and bioassays. J. Chem. Ecol. **13**:1695–1706.

White, P.R. & J. Chambers. 1989. Saw-toothed grain beetle *Oryzaephilus surinamensis* (L.) (Coleoptera: Silvanidae)—antennal and behavioral responses to individual components and blends of aggregation pheromone. J. Chem. Ecol. **15**:1015–1031.

White, P.R., J. Chambers, C.M. Walter, J.P.G. Wilkins & J.G. Millar. 1989. Saw-toothed grain beetle *Oryzaephilus surinamensis* (L.) (Coleoptera: Silvanidae)—collection, identification, and bioassay of attractive volatiles from beetles and oats. J. Chem. Ecol. **15**:999–1013.

Appendix: List of Names and Addresses of Suppliers

Company	Address	FAX, Website, or E-mail
Acros Organics	711 Forbes Avenue	800-926-1166
	Pittsburg, PA 15219	
Advanced Separation Technologies, see ASTEC		
Aldrich Chemical	P.O. Box 2060	aldrich@sial.com
	Milwaukee, WI 53201	800-962-9591
Alfa-AESAR	30 Bond Street	800-322-4757
	Ward Hill, MA 01835	http://www.alfa.com
Alltech Assoc.	2051 Waukegan Road	847-948-1078
	Deerfield, IL 60015	http://alltechweb.com
Analtech	75 Blue Hen Drive	302-737-7115
	P.O. Box 7558	http:/www.analtech.com
	Newark, DE 19714	
Analytical Research	P.O. Box 140218	352-466-0055
Systems	Gainesville, FL 3261	74747.3151@compuserve.com
Analytichem International, see Varian		
ASTEC	37 Leslie Court	210-428-0152
	P.O. Box 297	
	Whippany, NJ 07981	
Baxter Scientific	1430 Waukegan Road	Phone: 708-689-8410
	McGaw Park, IL 60085	
Bedoukian Research	21 Finance Drive	Phone: 203-830-4000
	Danbury, CT 06810	
Bio-Rad Labs	3300 Regatta Boulevard	800-227-5589
	Richmond, CA 94804	
Biotage	1500 Avon Street Extension	804-979-4743
	Charlottesville, VA 22902	
Brechbuhler AG	Steinwiesenstrasse 3	41-1-730-48-25
	Ch-8952 Schlieren/Zurich	
	Switzerland	
Bruker Instruments	19 Fortune Drive	508-667-3954
	Manning Park	
	Billerica, MA 01821	

Carlo Erba	See ThermoQuest Corp.	
Chromapon, Inc.	P.O. Box 4131	310-693-6188
	Whittier, CA 90607	
Chrompack	1130 Route 202	908-722-8365
	Raritan, NJ 08896	
Daicel Chemical Industries	8-1, Kasumigaseki 3-chrome	03-507-3193
	Chiyoda-ku, Tokyo 100	
	Japan	
EM Separations	480 Democrat Road	609-423-4389
	Gibbstown, NJ 08027-1297	
Finnigan Corp.	355 River Oaks Parkway	408-433-4823
	San Jose, CA 95134	http://www/finnigan.com
Fisher Scientific	711 Forbes Avenue	800-926-1166
	Pittsburgh, PA 15219	
Fluid Metering, Inc.	29 Orchard Street	516-624-8261
	P.O. Box 179	http://www.fmipump.com
	Oyster Bay, NY 11771	
Fluka	980 South Second Street	516-467-0663
	Ronkonkoma, NY 11779	
Hewlett-Packard	HP Chemical Analysis Group	302-633-8954
	2850 Centerville Road	http://www.hp.com/go.chem
	Wilmington, DE 19808-1610	
Isco	Separations Instruments	402-464-0318
	Division	
	4700 Superior	
	Lincoln, NB 68504	
J&W Scientific	91 Blue Ravin Road	916-985-1101
	Folsom, CA 95630-4714	http://www.JandW.com
Mattson Instruments	1001 Fourier Court	608-831-2093
	Madison, WI 53717	
Micromass Ltd.	Floats Road, Wythenshawe	44-161-998-8915
	Manchester M23 9LZ	http://www.vgorganic.co.uk
	United Kingdom	
Narishige, Inc.	1 Plaza Road	516-621-5081
	Greenvale, NY 11548	
P.C., Inc.	8106 Appalachian Way	301-299-2978
	Potomac, MD 20854	
Perkin-Elmer Corp.	761 Main Avenue	203-762-4222
	Norwalk, CT 06859	http://www.perkin-elmer.com
Pharmacia	800 Centennial Avenue	800-329-3593
	P.O. Box 1327	http://www.biotech.pharmacia.se
	Piscataway, NJ 08855	
Phenomenex	2320 West 205th Street	310-328-7768
	Torrance, CA 90501	
Pierce Chemical Co.	P.O. Box 117	815-968-7316
	3747 North Meridian Road	
	Rockford, IL 61105	
Quadrex	P.O. Box 3881	203-393-0391
	New Haven, CT 06525	
Regis	P.O. Box 519	708-967-5876
	Morton Grove, IL 60053	

Restek	110 Benner Circle	814-353-1309
	Bellefonte, PA 16823-8812	http://www.restekcorp.com
Sadtler Research Labs.	3316 Spring Garden Street	215-662-0585
	Philadelphia, PA 19104	
Sanki Engineering	520 Fellowship Road	609-231-4599
	Suite D-406	
	Mt. Laurel, NJ 08054	
Scientific Glass Engineering	see SGE	
Scientific Instrument Services	1027 Old York Road	908-806-6631
	Ringoes, NJ 08551	http://www.sisweb.com
Sensidyne	16333 Bay Vista Drive	813-539-0550
	Clearwater, FL 34620	
SGE	2007 Kramer Lane	512-836-9159
	Austin, TX 78758	
Sigma Chemical	P.O. Box 14508	800-325-5052
	St. Louis, MO 63178	http://www.sigma.sial.com
Supelco	Supelco Park	800-447-3044
	Bellefonte, PA 16823-0048	http://www.supelco.sial.com/supelco.html
Sutter Instrument Co.	40 Leveroni Court	415-883-0128
	Novato, CA 94949	
Tekmar	P.O. Box 429576	513-247-7050
	Cincinnati, OH 45242	http://www.tekmar.com
ThermoQuest Corp.	355 River Oaks Parkway	408-526-9810
	San Jose, CA 95134	
Thomas Scientific	P.O. Box 99	609-467-3087
	Swedesboro, NJ 08085	
Tosohaas	156 Keystone Drive	215-283-5035
	Mongomeryville, PA 18936	
Upchurch Scientific	P.O. Box 1529	800-359-3460
	Oak Harbor, WA 98277	http://www.upchurch.com
Varian	811 Hansen Way M/S B-111	415-858-0480
	Palo Alto, CA 94304	http://www.varian.com
Waters	34 Maple Street	508-872-1990
	Milford, MA 01757	http://www.waters.com
Wheaton Science Products	1501 North 10th Street	609-825-1100
	Milville, NJ 08332	
Wilmad Glass Co.	Route 40 and Oak Road	800-220-1081
	Buena, NJ 08310	

Index